Technology versus Ecology:

Human Superiority and the Ongoing Conflict with Nature

Robert A. Schultz
Woodbury University, USA

Information Science
REFERENCE
An Imprint of IGI Global

Managing Director: Lindsay Johnston
Editorial Director: Joel Gamon
Production Manager: Jennifer Yoder
Publishing Systems Analyst: Adrienne Freeland
Development Editor: Austin DeMarco
Assistant Acquisitions Editor: Kayla Wolfe
Typesetter: Travis Gundrum
Cover Design: Jason Mull

Published in the United States of America by
Information Science Reference (an imprint of IGI Global)
701 E. Chocolate Avenue
Hershey PA 17033
Tel: 717-533-8845
Fax: 717-533-8661
E-mail: cust@igi-global.com
Web site: http://www.igi-global.com

Library of Congress Cataloging-in-Publication Data

Schultz, Robert A., 1942-
Technology versus ecology : human superiority and the ongoing conflict with nature / by Robert A. Schultz.
pages cm
Includes bibliographical references and index.
ISBN 978-1-4666-4586-8 (hardcover) -- ISBN 978-1-4666-4587-5 (ebook) -- ISBN 978-1-4666-4588-2 (print & perpetual access) 1. Nature--Effect of human beings on. 2. Human ecology. 3. Technology--Social aspects. I. Title.
GF75.S38 2014
304.2--dc23
2013020698

British Cataloguing in Publication Data
A Cataloguing in Publication record for this book is available from the British Library.

Table of Contents

Section 3
Possible Outcomes

Foreword

"Our goal as a species is probably to find out what our role in the universe is, as a knowledge-gathering species," writes Dr. Robert Schultz, author of *Technology versus Ecology*. "The role of our species is to try to understand the role of self-aware species in a non-self-aware universe," writes Zog, Schultz's mythical friend, who visits the earth every 20,000 years. Schultz and Zog agree that understanding the role of the only fully self-aware, knowledge-gathering species on earth is of the greatest importance. Schultz argues that we need this understanding in order to take purposeful actions that will allow humans to survive. The question at the heart of this concern, as in all Schultz's writings, is ethics.

Ethics involves the regulation of "cooperative benefits" or "cooperative principles." In terms of our ecosystem, Schultz regards those ethics as involving "actions and policies necessary for the long-term survival of humanity." Eco-ethics also involves "the survival of the ecosystem of which human beings are a part" and on which human survival depends.

On the road to discovery, Schultz describes the long narrow path of self-interest followed by a species for which survival and the cooperative spirit appear to have been less important than the present enjoyment of the world one was in or the future enjoyment of a world that was not here. In the former case, nature was a resource that technologies increasingly exploited for pleasure and monetary gains, often interlinked ends. In the latter, nature was often despised and therefore ignored.

Ultimately, however, it is not classical technology in itself that is to blame for environmental degradation. It instead is the impulse behind the technology that leads humans away from nature. For in recognizing their difference, humans try to create a space apart from, rather than with, nature. The power derived from such separation cannot be ignored. It is seductive. Technologies that originate somewhat in harmony with nature, such as early farming, move from providing a way to mediate between the natural world and the social world to a means of using the natural world to satisfy social needs that can be monetized.

In this way technologies increasingly disrupt the natural order—especially modern technology, which is disruptive in its very conception. This disruption is manifest not just in the technologies but also in the very human disposition to possess, store, and hoard. That impulse, natural in humans, runs contrary to the rest of nature, which is bound and responsive to the current conditions of its existence. In identifying the human urge to preserve as a cause of ecological disruption, Schultz's argument might even be used to deconstruct the mantra of "Reduce. Recycle. Reuse," which simply reinstates the human attachment to "saving" in a more palatable form. Enviro-chic or green luxury is just another form of conspicuous consumption, of storing, using, and monetizing.

We need a different approach. The fundamental question that Schultz poses is "How can the strategies that got us into this mess possibly get us out of this mess?" We love to distinguish ourselves through various types of capital—cultural, financial, even environmental. The motility of the forms that this love of distinction takes presents an obstacle. Our fondness for what we do is so enormous that effective change seems impossible.

Enter Zog, the mythical traveler from other times and places, who much like the citizen of the world or the Persian traveler of the Enlightenment, surveys human activity and, with a critical detachment borne of his people's fascination with the conundrum humans present, helps us see clearly the choices between what will give life and what will exterminate it.

Ultimately, rooting for humans to survive, being on Team Human is not a bad thing as long as it does not mean a blanket validation of all human action. We may not root against Team Animal or Team Planet. Taking effective action requires us to both care about the fate of our species and care about the earth household that sustains our lives.

Schultz's work flows out of his training as a philosopher and his work as a technologist. Robert Schultz has followed up his books on ethics and technology in *Contemporary Issues in Ethics and Information Technology* (2005) and *Information Technology and the Ethics of Globalization* (2009) with a wider and more penetrating look at the ethical dimension of technological and ecological intersections. The current of his ideas flows through his two previous volumes. One will find the same exhilarating coming together of technology and ethics here that one finds in those. But here the stakes are higher, because Schultz has used his breathtaking knowledge and style to deal with the central issue of our day—human survival.

Of course, Dr. Schultz is not the first to see how industry arose from the alienation of humans from their environment, and how in so arising industry led to even greater alienation, which led to greater degradation of the environment and long human pain and suffering. John Milton, living at the beginning of industrialism, saw the change even with his blind eyes and recorded it in *Paradise Lost*. For Milton, the primal and pre-lapsed state of Adam and Eve showed the true relationship of humans to the planet, but in his epic, harmony with nature gives way to human desire and human need for power over others. Whereas before the fall, all the world was home, after the fall, humans required habitation, habiliment, technology, and death. One hundred years after Milton's passing, another poet, William Blake, saw Milton's prophecy fulfilled. One hundred years after Blake, Schultz notes the coming of "postmodern technology," technology reliant on petroleum-based products that is "incompatible with the ecosystem. These products replace products compatible with the ecosystem."

There is a story in the Talmud about a farmer who makes his land arable by casting the stones from it into the public way. A passerby stops as the farmer works and asks why the man takes what he hates from what is not his and throws that into what is his. The farmer, of course, is angry with the meddlesome stranger. However, when the farmer loses his property and stumbles over the stones he has cast into the highway, he discovers that individual human possession is fleeting, while the earth, which is the possession of all, including the individual, endures.

Humans' sense of separateness may lead to their sense of being a special species, and it may be that this sense of difference expresses itself in the idea of dominion. Still, both human difference and human dominion are shared, and it is important that we attend in an ethically rigorous way to understanding what actions will produce "cooperative benefit" for a "self-aware species" dependent on "a non-self-aware universe." Schultz's work calls us to that task. It is a sobering and energizing call, one to be heeded and acted upon.

David Rosen
Kendall College of Art and Design, USA

David Matthew Rosen *is the president of Kendall College of Art and Design, which houses the Wege Center for Sustainable Design. Rosen holds a B.A. from Haverford College and a Ph.D. from the Johns Hopkins University. He served as Interim President at the University of Maine at Machias, Senior Vice President at Woodbury University in Burbank, California, and Board Chair of the Maine Humanities Council. Dr. Rosen held a John Carter Brown Library fellowship at Brown University. His publications include The Changing Fictions of Masculinity.*

Preface

If we continue to operate in terms of a Cartesian dualism of mind versus matter, we shall probably also continue to see the world in terms of . . . man versus environment. It is doubtful whether a species having both an advanced technology and this strange way of looking at its world can endure. —Gregory Bateson (1971)

This book is about conflicts between technology and ecology.[1] Technology, ecology, and their relation were dealt with in two of my previous books. Those books started from the philosopher Martin Heidegger's view that modern technology is a new way of bringing things into being which is incompatible with older ways. Hence, there is a built-in conflict between technology and nonhuman ecological systems. In my earlier books, it was observed that since humans require ecological systems to survive, ecology must take precedence over technology. I now feel that this observation is not enough to resolve the conflict.

Virtually all human beings seem to share the attitude that they are in charge of the rest of nature. This attitude is shared both by those who would give priority to technology and by those who advocate living in harmony with nature. This attitude is probably the source of the conflict between technology and ecology. For how can we at the same time be masters of nature and yet subject to nature, as we clearly are? We need to explore how an animal among animals came to have this attitude. It may begin not with tool use but with storing tools, thus creating for the first time a purely human space distinct from the rest of the world. Although my investigation respects historical fact as determined by science, it is primarily conceptual rather than historical.

Before conceiving of this book, I wrote about issues of what human beings owed to the rest of life on the planet and to the planet itself.[2] One view is that principles for our relations with animals and the ecosystem are to be determined by human benefit. Impacts on the ecosystem are considered solely in terms of benefits to *us,* to humans. This is probably the view of many people. An unthinking version gives no weight at all to the well-being of nonhuman nature. If sea otters are eating so many clams that the welfare of human clam diggers is affected, kill a bunch of otters. If wolves are depredating your livestock, kill the wolves and poison their pups. If coyotes are eating your pet Chihuahuas, exterminate the coyotes.

A more thoughtful version would consider the impact of nonhuman nature on human beings. Thus, removing predators from an area could result in an explosion of deer and consequent destructions of trees and plants, both wild and domestic. Cutting down forests to plant crops could have the effect of damaging the fertility of the land. Concerns such as these recognize that living things are organized interdependently into what are called *ecosystems*. When these ecosystems are damaged, there are negative consequences for human beings. Human beings are, after all, part of these ecosystems.

This was my view in my previous writings on the subject. In those writings, the preservation of the ecosystem was tied to human welfare, and we cannot have human welfare without the ecosystem. At least this formulation does consider actions involving animals and the environment in a broader context. We do not exterminate wolves just because they kill our sheep, but rather look to their place in the ecosystem and consider the damage done by extra deer if the deer population is not kept in check by enough wolves.

What I finally realized, and what led me to write this book, is that we humans assume a privileged place in nature, which is not justified. This privilege has both negative and positive aspects. Negatively, we speak of *intervening* in nature when, say, we cull an elephant herd. Yet we do not speak of an elephant intervening in nature when it kills its trainer. Not only that, but we assume we can care for the environment in ways the environment cannot care for itself. Even this positive aspect needs to be questioned. When environmentalists demand that coyotes rather than human hunters reduce excessive deer population, they thereby assume that human beings are not part of nature! (Associated Press, 2010).

But then what is the appropriate point of view for human beings to use as the basis of sound principles for treating animals and the rest of the world? We need to think through what makes us human beings "too distantly a part" of the world we live in:

It is the human that is the alien,
The human that has no cousin in the moon.

It is the human that demands his speech
From beasts or the incommunicable mass.

If there must be a god in the house, let him be one
That will not hear us when we speak: a coolness,

A vermilioned nothingness, any stick of the mass
Of which we are too distantly a part.
("Less and Less Human, O Savage Spirit," Stevens, 1959)

These words of the poet Wallace Stevens pose the conundrum this book is an attempt to resolve. Of course it is better for us to take charge of the ecosystem than to trash it, but taking charge still puts us in a false position. Our true position is that we are in fact a part of the ecosystem along with the rest of nature. We have to be able to express our true position somehow. This is not just a philosophical exercise. Unless we can assume our true position and adopt that point of view, we stand an excellent chance of extinction. We are now in a race against time to understand and act on our true position in nature.

Man's ability to manipulate the environment goes way beyond survival, to the point of threatening the entire current ecosystem. Will this animal shortly make itself extinct and maybe leave room for another attempt not so imbalanced? Humanity's threatening the entire ecosystem means that the point of view of man as superior is *actually incorrect.* The question is, what is a correct alternative view?

Chimpanzees that make tools are intelligent. However, their ability to make tools does not divorce them from the ecosystem. It is making *and storing* tools. Note what the storing of tools accomplishes. It sets aside a part of the world as different from the rest of the world and as essentially human, and the purpose of that different part of the world is to change the rest of the world. This innocuous beginning of technology does not lead to our divorce from the world for quite a long time. Our separation from the world lies dormant and accumulates slowly. Finally, with modern technology, the entire world is turned into storage ("standing reserve" in Heidegger's terms), into items on hand for some further use (Heidegger 1955).

Other species with similar or possibly greater intelligence do not appropriate the world, for example dolphins. The $64,000 question is what do they use their intelligence for if not to appropriate the world? Our intelligence used for appropriation could easily lead to our extinction, taking many other species with us.

Many human beings regard the results of our separation from the ecosystem as a triumph rather than a mistake. That we are able to prevail over nature is thought to be a distinctive human strength. Climate change denial may be due to such beliefs. Climate change denial often stems from the belief that human economic activity takes precedence over the rest of nature. Indeed, climate change deniers apparently believe that human economic concerns (in particular increased profits for fossil fuel corporations) take precedence over the findings of science.

So are our obvious differences from other species ultimately a good thing or a bad thing? The answer is neither. The parallel question is whether globalization is a good or a bad thing. The answer is both or neither. Globalization is a human activity that has both good and bad aspects. So too, humanity's activity in nature has both good and bad aspects. The aim of this book is to sort these out. Could it be that an obviously powerful and intelligent species is too smart for its own good?

ORGANIZATION OF THE BOOK

This book is organized into three main sections. Section 1, "Where We Are Now," examines current aspects of humanity's relation to nature. Section 2, "The Origins of Technology and Human Superiority," starts from the time humans were part of nature and traces human development from there to contemporary times. Section 3, "Possible Outcomes," considers ways in which humans can be in equilibrium with their environment.

Chapter 1 examines several versions of human superiority. None give humans the right to do whatever we want to the rest of the environment. Although humans have physical superiority, that does not give us "dominion" over the rest of the world.

Chapter 2 examines the ways humans can seriously damage the ecosystem, our support system. Modern technological manufacturing regularly produces harmful side effects, either toxic substances or unexpected threats to the entire ecosystem such as ozone layer destruction or global warming. Human economic activity is leading to mass extinction of other species. Humanity is threatening the entire ecosystem. This means that the view of humans as superior is actually incorrect.

Chapter 3 examines how modern technology dramatically accelerates our conflict with the ecosystem because of its nature. Modern technology treats everything as nothing but resource. This explains why modern technology *must* conflict with the ecosystem. It does not recognize the ecosystem as anything other than a set of resources it can exploit. Corporations turn out to be the institution par excellence for advancing modern technology, since for them everything becomes a means to maximizing profits.

Chapter 4 deals with another type of separation of humans from the environment. Humans view themselves as not part of nature when they believe they are primarily nonphysical minds or souls or spirits, but we know that humans arose on this planet through natural selection. Consciousness, although it is in some ways nonphysical, is still very much a part of nature. The work of two contemporary philosophers, Richard Taylor and John Rowlands, provides an account of the relation of consciousness to the world.

In Section 2, the origins of the human separation from nature and human superiority are examined. Chapter 5 begins with our immediate ancestor *Homo erectus* who made tools complex enough to save and store, thus creating a purely human space for the first time. *Homo erectus* later tamed fire and built temporary shelters, again altering the relation of human beings to nature. When members of our species *Homo sapiens* appeared, they made and wore clothing, carried out ceremonial burials, began making representations (cave paintings), and exterminated nearly all animals larger than themselves.

Chapters 6 and 7 examine two developments that perhaps separated humans from the ecosystem the most: Farming and civilization. Both are fairly recent developments, farming dating from about 11,000 years ago (6% of the time of our species on earth) and civilization about 7,000 years ago (4% of our time on earth). Chapter 6 examines farming. Farming separates us from the ecosystem by replacing the ecosystem with plants and animals of our own choice, but farmers must still respect the environment which allows those plants and animals to thrive. Once humans began farming, forces arose which made it difficult to return to a hunter-gatherer lifestyle, even when that lifestyle may have been preferable.

Chapter 7 examines the origin of civilization. Civilization has its origin in farming, because farming produced a food surplus, so that a number of people were freed from work connected with producing food. This separation from the environment enables humans to develop a wide range of activities, pursuits, and occupations, accompanied by many technological innovations. These new activities and innovations are often regarded as demonstrating the superiority of humans, but since almost all these pursuits and innovations benefit humans only, it is only physical superiority that is once more demonstrated.

Chapter 8 places into a historical context the arrival of modern technology about 250 years ago. Modern technology exacerbates conflicts with the ecosystem. The products and by-products of modern technology are typically either toxic or cannot be incorporated into the ecosystem—they are not biodegradable, in other words. The largely petroleum-based products from about 1900 replace products compatible with the ecosystem, and conflicts with the ecosystem accelerate.

Chapter 9 examines a number of views of the causes of humanity's conflict with the ecosystem, which are often said to be at least one of the following: human institutions misusing technology, technology itself, population growth, economic growth, and human attitudes. The causes are modern technology together with attitudes of human superiority and human exceptionalism. Population growth and economic growth are secondary causes—they make environmental problems worse.

Section 3 looks at various aspects of the problem of humanity's achieving equilibrium with the environment.

Chapter 10 considers the extent to which environmentalism and sustainability can help with humanity's problems with the ecosystem. Sustainability is maintaining equilibrium with the ecosystem. Four accounts of how to achieve sustainability are evaluated: (1) unlimited economic and population growth; (2) protecting diversity through cooperation with corporations; (3) reducing population and dealing with climate change; and (4) humans' adopting the attitude that the ecosystem is primary.

Unlimited economic and population growth is a non-starter. It would have to be achieved through massive applications of modern technology, which would clearly create more environmental problems. Corporations can be trusted only to maximize their profits. Although reducing population and dealing with climate change will help, by themselves they do not address the continuing threat of modern technology to the ecosystem. The final alternative, that humans must come to realize that the ecosystem is primary, is correct, but this realization needs a firmer basis than emotional commitment.

In Chapter 11, the place of science and technology in the relation of humanity to the ecosystem are considered. The role of modern technology needs to be reduced because of the ongoing threat of catastrophic environmental consequences. Because modern technology needs to be employed to monitor the ecosystem and to deal with potential celestial collisions, some of it must be continued. Parts that do not contribute to the survival of the ecosystem need to be reduced as quickly as possible, without causing any more damage than necessary. Economic growth needs to be rethought with environmental costs properly included.

Chapter 12 examines issues concerning our relation to other species. If we are a species among species, how should we relate to other species? Some animal rights theorists go overboard and say it is unethical for humans to interfere with other animals at all. This view once more makes humans not a species among species. Since ethics is principles for social cooperation, only a restricted class of animals—work animals and pets—can have ethical standing. We need to find a middle way between no human interaction with animals and human superiority, which considers animals as mere objects for our use. In any case, eco-ethics requires us to treat animals humanely and not as objects.

Chapter 13 attempts to determine the role of humans as a species among species. Civilization may not have that much long-term value for our species because it fosters fairly complete separation from the ecosystem. Ultimately, our survival rests with our being able to use our knowledge to respect the ecosystem and to live within its limits. Of the various distinctive characteristics of the human species, only long childhood and menopause bear directly on our being a species among species. Both give us the opportunity to learn more before adulthood. We are therefore adapted to learn more about our surroundings and act on the basis of that knowledge. Indeed, our goal as a species may be to find out what is our role in the universe as a knowledge-gathering species.

Chapter 14 presents some considerations about the extinction of the human race. Sudden and untimely extinction of the human race is what we hope we can prevent, but eventually the human race will become extinct, like any other species. Ultimately, we should be concerned with what we can do toward our species goals within the possible one-to-two million-year lifespan of our species. Our individual lifespans are not meaningless even though we know they will end. Therefore, what we do as a species is not meaningless just because it will come to an end.

Chapter 15 summarizes the principal ideas of the book. The book closes with two stories about the future of the planet, one story with humans in it and one story without humans.

We may be able to avoid the sudden and untimely extinction of the human race if we treat the ecosystem with the respect it deserves. We cannot treat the ecosystem with respect if we continue to deploy modern technology and unlimited economic growth. To treat the ecosystem with respect is to realize that our role on this planet is limited and that, regardless of our physical power, the rest of the planet does not exist for our benefit. I hope this book will make it clear why human superiority and human exceptionalism are ideologies we need to discard.

Any belief in moral, ethical, or practical human superiority can only lead to the early extinction of humanity. Often, people think there are only two alternative ways of looking at the world: either humans rule or God rules. In this book, both these ways of looking at the world are shown to be incorrect. Evidence for a third view is presented: that humans are a species among species. My evidence includes findings from the sciences of biology, ecology, anthropology, paleontology, and psychology, as well as relevant results from both classical and contemporary philosophers.

Robert A. Schultz
Woodbury University, USA

REFERENCES

Associated Press. (2010, October 18). Coyotes, not hunters, proposed for Valley Forge. *Animal News*. Retrieved 10/20/10 from http://www.bornfreeusa.org/news.php?p=2516&more=1

Bateson, G. (1971). The cybernetics of self. *Psychiatry*, *34*(1).

Diamond, J. (2005). *Collapse: How societies choose to fail or succeed*. New York: Viking Books.

Heidegger, M. (1955). The question concerning technology. In *The Question Concerning Technology and Other Essays*. New York: Harper & Row.

Schultz, R. (2006). *Contemporary issues in ethics and information technology*. Hershey, PA: IGI Global.

Schultz, R. (2009). *Information technology and the ethics of globalization*. Hershey, PA: IGI Global. doi:10.4018/978-1-60566-922-9.

Stevens, W. (1959). *Collected poems*. New York: Vintage Books.

Tainter, J. (1988). *The collapse of complex civilizations*. Cambridge, UK: Cambridge University Press.

ENDNOTES

1 I am indebted to David Rosen for important clarification on the role of technology in my enterprise.

2 Schultz (2006) Chapters 12 and 13. Schultz (2010) Chapters 14 and 15.

Acknowledgment

My special thanks to David Rosen, not only for his Foreword, but also for his encouragement when I was deciding whether to write this book. He suggested that my idea of going back to a time when humans did not think of themselves as in control of nature and then looking at what happened later was not crazy. Jane Whitehill deserves special mention for her careful and thorough reading of Chapters 1-4. Her comments resulted in some major improvements to those chapters. Daniel Quinn and Sarah Mallery supplied helpful comments and criticism. My daughter Katie Schultz supplied some valuable firsthand information about leaving a vegetarian diet. Neil Anapol introduced me to Heidegger's writings on technology many years ago. Mike Gladders and Dorian Abbot of the University of Chicago Astronomy Department provided help with the intricacies of "runaway greenhouse," as did Mike Guidry of the University of Tennessee Astronomy and Physics Department. Finally, Monica Speca and Joel Gamon of the IGI Global editorial staff provided prompt and clear answers to all my questions. Also my thanks to John Dittmeier, Andrew Ross, Rebecca Schultz, and the Woodbury University Library Staff.

Robert A. Schultz
Woodbury University, USA

Section 1
Where We Are Now

Chapter 1
Humans as Masters of the World

ABSTRACT

Several versions of human superiority are characterized and analyzed. None give humans the right to do whatever they want to the rest of the environment. Although humans have physical superiority, that does not give humans "dominion" over the rest of the world.

INTRODUCTION

Zog is an infrequent visitor to the planet earth. He drops by every 20,000 years or so. During his recent visit, he was astonished at the changes that had happened since his last visit 20,000 years ago.[1] During his last visit, he saw small bands of accomplished human hunters with very effective stone tools. Their diet included tasty stews of cooked meat and vegetables. They lived in temporary settlements, often utilizing sites with natural protection from the elements such as caves. They had been producing beautiful artwork in these caves for some time.[2] Although their impact on the landscape was not dramatic, Zog noted that species of large animals had often disappeared soon after humans occupied an area.[3] He also noted that only two of the three human species present during his visit of 40,000 years ago had survived. *Homo sapiens* was going strong, *Homo sapiens neanderthalensis* had vanished completely, and *Homo floresiensis* was present only on a few very isolated islands. (Zog would have called them hobbits if he had been able to read *The Lord of the Rings.*)[4,5].

DOI: 10.4018/978-1-4666-4586-8.ch001

On this visit, Zog noted that substantial areas of the planet had been converted to human use. Other species—such as dogs, cats, and horses—were present in those areas mostly only under the control of humans. Zog was relieved to see some of his distant cousins doing very well in spite of this (Zog's distant relatives were very much like our cockroaches.). From space, at night, large areas of the planet were artificially illuminated. But Zog was concerned that humanization was resulting in the loss of many species. He was concerned that humans did not seem fully conscious of the extent of changes they were making to their environment. But, being a politically correct alien, he did not want to be negative about what humans were doing to the planet. So in his report he stated that human beings had apparently become masters of the planet, in almost total control of its life forms and their environment.

Zog found that human beings were very much aware of being in control of the planet and its resources. In fact, with very few exceptions, humans either took their superiority for granted or thought their superiority was a good thing. Even people concerned about the negative consequences of human actions for the environment for the most part still supposed that humans were in control. These *environmentalists* saw serious problems with the consequences of human activity. But most humans thought more human activity would solve any such problems.

We now leave Zog. He will return at the end of this book and we will see what he might find on his next visit 20,000 years from now.

VERSIONS OF SUPERIORITY

The purpose of this chapter is to examine four related versions of the assumption of human superiority. These are:

1. Human superiority
2. Human exceptionalism
3. Cheerleader superiority
4. Physical superiority

1. **Human Superiority:** We humans are in charge of what happens to the environment, and our agency is the only agency humans need to consider.

This version was suggested by a 2010 news story from the Philadelphia area. In Pennsylvania, as in many parts of the northeastern U.S., deer populations have exploded. One cause is that humans have removed deer predators such as wolves from the environment. Large deer populations eat any and all available vegetation. With an unchecked deer population, vegetable and flower gardens vanish and even trees and shrubs are vulnerable. A Pennsylvania agency decided to employ some riflemen to kill excess deer. An environmental group objected and proposed instead that coyotes be imported and used to kill the deer.

The agency objected, possibly correctly, that there were insurmountable difficulties with using coyotes. But that is not what caught my attention. What caught my attention is that *the environmentalists were still assuming that human beings were in charge.* It was up to *us* to import the coyotes and it was up to *us* to decide on the method of reducing the deer population. In fact, in making their proposal, the environmentalists were assuming that human beings are not part of nature![6] That is, the riflemen are *not* part of nature but the wolves *are.*[7]

This is the underlying attitude that will be examined in this book. It is not only that human beings are *better* than other animals. Rather, human beings are *entirely distinct* from other animals and the rest of nature. This view could be called *human exceptionalism.* Human exceptionalism accepts the human superiority assumption above but adds the following:

2. **Human Exceptionalism:** We humans are not part of nature. Exceptionalism is often combined with human superiority: We are also in charge of nature.

Surprisingly, many environmentalists tacitly accept exceptionalism. Bill McKibben is an influential environmental activist. In his best-selling *The End of Nature* (1989), he tells us that human beings are no longer part of nature, they have *ended* nature. This may be an attention-getting way of describing the situation. But it is not correct. There is a problem precisely because we *are* still part of nature. Injecting large amounts of greenhouse gases into the atmosphere has consequences because of the effects of increased concentrations of gas in the atmosphere. This is a part of nature over which we have very limited control once the gases are in the atmosphere.

McKibben clearly thinks of nature as *non-human* nature. And he has primarily aesthetic reasons for wanting to preserve and restore such nature. There is nothing wrong with such goals, but they leave completely untouched the question of the proper place of human beings in the rest of nature. In the world of McKibben's proposed solution, human beings essentially stand outside the world and preserve and protect it. Again, a good goal, but how did we get into this position?[8]

Versions of these two views—superiority and exceptionalism—appear in many forms in recorded history, beginning with statements of superiority in the Old Testament (Genesis 1:28), and are supported by religious views of human beings as being unique in having souls. Related views are also asserted by biologists in the last few centuries. The biologist Julian Huxley asserted that humans have left the realm of biological evolution and are now subject only to cultural evolution, which is under their control (Huxley 1953). The rest of this chapter examines various versions of these views.

These views are strikingly implausible. Human beings arose from nature in ways science has been able to describe. That human beings should now *not* be part of nature is completely unbelievable. Of course, much of the support for human exceptionalism comes from supernatural faith-based beliefs inculcated by various religions. I think such support is not worth much. Worse, some religions reject major parts of science, including evolution. I believe this rejection of science should be viewed as a denial of reality. In this book only naturalistic or scientific support for human superiority and exceptionalism will be considered as valid. Spiritual bases for superiority and exceptionalism will be discussed further in Chapter 4, *Mind/Body/Consciousness*. However, not all religions conflict with science. The Dalai Lama has said that if any part of Buddhism conflicted with science, he would have to give up that part of Buddhism (Dalai Lama, 2005, p. 3).

Superiority and Exceptionalism as Connected in Religion

Genesis, the first book of the Judeo-Christian Bible, has a clear statement of superiority: "God blessed them and said to them, 'Be fruitful and increase in number; fill the earth and subdue it. Rule over the fish in the sea and the birds in the sky and over every living creature that moves on the ground" (Genesis 1:28). So far as I can determine, no other culture's creation or foundation myths include dominion over the earth.[9] Domination of animals is explicit in this verse, but the verse also says that the remaining portions of the earth are to be filled and under the control of human beings. So these attitudes are explicit in the origin myths of Judaism and, by adoption, Christianity and western civilization.

However, some Christian thinkers detect a different attitude in Genesis. Theodore Hiebert first notes further corroboration of human superiority and authority:

This conception that humans have a special status above the rest of creation appears to be corroborated by other details of the creation account in Genesis 1. While all other life originates within the sphere of the earth to which it is related—according to the divine command, "Let the waters bring forth swarms of living creatures" (Gen. 1:20), or "Let the earth bring forth living creatures of every kind" (1:24; cf. 1:11-12)—humans do not originate in this way. They are created by God alone (1:26). Furthermore, and even more significant[ly], humans alone among all forms of life are made in God's image (Hiebert, 1996, p. 3).

Hiebert remarks that a more recent interpretation of Genesis is that humans are God's stewards rather than wielders of "unrestrained and exploitative dominion." However, Hiebert finds this stewardship interpretation lacking because it ignores two important features of the historical context of genesis: First, that farming in ancient Israel was "adversarial" to nature because of the harshness of the land; second, that priests transferred their own absolute authority to humanity (Hiebert, 1996, p. 4). These are not very persuasive reasons to restore a stronger "dominion" reading of Genesis. There were many cultures with harsh environments which did not proclaim dominion over the planet. Similarly, neither did any of the many cultures with powerful priesthoods.

Since the Bible is not my starting point, these historical textual questions are more or less beside the point. However, it is reasonable to think that the attitude expressed in Genesis 1:28 was not expressed there for the first time, and that Genesis 1:28 reflected attitudes present in Hebrew culture when Genesis was produced.

Looking more closely at the notion of stewardship, an important question is whether it requires a creator God. A *steward* is someone charged with caring for and maintaining property or other affairs for someone else. If we are stewards of the environment, who is this 'someone else'? A creator God with the usual properties is the obvious answer—if we believe in such a God.[10] But because the properties, beliefs and requirements of God are highly contested, theistic stewardship could not be a basis for policy or action or attitude toward the environment. Could nonhuman nature itself be the 'someone else' for whom we are stewards? This question involves the nature of an ethical relation of humans to the environment, and we will return to it below.

Also involved with the notion of stewardship is the idea of *responsibility* (Swenson 2010). This again places human beings in a superior position with respect to the environment. It is certainly better for human beings to be responsible with respect to the environment instead of uncaring and exploitive. But being responsible still means regarding yourself as able to dictate and control the rest of nature.

According to Hiebert, Genesis 2 exemplifies a different attitude than human dominion. In Genesis 2, humans are (correctly) characterized as living beings among living beings:

. . . the archetypal human is made not in the image of God but out of topsoil, out of the arable land that was cultivated by Israelite farmers (Gen. 2:7). As a result of this kind of creation, humans hold no distinctive position among living beings, since plants and animals also were produced from this same arable soil (2:9, 19) (Hiebert, 1996, p. 6).

The critical question is then how did humans come to think otherwise?

Hiebert concludes with the more important question of which view of the role of humanity is to be followed today:

On the other hand, we know ourselves as humans . . . to be only a single species in a large and complex Web of life we do not entirely understand and can never really control. The Web of life functioned beautifully without us before we arrived as humans in the very last fraction of creative time, and it would do fine—much, much better in the last 100 years—without us. Our only hope of survival, in fact, is in recognizing our dependence on this Web of life and adapting our behavior to conform to the process created into it and to the demands it makes upon us. In such a context, it seems reasonable—even necessary—to . . . conceive our responsibility as one of service within the world of God's creation (Hiebert, 1996, p. 7).

I have absolutely no quarrel with these sentiments, except with the efficacy of the last six words. We need to find out how human beings got to be so different from other animals that they often think they are not animals at all. My intention in this book is to present a plausible account of how this happened.

3. Cheerleader Superiority

Another version of human superiority could be called "cheerleader superiority" because it is common in connection with sports teams.[11] The preferred sports team need not really be superior, but the idea is that enthusiastic fan support helps the team do better and thus become more superior. Occasionally the enthusiastic fan support becomes destructive, which does not help the chosen team at all.[12]

This kind of cheerleading support is also common for nations. Again, there may be no objective evidence that the nation deserves support. And indeed there can be objective evidence that a nation deserves nonexistence—consider Germany when it was committing genocide in the 1930s and 1940s. But this kind of unreasoned loyalty is probably necessary for citizens to die to protect their nation. And the willingness for its citizens to make this ultimate sacrifice may be necessary for the continued existence of the nation.

Human superiority may be a generalization of the superiority of more limited groups. Cheerleader superiority can be generalized to the entire human race. It can be expressed as "man's intellect can solve any problem."[13] There are reasons to believe this view is a form of cheerleading. One reason is somewhat technical. The mathematician Kurt Gödel proved in 1933 by unimpeachable methods that there are problems in mathematics that are unsolvable by any amount of human ingenuity. Since then, a number of mathematical problems have been proved to be unsolvable (Quine, 1982, pp. 216-217).

There are other reasons for thinking that the human intellect has limits. The philosopher Immanuel Kant's *Critique of Pure Reason* (1781) has precisely the aim of determining the limits of human intellect. It would probably be helpful if scientists, especially theoretical physicists, had some familiarity with Kant's work. Kant shows the limits to certain patterns of scientific explanation with great clarity. Yet even Nobel prizewinning physicists seem to have no inkling of the relevance of Kant's findings.

For example, one procedure in physics is to explain what something is by saying what it is composed of. Thus, we explain what matter is by saying it is composed of molecules of certain types. We explain what molecules are by saying they are composed of atoms. We explain what atoms are by saying they are composed of electrons, neutrons, and protons. We explain . . . At this point we can pause to consider Kant's point about this pattern of explanation. First of all, note that the meaning of the Greek word 'atom' is 'without parts.' So atoms were clearly misnamed because we now think they have parts. But note that at each stage of this explanatory pattern, there are two possibilities: Either the entities at that stage cannot be further decomposed—meaning they cannot be explained—or the entities at that stage can be further decomposed meaning we never reach any entities that cannot be decomposed. In this case, the explanation is never completed. Also in this case, theoretical physicists have infinite job security.

What would it mean to have an "unlimited" intellect? As Kant pointed out in his *Critique of Pure Reason* (1781) this would mean that human beings could apprehend the universe directly and completely, in unmediated form. This kind of apprehension, called *intellectual intuition*[14] by Kant, would be available to a supreme being. It is not available to humans who have clear limitations in how they relate to the world. Human beings are in space and time and they therefore cannot obtain knowledge of things in themselves, only of things as they present themselves to us. We have to remember that human beings are animals, and their initial presence is as a point of view in a world.[15] Our idea of space, as the place where we are present together with others and the rest of reality, is a construct. Animals exhibiting extensive coordinated purposive cooperative behavior may for that reason have this concept of space.[16] The construct is an unfilled space of indefinite extension, available for entities to appear in it.

The observer reappears at the outer edges of physical theory: In the theory of relativity of Einstein and in quantum mechanics. The theory of relativity is based on the premise that reality must be described from within the frame of reference of the observer (Einstein 1956). Quantum mechanics similarly notes that observation of the subatomic world cannot transcend the observer. We can determine either the position or direction of movement of subatomic particles, not both. And the limitations of our observation are intrinsic and cannot be overcome by better apparatus (Cassidy, 1998). We can conclude that human beings cannot have an account of the world, of reality, that is observer-independent. Therefore, the human intellect is intrinsically limited because it cannot transcend its basis in individual animals.

4. Physical Superiority

Besides intellectual superiority, there is also physical superiority. So far human superiority as an *attitude* has been discussed. However, human beings have been able to "be fruitful and multiply" beyond the wildest dreams of earlier times. In 1800, there were about 1 billion humans on the planet. In 1900, there were 1.6 billion. In 1999, there were 6 billion. With trends toward decreased fertility (Italy and other European countries as well as Thailand are already not replacing their populations), human population is expected to peak at between 9 and 10 billion in the late 21st century (Wilson, 2002, pp. 28-31).

Humans have been also able to convert increasingly large areas of the earth for their own use, whether for agriculture or for settlement as civilization. There is no question about the power of human beings. Animals and plants not appropriated for human use do not stand a chance in any conflict with human beings. These "wild" animals and plants for the most part owe their continued existence to the forbearance of human beings. As I mentioned at the beginning of this chapter, humans going back at least 60,000 years have been responsible for the extinction of almost all large terrestrial animals unlucky enough to have humans arrive in their neighborhood.

Even without considering human technology, humans have had and continue to have great power. Technology will receive a separate discussion in Chapter 3, "Science, Modern Technology, and Corporations." For now, it is enough to point out that technology makes possible a huge increase in the power of human beings. In the amazingly short space of 200-300 years, we have gone from a human environment not much different from that at the beginning of civilization, to trains, automobiles, and planes. From candles and oil lamps to electric light and electronic equipment. From ink on paper to electronic screens. Not to mention hot water, refrigerators, freezers, air conditioning, and indoor plumbing. Or advances in health and control of disease. Even the term 'world' usually means human beings. When back in the 1960s counter-culturalists shouted "The whole world is watching," they did not mean that whales were watching, nor buffalo.

But does having enormous power, *make* us superior? Or does it only make us enormously powerful? Hurricanes and tornados are very powerful but that does not make them superior. More plausibly, it is our ability to accumulate power that might make us superior. And this ability is tied to the ability to innovate, create and take chances on new things. One striking comparison between us, *Homo sapiens,* and our previous ancestors, is that we spread all across the globe. *Homo habilis* and *Homo neanderthalensis* usually remained in the areas they evolved in. Most striking is that to settle Australia and Polynesia, early humans had to set out in small boats in the open ocean without knowing there was going to be any land. This may not have been beyond the daring of our predecessors,[17] and perhaps not really rational on the part of *Homo sapiens.*

Humans as Masters of the World

Does our ability to make dramatic changes to the earth make us superior? Claims to superiority are not very interesting if they are merely *our* claims, from *our* point of view. A more interesting claim to superiority is that our making these changes to the planet are *better* from a more overall point of view than just that of humanity. And it is far from clear that this could be true. Certainly none of the myriad species we have made extinct would agree, if they could. It is not clear that our changes to the environment are better even for us, an issue we will explore later. Actually, there is one version of exceptionalism that seems correct: We are the only animal that generally believes it has the right to manipulate the environment and any living things in it, in any way we wish. But there is no obvious justification for this right.

Could it be that other animals have attitudes of superiority and exceptionalism and just keep them to themselves? This hypothesis would make no sense unless there was evidence of other animals asserting their dominion. And there is no such evidence. First of all, no other animal manipulates the environment to anywhere near the extent of human beings. Insects build hives and nests, mammals build dens, and beavers build dams. Birds build nests. But all of their constructions are quite local. So we have no rivals to our dominion in terms of being able to change the environment. Second, even animals with comparable brain size such as dolphins and whales seem to manipulate the environment only in transient ways, for example, dolphins herding fish. Third, top-of-the food chain predators mostly see humans as prey, and this gives them a certain superiority over an unprotected human. But given that humans can protect themselves and end up having any predator for dinner, any predator's claim to superiority is not convincing.

Perhaps the only animal which makes concerted attacks on humans and human settlements is the elephant. It is well known that elephants in captivity even for some time can suddenly and unpredictably kill their trainers or keepers. But in recent years, the frequency of elephant attacks on people, groups of people, and human settlements has dramatically increased. The ultimate cause seems to be habitat loss, resulting in not enough space for elephants, then resulting in "culling," the practice of killing what humans take to be excess elephants. Elephants are known to have excellent memories (recently demonstrated by neurological examination) and strong family and social ties. They mourn and memorialize their dead, and young elephants stay close to their mothers for the first 8 years of their lives. Almost all the attacks—which destroy villages and kill numbers of people—are by elephants whose families have been brutally killed, either for "culling" or by poachers for ivory. Sometimes a young elephant will be chained to his dead mother until means of moving the older elephant are arranged (Siebert 2006).

Charles Siebert, in his extensive account of the background for these elephant attacks, describes the use of what amounts to therapy to aid these elephants. There is not much discussion in his account of how elephant killings and elephant family disruptions might be reduced. Siebert treats poaching for ivory as regrettable but makes no mention of attempts to stop it. There is also no discussion of the appropriateness of the practice of "culling" elephants. This practice is by no means universally endorsed. Richard Leakey, paleontologist and for some years Director of Kenya's Wildlife Services, has long objected to culling elephants. Because the practice yields valuable ivory, it can easily be equivalent to poaching. Leakey also finds there is no scientific evidence supporting the effectiveness of culling as a means of managing elephant populations (Leakey & Lewin, 1995, Chapter 11).

Elephants are basically no match for human beings, even though their strength and intelligence leads them to try to fight back. In the case of the elephants, we can see a repeat of two ways human superiority expresses itself: First is the propensity of human beings to appropriate any portion of the planet for their own uses. Second is the use of highly developed technology to gain control over almost all other animals and plants.

The propensity to appropriate the planet will be discussed further in Chapter 2, "*Conflicts with the Ecosystem.*" The place of technology will be discussed further in Chapter 3, "*Science, Modern Technology, and Corporations.*"

ETHICS AND ECO-ETHICS

Even if we accept the human superiority principle, constraints on how humans relate to nature are possible. Human beings accept constraints on their behavior with each other called ethical principles. I will first indicate the nature of these ethical principles and then their potential extension to the relation of human beings to nature.

In this book, I deal with *ethics* rather than morality. I understand *ethics* as principles which regulate cooperative benefits. *Morality* includes principles justified by religious[18] or cultural peculiarities which are not shared by everyone.

The basis of ethics as cooperative principles is the realization that rules limiting individual self-interest can often produce greater cooperative benefits.[19] Making and keeping agreements are a major part of ethics so conceived. But ethical principles allowing us cooperative benefits involve more than keeping agreements. The principle of benevolence--to give aid to others in need--holds without any agreement. We simply assume that human beings recognize each other as fellow human beings and give aid because in so doing they expect that they will receive aid when they are in trouble.

By contrast, morality has a large arbitrary element because of its basis in beliefs that are explicitly not shared by all, such as religious beliefs. The principle that one ought to kill one's daughter if she marries an infidel can hardly be based on anticipated cooperative benefits. It is a membership rule for a religious sect. Failure to appreciate the distinction between ethical principles that insure cooperative benefits and moral principles that reflect mainly arbitrary religious or cultural beliefs may be responsible for the attractiveness of relativism, the belief that ethical beliefs are true only for specific groups.[20]

Three plausible candidates for standards for cooperative ethical behavior are:

- **Intuitionism:** There are no overall standards, just a variety of principles we feel are correct by intuition.
- **Utilitarianism:** The right thing to do is what produces the greatest good for the greatest number.
- **Universal principle:** Act on principles that could be willed to be universal law.

Intuitionism is actually not a standard. It says that there is no good explanation of right and wrong, but we nevertheless have strong intuitive feelings about what is right and wrong. For the intuitionist, these feelings need no justification. The Ten Commandments, taken by themselves, are an intuitionist theory. The major difficulty with intuitionism is that when different principles of right action conflict, we have no principled way of resolving the conflict.

Utilitarianism can be stated: Act so as to produce the greatest amount of good for the greatest number. Utilitarianism has much plausibility. For how could it possibly be wrong to do the action that produces the greatest good? How could it possibly be right to do an action which produces less good when you could have done better? Although a plausible idea, utilitarianism suffers from two major difficulties. One is that if we consider actions in isolation from one another, it is easy for a utilitarian to break promises or fail to fulfill contracts when more good would be produced in that case. The trouble is that then institutions which allow cooperative benefits, to live and work together, would disintegrate. Important goods are not available unless we consider ourselves bound to follow certain non-utilitarian rules.

Utilitarianism can, however, achieve these goods if it is modified to apply to rules rather than individual acts. Then one is still bound by social rules governing the institutions of keeping agreements and fulfilling contracts even though more good might be done in the individual case by breaking the social rule. One does actions not because the individual actions produce the greatest amount of good, but because the right action is to follow social rules which produce the greatest amount of good. This theory is called *rule utilitarianism.*

But how do we tell which rules these are? The second major difficulty is that summing goodness over individuals in any precise way has been proved to be impossible. So the notion of the greatest good for the greatest number can only serve as a metaphor. It simply can't be made precise (Arrow 1951)[21].

Universal principle ethics is the major alternative to utilitarianism. Universal principle ethics insist that rightness is not just some sum of goodness. The philosopher Immanuel Kant (1785) developed universal principle ethics, founded on his *Categorical Imperative*: Act on principles that could be willed to be universal law. For example, making an agreement you have no intention of keeping could not be willed to be universal law because then no one would make or accept agreements. The biblical Golden Rule, "do unto others as you would have them do unto you," is a similar but less formal version of The Categorical Imperative.

As with intuitionism, universal principle ethics has little guidance for what to do when principles for right actions conflict. Some account is needed of ethical social rules, especially how they fit together into a system without conflicts. The 20th century philosopher John Rawls expanded a suggestion of Kant's about how to do this into a comprehensive theory of justice (Rawls, 1999a). Rawls' theory has had wide influence and is used extensively by lawyers, jurists, and politicians.

At the level of society, rule utilitarianism is widely used as a theory of justice for society, especially by economists dealing with public policy.[22] Rawls' alternative is a theory that bases principles of justice on a social contract (Rawls 1999a). Utilitarianism allows uneven distributions of value, justifying suffering of the less advantaged by greater overall advantage. By contrast, on a social contract view, the well-being of everyone, including the worst off, is taken into account.

Eco-Ethics

My reason for bringing up (human) ethics in this much detail is to be able to determine what applies and what does not apply when we consider the extension of ethics to humanity's relations to other living beings and to the environment. Ethical principles of cooperation are justified by universal principle considerations of reciprocity (for individuals) or by social contracts (for societies or other institutions). But it does not seem sensible to base principles for relations with animals or with nature on considerations of reciprocity or social contracts. I will therefore call the different principles for these relations, *eco-ethical principles*.

So, what is the nature of eco-ethics? Possible ethical principles for human relation to the environment seem to assume human superiority precisely because animals and nature cannot respond to us as human beings on an equal footing. For example, the view of many people is that our relations with animals and the ecosystem are to be determined by human benefit. Impacts on the ecosystem are considered solely in terms of benefits to *us,* to humans. An unthinking version gives no weight at all to the well-being of nonhuman nature. If wolves are depredating your livestock, kill the wolves and poison their cubs. If elephants destroy your crops, cull the elephants. A more thoughtful version would consider the impact of nonhuman nature on human beings. Thus removing predators from an area could result in an explosion of deer and consequent destructions of trees and plants, both wild and domestic. Cutting down forests to plant crops could have the effect of damaging the fertility of the land. Concerns such as these recognize that living things are organized interdependently into what are called *ecosystems*. And those living things become what they are through complex interaction with the rest of the ecosystem. When these ecosystems are damaged, there are negative consequences for human beings. Human beings are, after all part of these ecosystems. When we look at the environment for the basis of a higher-level principle, we see that the environment does not consist of objects provided for whatever use we want to make of them.[23] Thus, human survival requires the well-being of the ecosystem. The correct principle for eco-ethics is:

- **Ecosystem Principle:** Actions and policies necessary for the long-term survival of humanity take priority over other human actions and policies. Further, the survival of the ecosystem of which human beings are a part has higher priority over actions and policies for human survival.

Thus, this principle plus the various principles of justice provide a basis for the ethics of the environment. A brief statement of the priorities they encapsulate is this: Without ecosystem, no human social cooperation; without human social cooperation, no individual interests. The major levels of cooperation are global and national.

In my previous writings on eco-ethics, this was as far as I got.[24] In those writings the preservation of the ecosystem was tied to human welfare. And we can't have human welfare without the ecosystem. At least this formulation does consider actions involving animals and the environment in a broader context. We do not exterminate wolves just because they kill our sheep, but rather look to their place in the ecosystem and consider the damage done by extra deer if the deer population is not kept in check by enough wolves.

But even this improved formulation for eco-ethics still assumes some version of human superiority. This superiority has both negative and positive aspects. Negatively, we speak of *intervening* in nature when, say, we cull an elephant herd. Yet we do not speak of an elephant intervening in nature when it kills its trainer. We assume we can care for the environment in ways the environment cannot care for itself. Even this positive aspect needs to be questioned. When environmentalists demand that coyotes rather than human hunters reduce excessive deer population, they thereby assume that human beings are not part of nature![25]

But then what is the appropriate point of view for human beings to use as the basis of sound principles for treating animals and the rest of the world? We need to think through what makes us human beings "too distantly a part" of the world we live in. Of course it is better for us to take charge of the ecosystem than to trash it. But taking charge still puts us in a false position. Our true position is that we are in fact a part of the ecosystem along with the rest of nature. Unless we can assume this position and think in terms of it, we may stand an excellent chance of extinction. Man's ability to manipulate the environment goes way beyond survival, to the point of threatening the entire current ecosystem. We will see in the next chapter whether humanity threatens the entire ecosystem. If humanity does threaten the entire ecosystem, this means that the point of view of man as superior is *actually incorrect.*[26] The rest of this book is an attempt to discover a correct alternative view.

REFERENCES

Allan, D. (2011). *Population growth over human history*. Retrieved 8/31/11 from www.globalchange.umich.edu/globalchange2/current/lectures/human_pop/human_pop.html

Arrow, K. (1951). *Social choice and individual values*. New York: John Wiley & Sons.

Associated Press. (2010). *Coyotes, not hunters, proposed for Valley Forge*. Retrieved 10/20/10 from http://www.bornfreeusa.org/news.php?p=2516&more=1

Bradshaw Foundation. (2011). *Cave art of France*. Retrieved 7/24/2011 from www.bradshawfoundation.com/ france/index.php

Cassidy, D. (1998). Quantum mechanics 1925-1927: The uncertainty principle. *The American Institute of Physics*. Retrieved 9/02/11 from www.aip.org/history/heisenberg/p08.htm

Clottes, J. (2002). Chauvet cave (ca. 30,000B.C.). In *Heilbrunn Timeline of Art History*. New York: The Metropolitan Museum of Art. Retrieved 7/24/2011 from www.metmuseum.org/toah/hd/chav/hd_chav.htm

Dalai Lama. (2005). *The universe in a single atom: The convergence of science and spirituality*. New York: Morgan Road Books.

Einstein, A. (1956). *The meaning of relativity* (5th ed.). New York: Barnes & Noble.

Flannery, T. (2010). *Here on earth: A new beginning*. London: Allen Lane.

Gauthier, D. (1986). *Morals by agreement*. Oxford, UK: Clarendon Press.

Hiebert, T. (1996). *Rethinking dominion theology*. Winnipeg, Canada: Direction.

Huxley, J. (1953). *Evolution in action*. New York: Harper & Brothers.

Kant, I. (1781). *Critique of pure reason* (Smith, N. K., Trans.). London: Macmillan.

Kant, I. (1785). *Groundwork of the metaphysics of morals*. Retrieved 3/24/12 from www.earlymoderntexts.com/kgw.html

Leeming, D. A., & Leeming, M. A. (2009). *A dictionary of creation myths*. Oxford, UK: Oxford University Press.

Lynas, M. (2011). *The god species*. Washington, DC: National Geographic Society.

McKibben, B. (1989). *The end of nature*. New York: Random House.

McKibben, B. (2010). *EaArth*. New York: St. Martins Press.

Mill, J. S. (1863). *Utilitarianism*. Retrieved on 5/10/04 from //etext.library.adelaide.edu.au/m/mill/john_stuart/m645u/

Quine, W. V. (1982). *Methods of logic* (4th ed.). Cambridge, MA: Harvard University Press.

Rawls, J. (1999). *A theory of justice* (revised ed). Cambridge, MA: Harvard University Press.

Schultz, R. (2006). *Contemporary issues in ethics and information technology*. Hershey, PA: IRM Press.

Schultz, R. (2009). *Information technology and the ethics of globalization*. Hershey, PA: IGI Global Press. doi:10.4018/978-1-60566-922-9.

Siebert, C. (2006, October 8). An elephant crackup? *New York Times Magazine*. Retrieved 9/01/11 from www.nytimes.com/2006/10/08/magazine/08elephant.html?pagewanted=1

Singer, P. (2002). *Animal liberation*. New York: HarperCollins.

Swenson, K. (2010). *The bible and human 'dominion' over animals: Superiority or responsibility?* Retrieved 9/15/11 from www.huffingtonpost.com/kristin-m-swenson-phd/the-bible-and-human-domin_b_681363.html

Tolkein, J. R. R. (1954). *The lord of the rings*. London: Allen & Unwin.

Wilford, J. N. (2009, May 7). Homo floresiensis. *New York Times*. Retrieved 7/29/2011 from topics.nytimes.com/topics/news/science/topics/archaeology_and_anthropology/ homo_floresiensis/index.

Wilson, E. O. (2002). *The future of life*. New York: Vintage Books.

Winton, R. (2011, April 6). Giants fan had feared for safety. *Los Angeles Times*. Retrieved 10/14/2011 from articles.latimes.com/2011/apr/06/local/la-me-dodgers-beating-20110406

ENDNOTES

1 Most current popular methods of referring to the past seem to me to cause unnecessary effort in deciphering them. So I will use the completely transparent phrase 'years ago' instead of A.D., B.C., C.E., B.C.E., B.P., or ka.

2 The Chauvet cave paintings, so far the oldest discovered, date from about 30,000 years ago. See Clottes (2002). See also the Bradshaw Foundation's collection of cave paintings around the world.

3 Tim Flannery (2010), Chapter 7, "New Worlds."

4 Tolkein (1954).

5 There is still some dispute about whether the small skeletons found on the island of Flores are a different species. See Wilford (2009) "Homo Floresiensis."

6 Associated Press (2010). "Coyotes, not hunters, proposed for Valley Forge."

7 There are recent denials of human superiority such as those by Peter Singer (2002) and animal rights activists. I believe their positions ultimately have similar problems. The issues are complex and will be dealt with in Chapter 13, "Regaining Our Place as a Species among Species."

8 McKibben's more recent book, *EaArth* (2010), continues his theme of separation. It also seems to espouse a doctrine, not of human superiority, but of *Vermont* human superiority. My criticisms are of McKibben's theoretical positions. I applaud his practical work in ameliorating climate change through his organization 350.org.

9 Leeming (2009) has a comprehensive compilation of origin myths from almost all cultures. None but Judeo-Christian origin myths say that humans are in charge of the rest of the earth.

10 Some religions, notably Buddhism, are not theistic. That is, they hold that there was not a (benevolent) creator god. This doctrine was due to the Buddha himself.

11 Cheerleading is a manifestation of what the Greeks called *thumos* (θυμος). *Thumos* is translated as "spirit," not as in ghost but as in "that's the spirit" or as in "a spirited horse" or as in "school spirit."

12 In a widely publicized case, a San Francisco Giants fan was beaten into a coma by two Los Angeles Dodgers fans in Dodger Stadium (Winton 2011).

13 This formulation is due to Dr. William Goldsmith.

14 In the original German, *intellektuelle Anschauung*

15 This is *not* the view that human beings start with unorganized sensations. The place of human beings in the world will be discussed in more fully in Chapter 4, "Mind Body Consciousness."

16 Predators hunting in packs whose behavior must be consciously tailored to the behavior of others may be the best example: Lions, wild dogs, wolves, even dolphins herding schools of fish. Defensive behavior such as the schooling of fish may not require awareness of the individual animal's place in the cooperative activity. Similarly with social insects such as ants and bees.

17 Some paleontologists belief *Homo erectus* built rafts and crossed ocean to populate what are now Indonesian islands.

18 Most religions say their special beliefs require faith. However, since there are a number of seriously conflicting faith-based beliefs, the chances that any are correct is very small.

19 The payoff matrix called the *Prisoners Dilemma* applies in most situations in which there is an ethical principle providing cooperative benefits, and the choice is to observe that principle or not to observe and act on self-interest instead (see Appendix A). What the payoff matrix reflects is that one can always do better from a selfish or self-interested point of view if everyone else obeys the (cooperative ethical) principle but you do not. If everyone acts selfishly (that is, disobeys), the cooperative principle with its cooperative benefits is no longer available—we are at alternative (3,3), which means everyone is *collectively* worse off than if everyone obeyed (2,2). Therefore, the only way we can have ethical principles is if we treat principles which are cooperatively rational (produce 2,2 as opposed to 3,3) as of higher priority than considerations of self-interest (Schultz, 2010, Chapter 4, "The Basis of Ethical Principles," "The Rational Basis of Ethics."). Recent authors using game theory and the prisoner's dilemma as foundations for ethics include David Gauthier (1986).

20 My distinction between ethics and morals is based on a distinction drawn by the philosopher John Rawls in developing his theory of justice. Rawls distinguishes principles of justice regulating cooperative behavior and comprehensive doctrines which are not allowed to affect the (cooperative) social contract (Rawls 1999). For further discussion, see Schultz (2010), Chapter 4, "The Basis of Ethical Principles."

21 Kenneth Arrow won the Nobel Prize in 1972 by proving in his "general possibility theorem" that a consistent and very minimally just amalgamation of individual preferences is impossible. Such an amalgamation is called a "social choice." Utililitarianism as a usable theory would need to make such impossible social choices. Arrow's proof uses fairly abstract mathematics (theory of partial

orderings) and is not accessible to non-mathematicians. For a brief (but still technical) account, see encyclopedia.thefreedictionary.com/Arrow's+theorem.

22 Mill (1868) is the classic reference.

23 This observation is the basis for a correct ethical placement of technology, discussed in Chapter 3, "Science, Modern Technology, and Corporations."

24 Schultz (2006), Chapters 12 and 13, and Schultz (2010), Chapters 14 and 15.

25 Associated Press (2010). "Coyotes, not hunters, proposed for Valley Forge."

26 Mark Lynas, in his *The God Species* (2011, Introduction) cogently describes the damage to the ecosystem wrought by human beings. He also describes an experiment, which shows that natural selection can lead to the destruction of an ecosystem when one species takes over. Then he inexplicably concludes that because human powers are the result of natural selection, that humans are exempt from these bad results of natural selection.

APPENDIX

Table 1. Cooperative Benefits Payoff Matrix

	Person A Obeys Principle	Person A Disobeys (Acts Selfishly)
Person B obeys principle	(2,2)	(4,1)
Person B disobeys (acts selfishly)	(1,4)	(3,3)

Chapter 2
Conflicts with the Ecosystem

ABSTRACT

Although human beings are part of nature, they are seriously damaging the ecosystem, our support system. Modern technological manufacturing regularly produces harmful side effects, either toxic substances or unexpected threats to the entire ecosystem such as ozone layer destruction or global warming. Human economic activity is leading to the mass extinction of other species. Humanity is threatening the entire ecosystem. This means that the view that humans are superior to nature is actually incorrect.

INTRODUCTION

If humans are ultimately are part of nature, animals among animals, how can we be in conflict with nature? Everything we do is part of nature, in the final analysis. But in this final analysis, we would not be in conflict with nature if we turned the entire planet into a lifeless desert. "Nature" in this sense is simply anything that happens. The lifeless desert outcome is now unfortunately a distinct possibility. That would be a conflict with nature, when "nature' is understood in a more relevant way. The part of nature which is relevant to us as human beings is the *ecosystem.* The ecosystem is the support system for all living things on the planet. It is a complex web of interdependent living things. Our air, water, and food are supplied by the ecosystem. Oxygen is not a static component of the air. Plants are constantly

DOI: 10.4018/978-1-4666-4586-8.ch002

generating it, and without it, we could not survive. We also need water to drink, preferably clean water if we do not want to die prematurely of disease. Components of the ecosystem recycle and purify water. Humans eat both plants and animals as food. Both require support from the ecosystem to be available. Plants need water, nutrients from the soil, carbon dioxide, and sunlight in order to grow by means of photosynthesis.[1] Animals need to eat plants or other animals in order to survive.[2]

Human beings have conflicts with the ecosystem because they are powerful enough to cause serious damage to the ecosystem. Often, when ecosystem damage causes harm to human beings, there is a consensus that the damage needs to be eliminated. Two examples would be sewage treatment plants and various steps taken to reduce air pollution. But sometimes, the pursuit of purely human ends such as profit takes priority and the damage continues to accumulate. Industrial pollution is an example. When ecosystem damage only indirectly harms humans, it is harder to get a consensus on action. Carcinogenic substances are examples. It is even harder to get a consensus when ecosystem damage is primarily to nonhuman parts of the ecosystem. Species extinction and endangerment are examples. We also need to consider destruction of the nonliving environment. Often the harm caused by such destruction, as for example in open pit mining, is immediate or indirect harm to humans or destruction of animals, plants, or their habitat. But in addition irreplaceable landscape can be lost. Mountaintop removal mining in West Virginia obliterates the original mountain landscape and replaces it with a barren moonscape. (Mountain Justice 2012) We need to consider the eco-ethical basis for concern about destruction of landscape as well as the harm caused to living things.[3]

Each of these types of harm will be considered in turn. I want to make clear what makes them conflicts with the ecosystem. In particular, what is the role played by assumptions about human superiority and human exceptionalism in generating and maintaining these conflicts. As we saw in Chapter 1, there is no obvious justification for assuming human superiority as an ethical principle or principle of action.

The types of harm caused by conflicts with the ecosystem are:

- Direct harm to humans
- Indirect harm to humans
- Harm to other species
- Destruction of nonliving environment

These categories can overlap. Toxic substances which cause direct harm to humans usually also cause harm to other animals. Indeed, animal testing is one way to determine potential harm to humans. And the destruction of nonliving environment will also often be habitat loss and consequent harm to nonhuman species. When there is harm in multiple categories, the harm in both categories will be considered. Thus when there is harm to humans and other species, both types of harm will be addressed.

DIRECT HARM TO HUMANS

Pollution is the first direct harm considered. The Biology Online Dictionary gives a usable definition of pollution:

The change in the environment caused by natural or artificial input of harmful contaminants into the environment, and may cause instability,[4] disruption or harmful effects to the ecosystem.

The different forms of pollution are as follows: Air pollution, water pollution, soil contamination, radioactive contamination, noise pollution, light pollution, thermal pollution (Biology Online 2009).

The core of this definition is this: Pollution is additions to the ecosystem that cause harm. The definition does not state that it applies only to human beings, but this is usually our practice. We normally only think of humans as polluters. Algae produced a surge of domoic acid in Monterey Bay, which resulted in a massive wave of sea mammal deaths starting in the late 1990s. These algae were just being algae, not polluters (Northwest Fisheries Science Center 2009). Also, we sometimes do not consider nonhuman animals or plants to be victims of pollution. Although they can be harmed by additions to the environment, the cause (some harmful substance) is sometimes not considered to be pollution if it is not the result of human activity. But if insects had a voice in the matter, they would consider insecticides to be pollution. To be an insecticide, a substance has to be very harmful to insects. But insects don't have a voice, so insecticides would not be considered to be pollution because of their harm to insects. Obviously, they are considered to be pollution because of their harm to humans.[5]

The types of pollution mentioned in the definition fall under the four major states of matter: Air, water, earth and energy (which is fire generalized). Soil pollution is earth pollution. Noise pollution is a form of air pollution. And radioactive contamination, light pollution, and thermal pollution are forms of energy pollution.

It is ultimately more important to identify harm to the ecosystem than to determine whether something is strictly speaking pollution. So my next topic is human changes to the ecosystem which create direct harm to humans.

Do Wild Bears Pollute in the Woods?

When other animals produce waste, we usually do not consider it pollution. Of course, animal waste can indeed be pollution if it is part of a human project such as factory farming of animals. But can animals or living organisms or nature itself pollute without human interference? Volcanoes are not air polluters, even though volcanoes produce harmful toxic gases. And when wild animal waste causes pollution, it is almost always the result of human-caused habitat change. David Fahrenthold reported that scientists in Virginia and Maryland recently found that wild animal dung was causing serious water pollution problems. But the cause is "unnaturally high populations of deer, geese and raccoons living in modern suburbs and depositing their waste there." And the cause of "unnaturally high" populations of what are in fact prey animals are the absence of predators, which humans are responsible for. Scientists determined it would be neither feasible nor popular to kill or relocate enough animals to significantly reduce this pollution. Fahrenthold remarks that "It could be the ultimate irony of people's impact on nature that the entire system has changed so radically that wild animals now degrade their own environment" (Fahrenthold, 2006).

Human waste is often pollution. 'Waste' is output from an animal, plant, or human process that has no further use. Animal excrement often is used as fertilizer for plants. So used, it is not waste. Human waste is used as fertilizer in China, and is called "night soil." So used, it is neither waste nor pollution.

Now, why, if humans are animals among animals, do their individual and social by-products require special treatment? It was not so long ago in European cities—until the 1800s—that chamber pots containing human waste were simply dumped into the street. The city of Paris still flushes its streets every night, but no longer to remove excrement. The reason for the necessity of treatment is that people are more concentrated.[6] With enough space, we could behave as bears or elephants, and our waste would be recycled back into the environment with no extra effort.[7] The presence of human waste becomes pollution when it occurs in large enough quantities to be a threat to the normal functioning of the ecosystem of the area. Indeed, when wild animals cause pollution, as just discussed in Virginia and Maryland, it is for the same reason. And ultimately humans are responsible.

So, is it only the greater density of settled and civilized human beings that causes us to be polluters? This might be true of we considered only waste produced by individual human beings by bodily processes.[8,9] But human beings also pollute in other ways.

Toxic Substances

Human-made toxic substances create an important kind of conflict with the ecosystem. 'Toxic' means 'harmful to humans or other living creatures.' Even though non-human activity can produce a damaging amount of toxic substances in one area—as in the case of the algae producing domoic acid—we do not usually consider it pollution unless produced by humans or by living things under the control of humans. But we do occasionally consider it pollution when non-human animals and plants are affected by human pollution, even when humans are not directly affected.

Modern human manufacturing processes often have toxic by-products. Why this should be so is discussed in the next chapter. The "Safe Chemicals Act of 2010" of Sen. Frank Lautenberg attempts to deal with this problem. The Act would require safety testing of all industrial chemicals, and puts the burden on industry to prove that chemicals are safe in order stay on the market. Under current policy, the EPA can only call for safety testing after evidence surfaces demonstrating a chemical is dangerous. As a result, EPA has been able to require testing for just 200 of the more than 80,000 chemicals currently registered in the United States and has been able to ban only five dangerous substances (Lautenberg Press Office, 2010).

The Safe Chemicals Act is pending legislation as of late 2011 and attempts to get a status update from Sen. Lautenberg's office were not successful. It is still the case that few new chemicals are tested for safety.

The case of Polychlorinated Biphenyls (PCBs) illustrates the need for such testing. PCBs were first commercially produced in 1927 and ultimately banned in 1979. PCBs are human-made substances which can be either liquid or a waxy solid. They were used in hundreds of industrial and electrical applications because of their non-flammability and insulating properties.

Within four years after production began, virtually all workers involved with PCB production were exhibiting symptoms of PCB exposure, including a particular variety of severe acne called chloracne, which is caused by exposure to toxic chlorine compounds. Other symptoms were loss of energy, appetite, and libido. An examining physician diagnosed systemic poisoning. (Katers, 2009) By 1937, a number of companies using PCBs became concerned about the effects, including horrible skin rashes on their workers and even their customers. General Electric decided to continue using them anyway because otherwise they would have to give up a huge part of their business. At the time GE regarded its decision as unethical but necessary for business. Monsanto, which by then had become the manufacturer of PCBs, claimed there was no proof that their product was causing any problems and continued this claim for years in the face of several cases of multiple fatalities[10] (Katers 2009). For example, in Japan in 1968 several hundred were sickened by vegetable oil contaminated with PCBs. 50 eventually died. In 1970, Monsanto oversaw tests, which found PCBs to be safe. These test were conducted by a lab (Industrial Bio-Test Labs) later found to be criminally fraudulent in its environmental safety testing. However, the fraudulent report resulted in PCBs continuing to be deposited in the environment until 1976, when they were specifically prohibited by the Toxic Substances Control Act (Katers, 2009).

The toxic effects of PCBs are compounded by the fact that these chemicals are not water-soluble but instead fat-soluble. As a result, they persist in the environment for long periods of time. Also, their concentrations increase in animals higher up in the food chain such as eagles and humans.[11] Serious concerns were raised by researchers in the mid-60s who found that high concentrations of PCBs were found in top-of-the food chain animals everywhere on the planet (Katers 2009).

There are several ways of dealing with toxic substances. One way or another, they must be removed from the environment: They can be reprocessed to reduce their toxicity, either chemically or by entombment in cement; or they can be destroyed by subjecting them to high temperatures, either by incineration or by pyrolysis (Pyrolysis is destruction by high voltage electric arc, used when regular incineration would produce significant air pollution, as with PCBs.). Finally, toxic substances can be sequestered from the rest of the environment if they cannot be made nontoxic. This tactic does not work well for radioactive waste, which remains toxic and harmful for long periods of time—more than 30,000 years for waste containing plutonium. The longevity of radioactive waste reveals the problem with the sequestration of other toxic wastes: Sequestration is not a permanent solution, and we must hope that the toxic substances degrade over time.

In the U.S., what is called the Superfund was created in 1980 to deal with toxic wastes. The superfund is administered by the EPA (Environmental Protection Agency). The federal law creating it was the Comprehensive Environmental Response, Compensation, and Liability Act of 1980 (CERCLA). The Superfund has had limited success, cleaning up about 320 out of about 1600 hazardous waste sites (Sapien, 2007). The Superfund portion of the EPA Website reveals numerous bureaucratic outgrowths having little to do with its main focus. The site has no information about how it is determined that waste is hazardous.

Government inefficiency is an important stumbling block in dealing with damaging conflicts with the environment. But more important is corporate resistance even to admitting there is a conflict. Monsanto's behavior in the case of PCBs is typical of a corporation whose profits are threatened by negative results from its products. A proper diagnosis of this behavior is critical. It is important not to blame capitalism as the sole cause, because similar behavior has occurred under socialism.[12] Many man-made chemicals are incompatible with the environment. They need either to be sequestered from the ecosystem or to be detoxified. Especially since the advent of modern technology, human processes which replace naturally occurring products tend to produce waste which is often toxic. In the next chapter, we will see why modern technology tends to have these effects. In addition, it is important to come to understand how managers can come to discount the impact of their decisions on the world we and they both live in.

INDIRECT HARM TO HUMANS

We have seen how toxic substances cause harm. But other man-made substances which are not toxic can also cause harm. Perhaps from the point of view of the ecosystem, the worst of these products are non-toxic plastics, derived from fossil fuels such as petroleum. Because most such products are non-biodegradable, they do not participate in natural recycling. Plastics disintegrate into smaller pieces, right down to the level of molecules. These smaller pieces are non-biodegradable, they remain in the environment. Natural components recycled into the ecosystem instead become mostly water and oxygen. The area known as the Great Pacific Garbage Patch in the northern Pacific Ocean has the highest levels of decomposed plastic on the planet. Before it disintegrates, plastic kills as many as 100,000 marine animals who mistake it for food. When decomposed, plastic still enters the food chain and accumulates in animals higher in the food chain (Amaral, 2003). A similar patch has appeared in the North Atlantic.

On land, plastics—single-use bags and packaging—are a major component of landfills (Berkeley Plastics Task Force, 2009). The bags will eventually break down—in 1000 years. Even then, the molecular components will persist indefinitely. "Every year, more than 500 billion plastic bags are distributed, and less than 3% of those bags are recycled." Even when recycled, they become lower grade plastic not reusable as bags. Hence recycling does nothing to reduce the production of plastic bags[13] (Ecology Center, 2011).

Petroleum-based fertilizers are in some ways similar to plastics. They are not especially toxic, and are beneficial when used for their intended purpose of enabling plant growth. Indeed, these fertilizers are credited with dramatically increasing agricultural yields. However, they invariably are washed into nearby streams and rivers. There they make those rivers and streams unusable for drinking and swimming.[14] They are a major cause of lake eutrophication[15] and they eventually create dead zones in the ocean at the mouths of rivers. Because of the increased yields, farmers and especially agribusinesses treat the fertilizer pollution as a regrettable side effect.

When synthetic laundry detergents replaced soap in the 1940s and 1950s, an increase in water problems began to emerge (Knud-Hansen 1993). Phosphates added to laundry detergent dramatically improved the cleaning efficacy of detergents. But this ingredient, phosphate, causes severe eutrophication.[15] It has been limited or banned from laundry detergents in many states in the US, Canada, the EU, and Japan. Substitute ingredients have been found which have similar cleaning ability but which do not cause eutrophication. However, phosphates are still present in dishwasher detergents because an effective substitute has not been found (MacIntosh, 2007).

Genetic engineering is in some ways a special case of indirect harm, but is similar in corporate attitudes toward its possible harm. Genetic engineering (genetic manipulation) has unforeseen consequences which do not play out within individual life spans. So how can "genetic engineering" be safe? The long-range effects of such genetic manipulation as producing a more frost-resistant strawberry cannot be predicted. In nature, genetic changes are honed over time against the existing environment by natural selection.[16] By making changes out of context, we are asking for a disaster as extensive as the disappearance of the ozone layer.

Current uses of genetic engineering show that this technology does not fully understand the science at its basis. Genetic engineering does not take account of its own relation to the ecosystem. We already have cases in which genetically altered individuals cannot be safely released into the environment. Salmon engineered for increased size are also sterile. Wild female salmon prefer to mate with the larger engineered salmon and thus will produce no offspring. If engineered salmon are introduced into the wild, salmon extinction would be highly probable. Genetically engineered changes are thus being judged only in the context of current benefit in terms of limited human aims. From the point of view of natural selection in the ecosystem, human genetic manipulation will impact gene pools in the same destructive way as strong bursts of radiation.

The story is the same as with many of the ultimately harmful chemicals developed by technology: A technological application with superior properties turns out to have harmful side effects on the wider environment, usually not predictable until the harms have begun to be realized. When genetic engineering repeats this story, any bad consequences will be infinitely worse. The bad consequences will turn up, as in the example of the salmon, in changes to the genetic material, which make the changed organisms less able to survive through natural selection. And, because of the very real possibility of interbreeding with non-engineered stock, the parameters for natural selection will be reset in a way that has very little to do with what natural selection would have done. All it will take is for engineered major food crops to be selected out with no ready way to reverse the changes.

Possibly even worse, some genetically engineered changes are essentially frivolous. Monsanto genetically engineered flaxseed to be resistant to its proprietary insecticide Roundup. The genetically engineered seeds crossbred by natural means with regular flaxseed in adjacent farms. Monsanto then sued a number of farmers for patent infringement and won at least some of these suits. A number of other types of seeds have been genetically engineered in the same way. Monsanto is tampering with the very basis of food production with no way to estimate possible harmful consequences. In this case, Monsanto's profits are foremost, and any possible harm to the ecosystem and its genetic basis are completely irrelevant for them.

This persistent feature of modern technological production, namely the adoption of new technologies without adequate assessment of ecological side effects, is a critical flaw. Its causes need to be thoroughly understood. At worst, if persisted in it could mean the destruction of all life. Understanding this flaw is a major concern of the next chapter and Chapter 8, "More about Modern Technology."

GLOBAL HARM TO THE ECOSYSTEM

It is usually easy to recognize the harm from local conflicts with the ecosystem. But when there is no local harm, it can be difficult to recognize the indirect harm when the conflict with the ecosystem is global. Two examples: The destruction of the ozone layer by chlorofluorocarbons; and global warming.

A "safe" synthetic compound, Chlorofluorocarbon (CFC), developed in the 1940s for use in air conditioners and aerosol cans turned out to be inert except in the upper atmosphere, where it destroys the ozone layer which protects us from ultraviolet radiation. The ozone layer protects all life against excessive ultraviolet radiation, and when it is damaged or destroyed, the surface of the planet on which we and all animals and plants live, is subjected to dangerous levels of ultraviolet. There are two disturbing implications: (1) there does not seem to be any way that this result of normal chemical engineering could have been predicted. (2) All life has evolved under the protection of the ozone layer; this sudden a change has unpredictable consequences for all life forms in the ecosystem (Asimov & Pohl, 1991, pp. 91-110). One reassuring development: All nations have recognized the threat and accepted the Montreal Protocol. This protocol bans the use of chlorofluorocarbons, and the ozone layer is expected to recover fully by about 2050.

However, in 2012 an entirely different threat to the ozone layer emerged: More severe thunderstorms caused by global warming were found to be depositing ozone-destroying material in the ozone layer (Fountain, 2012).

Paradoxically, sometimes local pollution can be a necessity for life at the global level. Locally ozone is an air pollutant in cities caused by vehicle emissions. But the presence of ozone in the upper atmosphere is a necessary shield against excessive ultraviolet radiation.

Likewise, global warming is an overall effect caused by the increase of levels of so-called "greenhouse gases," of which carbon dioxide is the most important. They are called "greenhouse" gases because they increase the ability of the earth's atmosphere to retain heat, just like the panes of glass in a greenhouse. Yet carbon dioxide at the local level is nontoxic and the levels which produce global warming have no impact on any kind of life.[17] The increased levels of greenhouse gases are the by-product of industrial technology. Carbon dioxide is the most prevalent gas, with much of it coming from the burning of fossil fuels in the internal combustion engines of cars, buses, and trucks. The size of temperature increase produced by the increased carbon dioxide is difficult to predict, but the size of the increase would normally happen over tens of thousands of years. As of 2011, the average air temperature has increased 3 degrees Fahrenheit since 1970. One current consequence is storms of increasing severity. Warmer atmosphere holds more water vapor, which supplies more energy for storms and rainfall. Hundred-year storms and floods have become common (Hansen, 2009, p. 253).

Severe strains on plants and animals are also happening. Polar bears are famously losing essential parts of their environment, that is, frozen ocean. Some animals can move to more congenial colder places when the atmosphere warms, but trees cannot move with anywhere near the speed necessary to cope with global warming.[18] And some animals and plants on mountains run out of space—they are at the top of the mountain with no further higher cooler places to move to (Wilson, 2002, p. 69).

Human-caused global warming, because it directly affects human economics, has aroused contrary argumentation based on the premise that human economic activity cannot possibly affect the entire environment. This is probably wishful thinking. Science tells us that our atmosphere's composition is not a given; the amount of carbon actually in the atmosphere is about the same as that present in the totality of living beings (the "biomass"). Consequently, adding the carbon locked up in fossils in large quantities is likely to have an impact. In Figure 1, the black arrows indicate carbon exchanges exclusive of human beings. The red arrows are carbon added by human activity. As you will notice in Figure 1, the amount of carbon dioxide in the atmosphere (597 billion tons) is one quarter of the amount contained in soil and vegetation (2300 billion tons). But soil contains about 1600 billion tons, leaving 610 billion tons contained in vegetation, about the same as that in the atmosphere. Consequently, human activity adding 6.4 billion tons per year will make an impact. In 20 years, that would be an extra 128 billion tons of carbon dioxide.

The issue of human-caused climate change has caused corporations whose (short-term) interests are affected to distribute not only false information about a lack of scientific consensus but also false scientific information.[19] This has also occurred in the case of other conflicts with the ecosystem. I call it "knowledge pollution" and will discuss it further in the following paragraph. There is consensus that there is a relatively short window to reduce carbon emissions before drastic effects occur, so this knowledge pollution could be damaging. Recent credible projections of the result of lack of rapid drastic action is an average temperature increase of about 10°F by 2050. This change alone will be incredibly disruptive to

Figure 1. Carbon cycle (From IPCC AR4 WG1)

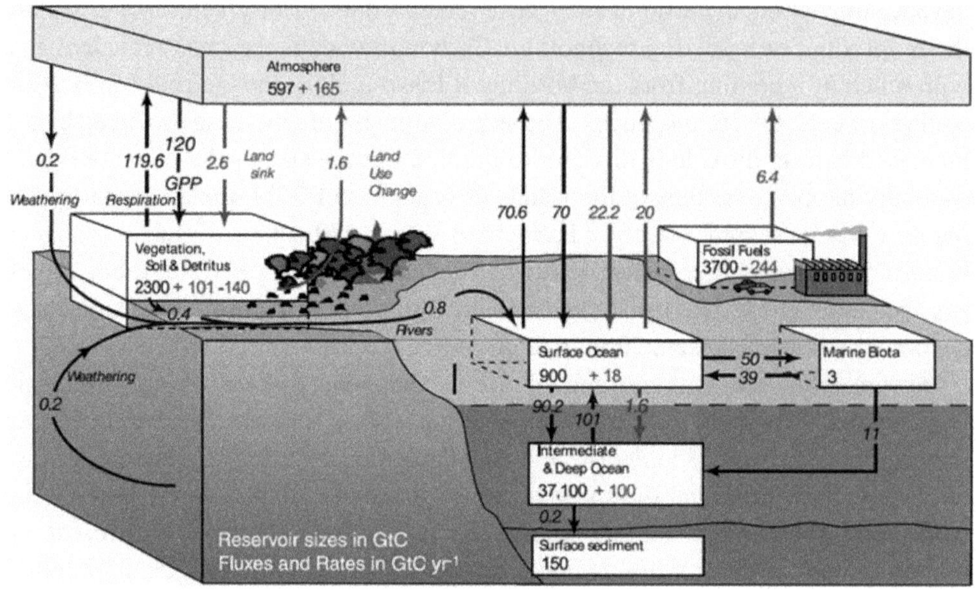

all life, but will also cause great weather and climate change. For comparison purposes, a 10 degree (Fahrenheit) decrease was enough to cause an ice layer 4000 feet thick over Wisconsin. (Co2gether, 2012) There is also a real but smaller possibility of what is called "runaway greenhouse," in which the earth's temperature becomes like Venus' surface temperature of 800° F.

A distinct but important form of indirect harm is "knowledge pollution." It is almost always practiced by corporations or other large institutions because smaller institutions don't have the resources or media access to pollute knowledge successfully. In the US, the two largest corporations Exxon Mobil and Koch Industries—both fossil fuel companies—have spent on the order of $100 million to fund climate change denial organizations since 1997. In addition, Koch Industries spends just under $40 million for lobbyists in congress (Baxter 2011). *Time* reporter Bryan Walsh found a concerted effort by large conservative corporations and other organizations to discredit what is in fact a scientific consensus. He observed that the tactics were similar to those used by tobacco companies against claims that smoking causes lung cancer (Walsh 2011).

An example of these tactics: A coworker told me of a poll purporting to show that scientists overwhelmingly rejected the idea of human-based climate change. It was conducted by an organization called the Oregon Institute of Science and Medicine, which upon examination turned out to be a mailing address and Internet site at a rural Oregon farmhouse. Its sole function seemed to be to conduct this poll. Their Website indicated anyone with any graduate degree could participate. Participation was online with no attempt to check credentials, and thus no attempt was made to verify whether anyone's credentials were relevant. A Web search will also yield a bewildering variety of histories of carbon dioxide levels. More than a few feature diagrams show the actual historical correlation between rising carbon dioxide levels and rising temperatures, only to show an opposite recent correlation between rising carbon dioxide levels and falling temperatures. Again, we observe the rejection, not of all science, but rather of science not in the apparent economic interest of these corporations.

Knowledge pollution is also practiced with regard to food. A recent TV commercial sponsored by the Corn Refiners Association (www.sweetsurprise.com) claims that high-fructose corn syrup is nutritionally the same as sugar. However, responsible studies at the University of California San Francisco medical school have shown that the effects on the body of high-fructose corn syrup are almost identical to the well-known negative effects of alcohol (ethanol), such as damage to such organs as the liver, heart, brain, and lungs. Except high-fructose corn syrup does not make you drunk (Norris, 2009).

Knowledge pollution is also practiced in nondemocratic regimes. In such regimes—China, for example—the state has the power to enforce false beliefs. When knowledge pollution is practiced, it is not so much that state or corporations actually know that what they are imposing is false. Rather, they have a *pragmatic* view of truth: Truth is what works, that is, truth is what produces the desired results. Corporate and state knowledge pollution is a decisive objection to this pragmatic theory of truth. [20]

HARM TO OTHER SPECIES

Ecological harm to other species is often a side effect of products or procedures intended to benefit humans. Insecticides are an interesting case. They are usually considered toxic to bugs and not to people, but it is not easy to see why. The insecticide industry got its start after World War II by repurposing products developed by the Nazis for use as chemical weapons against humans. In other words, instead of these products killing humans, they were used to kill insects. Insecticides such as DDT (an organochlorine) were derived from nerve gases. Insecticides such as malathion (an organophosphate) are related to the banned chemical weapon Sarin (Flannery 2010).

Belief in human exceptionalism and separation from the rest of the environment plays a part in attitudes toward insecticides. A product originally designed to kill humans is bound to have toxic effects on humans, animals, and plants. So it was actually wishful thinking that harm would come only to insects but not to humans. The organochlorines such as DDT are long lasting. They are not water- but fat-soluble so they tend to accumulate in the body as one moves up the food chain.[21] The results of their use were the extermination of fish from a number of rivers and the decimation of the bird population chronicled by Rachel Carson in her *Silent Spring* (1962). Both types of insecticides were banned from use in the US in 1973. They continue to be used in developing countries where in 2007 they caused about 220,000 deaths and severe effects for another 3 million people (Flannery 2010).

When sprayed indiscriminately, the insecticides tend to kill not only the targeted insects but also their predators, which ultimately leaves the situation unimproved. And the obvious problems with using insecticides continue to this day. Massive die-offs of honeybees have caused great concern since 2006. Called 'Colony Collapse Disorder' (CCD), up to 85% of bees vanished in Ireland. Bees are essential for pollinating crops and hence for crop growth. A number of causes have been suggested for CCD, without a smoking gun. Some observers think responsible "the widespread application of a relatively new class of systemic insecticides—the neonicotinoids [especially imidacloprid]—that are highly toxic to insects including bees at very low concentrations" (Cummins 2007). The US National Research Council report on the honeybee situation did not consider either insecticides or genetically modified crops as possible causes. "Any suggestion that GM crops and pesticides may be causing the decline of honeybees is met with heated denial from the proponents [such as Monsanto]" (Cummins 2007).

This corporate behavior was also present when bird deaths and human poisonings as a result of DDT spraying were first reported. The authorities said "The powder [DDT spraying residue] is harmless" (Flannery 2010). Similarly, Dupont described studies linking CFCs to ozone depletion as "garbage science" (CIESIN n.d.). Also tobacco companies long denied any link between smoking and lung cancer. And, also in our own day, the massive effort of the fossil fuel companies to discredit science demonstrating global warming and especially its causes by human activity.

Although it is natural for a company to defend its product, there is a clear pattern of the most vociferous opposition in cases where there is plenty of plain evidence—often excellent scientific evidence—that the corporation is wrong. In some cases, especially the global warming case, adopting the corporate view and putting economics ahead of science spells almost certain catastrophe for the rest of us—and also for them. This behavior is not confined to "bad" corporations. These corporations are functioning quite well as corporations. Rather, the global warming case alone shows that corporations as human institutions must either be abolished or radically restructured. Since corporations are a purely human activity, they would also vanish if human beings vanished or were greatly reduced in numbers. Therefore, they cannot be allowed to make this gamble. These issues will be discussed further in the next chapter.

How might these corporate decision makers be thinking? Either they distrust science and genuinely believe in other means of gaining knowledge or anything outside the successful operation of their corporations is irrelevant to them. The first way of thinking—the distrust of science—is not likely, since the reliability of scientific results often is critical for their operations (This is especially true of fossil fuel corporations.). The second way of thinking—the irrelevance of the nonhuman environment—is much more likely. We could also describe the second way of thinking in economic terms: The environment, organized as the ecosystem, is an externality.

How to deal with harm to other species raises important questions for eco-ethics. Attempts to preserve or rebuild endangered species sometimes conflict with human economic activity, and those who believe in unconstrained human superiority raise questions about why a dam can't be built even if a small fish goes extinct (the snail darter, in a famous U.S. case)[22].

Habitat loss is usually a conflict between humans and other species. The habitat of a living thing is where it usually or normally lives—and therefore *can* live. Although living things can change their habitat, the loss of a significant portion of a habitat can mean the extinction of a species. Both animals and plants can change their habitat. Rats and cockroaches are now adapted to human urban environments, but they existed before urban environments existed. So they added or changed habitat. Similarly for gingko trees and ailanthus trees. Gingko are now a cultivated species planted widely in urban environments because they are pollution resistant and accept confined growing spaces. Ailanthus (*Ailanthus altissima* or tree of heaven), on the other hand, has been cultivated in China throughout recorded history. The ailanthus spreads easily in temperate urban areas. It is now regarded as an invasive species in those areas, is hard to eradicate, and is derogatorily referred to as "ghetto palm" and "stink tree."

Although all these species were able to adapt to human environments, many if not most species suffer or die out if their original habitat ceases to exist. One common form of habitat loss for many species is forest or grassland converted by humans into farmland. The ecological damage caused is loss of animals and plants, or, more seriously, species extinction. Why this should concern us as humans is discussed below.

But the same human activity—cutting down forests to produce farmland—can also produce serious collateral damage directly to humans. The soil in most tropical rain forests is thin and poor, so they can be used as farmland for only a few years.

At that point, what is left is land with no further use for farming. The original forest with its rich supply of plants and animals is also gone.

Deforestation can also promote erosion, as happened in the ancient Mayan city-state of Copan, in present-day Honduras. Urban development and population increase caused the Mayans of Copan to clear steep hillsides to grow food. The resulting erosion buried the more productive lowlands with enough silt to reduce their productivity significantly. This contributed to the failure of the food supply for the city and its eventual collapse. (Diamond 2005, p. 170) Deforestation can also produce unwelcome changes in weather, such as dramatic decreases in rainfall. This happened in Australia, with the result that the large interior lakes vanished completely. In the Amazon, the forest produces 80 percent of the moisture for rain. So significant deforestation would dramatically reduce rainfall (Flannery 2010, p. 83).

Removing a forest obviously produces a dramatic change in habitats available for humans, animals, and plants. Using grassland for farm fields or pasture seems to be much less of a radical change. However, the consequences can sometimes be as bad or worse. In the early part of the 20th century, the Great Plains grasslands of the U.S. were plowed for agriculture. The result was the Great Dust Bowl of the 1930s. The original grass cover protected the land from drought and wind erosion. The resulting drought and wind erosion caused the destruction of human habitat, which forced the migration of an estimated 2.5 million people. Some areas such as Los Angeles regarded the migrants as undesirables and posted police at the state border to keep them out (Ganzel 2003).

Pasturing is also not necessarily ecologically neutral. The introduction of sheep in Australia contributed significantly to land damage because the hard hooves of the sheep severely damage the grass cover. It was suggested that farmers raise kangaroos instead, but the suggestion was not well received (Diamond 2005). Beef ranching in the U.S. West and elsewhere also has severe ecological problems which are not well publicized. Even traditional open range grazing has its problems. Henning Steinfeld, of the UN Food and Agricultural Organization (FAO) observes

Extensive livestock production plays a critical role in land degradation, climate change, water and biodiversity loss. For example, grazing occupies 26 percent of the Earth's terrestrial surface, and feed-crop production requires about a third of all arable land. Expansion of livestock grazing land is also a leading cause of deforestation, especially in Latin America. In the Amazon basin alone, about 70 percent of previously forested land is used as pasture, while feed crops cover a large part of the remainder (Stanford University 2007).

Factory farming of animals has especially severe ecological problems. These problems mainly stem from concentrating animals at high densities in confined spaces. Large amounts of urine and manure are stored in huge pools called 'lagoons' which leak into the rest of the environment from time to time. The Natural Resources Defense Council (NRDC) writer observes:

Some people hear the word "lagoon" and picture blue water, surrounded by palm trees, perhaps, or with mountains in the background. A visit to a factory farm would quickly erase this beautiful image from their minds. At factory farms, "lagoon" means an open-air pit filled with urine and manure. Lots of urine and manure—some lagoons are larger than seven acres and contain as much as 20 to 45 million gallons of wastewater (NRDC 2005).

In addition to the odor, toxic gases are emitted from the 'lagoon' into the nearby environment. Finally, since no attempt is made to keep the animal enclosures clean, antibiotics are routinely used to prevent infection, as well as to promote growth. Bacteria immune to antibiotics also grow. This is now a serious health threat. As we will see in the next chapter, factory farming employs the distinctive features of modern technology.

Factory fish farms (aquaculture) have other problems affecting the surrounding environment. The wild sea trout population near salmon cages collapsed in Iceland in 1989. The probable cause was overfeeding of the captive salmon—it is very difficult to control the amount of food received by the captive fish. But perhaps the most telling ecological problem is that the captive fish are fed with processed wild fish. So aquaculture may not be an ecological improvement (Emerson 1999).

The wild fish to feed captive fish come from ocean fishing. There are two respects in which it is practiced in an ecologically destructive way: One is the practice of bottom trawling, dragging nets along the bottom which results in the destruction of many other species in the course of netting the desired fish. This threat is clearly the result of exploiting a technological advance. The other practice is simply overfishing, taking so many fish that the remaining ones cannot replenish the supply. This state of affairs has been described as a "tragedy of the commons," after an important paper of this title by ecologist Garret Hardin (Hardin 1968). This tragedy occurs when a number of self-interested individuals sharing the use of a common resource, deplete the resource to the point where it is no longer usable.[23] In the present situation, fisheries of the North Atlantic (northern cod), the Black Sea (blue fish, starry sturgeon, gray mullet, and Black Sea mackerel), and other areas have collapsed (UN Food and Agricultural Organization 2005).

Even industrial plant farming has its ecological problems. Harold Mooney, a professor of biological sciences at Stanford University, observes:

We are seeing land once farmed locally being transformed to cropland for industrialized feed production, with grasslands and tropical forests being destroyed in these land use changes, with resources feeding livestock rather than the humans who previously depended on those lands (Stanford University 2007).

Besides ecological problems, factory farming of animals raises issues of unnecessary cruelty to animals. Peter Singer's treatment in his *Animal Liberation* (2009, ch. 3) will be basis for discussion of this issue in Chapter 12, "Human Animal Relations."

One extreme form of habitat loss which would drastically affect both humans and nonhumans would be the use of nuclear weapons. Both Einstein and the French philosopher Teilhard de Chardin were impressed with the fact that nuclear weapons gave humankind the capability of extinguishing all life. Robert Oppenheimer, the head of the project to produce the first nuclear weapon, famously said on the success of a test of the weapon, "I am become Shiva, the destroyer of worlds." Teilhard felt that the emergence of such power demonstrated human superiority (Teilhard de Chardin 1964, pp. 145-153). This is clearly *physical* superiority but humanity deserves some credit for not having used such weapons in war since their first use in 1945. As we have seen, in many instances human groups have not been all that good in looking beyond immediate self-interest in employing technology. In this area, they have perhaps done better.

SPECIES EXTINCTION

The most dramatic consequence of habitat loss is species extinction. Habitat loss is the cause of extinction most directly as the result of human conflict with the ecosystem. Other causes of extinction tend to have other origins. They are: Hybridization of a rare species with an abundant one; invasive species; coextinction of a species dependent in some way on another species going extinct; and climate change extinction (Wikipedia 2011). Of course, if humans cause significant climate change, they will be more directly responsible for extinctions from this cause.

Even 60,000 years ago when humans were solely hunter-gatherers, they managed to drive the megafauna (large animals such as mammoths and saber-toothed tigers) of the Americas to extinction. Richard Leakey produces strong evidence that these megafauna extinctions in Australia and the Americas were caused in large part by human "overkill" (Leakey 1995, p. 194). The five previous mass extinctions were caused by environmental catastrophes, likely asteroid collisions. Human beings are currently causing a sixth mass extinction, all on their own. As much as 50% of all species may be extinct in 100 years (Leakey 1995, pp. 232-245).

Other living beings besides humans have caused mass extinctions by "polluting" the environment. About 3.5 billion years ago, an organism called a *stromatolite* became abundant. Stromatolites look roughly like large mushrooms. At this time, there was no or almost no oxygen in the atmosphere. Stromatolites exhaled oxygen, and over time the percentage of oxygen in the atmosphere eventually went up to our current 20 percent. This unfortunately caused a massive die off of organisms for whom oxygen was toxic. Oxygen literally oxidized their bodies and they died. So for these organisms oxygen was pollution. Of course, the stromatolites could tolerate oxygen so they became even more abundant (see Figure 2).

Human beings are currently causing a mass extinction on a large scale. But the important difference between us and the stromatolites is that the stromatolites' extinction-causing actions benefited them. Our extinction-causing actions may cause our own extinction.

The critical issue here is the same one brought up years ago by the snail-darter case. It is: Why should we care if a small species useless to us goes extinct? E. O. Wilson's *The Future of Life* (2002) is largely concerned with this issue. Wilson's answer is that biodiversity is important, that we should strive to preserve as many species as possible. Wilson cites the currently unknown potential for wild organisms

Figure 2. Present-day stromatolites in Shark Bay, Australia (Picture from Wikipedia Commons taken by Paul Harrison)

supplying us with drugs. About 40% of current pharmaceuticals stem from wild organisms, including nine of the ten leading prescription drugs[24] (2002, p. 119). Often it is far from obvious that a particular organism might be a drug source. The poison dart frog of Ecuador turned out to produce a poison which yielded a spectacularly effective nonaddictive pain killer. Although we have the substance, the frog's habitat has been replaced by banana plantations and this discovery almost did not happen (2002, p. 122). Searching for wild pharmaceuticals is one kind of what Wilson calls *bioprospecting*. Besides searching for new drugs, bioprospecting also involves searching in wild ecosystems for new food sources (humans depend on very few species of food plants, which creates a vulnerability in our food supply), new sources of fiber, and new sources of petroleum substitutes.

But usefulness to us is not a good answer to the question of why preserve biodiversity. If it turned out that the returns from bioprospecting diminished, then it could be concluded that biodiversity no longer mattered. A more important answer depends on the conclusions of the 20th-century science of ecology. Thus, this answer will be persuasive only to those familiar with those conclusions. The conclusion of ecology is that all life is interdependent in complex ways. One indication—and this is important for the preservation of as many species as possible—is that ecosystems cannot be recreated artificially by humans. When conditions are replicated as closely as possible, the same ecosystem does not replicate (Leakey 1995, pp. 167-170). So we do not know what is important in the ecosystem for our survival, and we do not know what we can remove without potentially disastrous consequences for us. This is why bioengineering is such a crapshoot.

I think the ecological answer about biodiversity is persuasive. But it ignores an important emotional consequence of evolution together with genetics and DNA. This consequence is that all life is related and humans are just another part of the Web of life. And this is *our* Web of life, unique and not reproducible. Therefore, we should not destroy or damage it. These reasons based on concern for life do not apply for not destroying the nonliving environment. We now turn to issues concerning that environment.

DESTRUCTION OF THE NONLIVING ENVIRONMENT

Human beings treasure natural wonders. These include dramatic geological features such as the Grand Canyon orYosemite; waterfalls such as Niagara or Iguazu Falls; and mountains such as Fujiyama and the Grand Tetons. Also treasured are dramatic living things such as redwood or sequioia trees. The kind of attraction large natural wonders have for us was analyzed by the philosopher Immanuel Kant in his *Critique of Judgment* (1790). He said such features exhibited the *sublime*. Sublime things are felt to be boundless and incomparable. We are not so much pleased by them as awed. We are moved to admiration and respect. I think Kant got it right when he said that these natural wonders move us because they call to mind the greater power of nature and the relative insignificance human beings. He got it wrong when he claims that these feelings themselves show the superiority of humanity over nature—because we can conceptualize nature in this way. According to Kant, it is our ability to conceptualize which makes recognition of the sublime a judgement rather than a reaction of fright or fear (Kant 1790, section 28). For one thing, this reasoning would apply to *any* judgement we made. And just being able to make judgements about a thing does not show our superiority to that thing. If I am concerned about a hurricane or erupting volcano, that does not show I am superior to those natural disasters. Kant may have been trying to avoid the conclusion that human beings are *not* superior to nature. The fact that his analysis is couched in terms of the term 'dominion' echoes the biblical declaration of human superiority.

Nevertheless, even though I believe Kant's analysis of the sublime is misguided, he did call attention to an important phenomenon in our relation to nature. Namely, that we are in *awe* of nature. *Awe* is not fear. Awe is rather great respect for superior power or strength. This attitude is not in conflict with nature. In fact, it leads us to preserve natural features even when that preservation is not useful to us. This is contrary to modern technology. As we will see, one of the hallmarks of modern technology is, according to the philosopher Martin Heidegger, the illusion that

. . . everything man encounters exists only insofar as it is his construct. And this illusion gives rise . . . to one final delusion . . .[That] man everywhere and always encounters only himself" (1955, p. 27).

Modern technology had just begun to hold sway by Kant's time at the end of the 18th century. Yet it is still striking to see him express one of its characteristic attitudes in his analysis of the sublime.

The attitude of awe toward natural wonders is itself in conflict with usefulness to humans. And there are other natural objects which we prize regardless of their usefulness. E.O. Wilson maintains that pastoral landscapes are preferred by humans because they mimic the savannah landscape in which humans first evolved. These landscapes are mainly grass with occasional trees or groups of trees with a body of water nearby (Wilson 2002, p. 134). In a way the evolutionary component of this preference is not that important. It is merely a fact about individual preferences. The fact that there may be an evolutionary cause for the preference is no reason for anyone to give special weight to the preference. I have a hereditary preference for coffee ice cream over chocolate ice cream.[25] That is a reason for me to choose coffee ice cream when I have the choice. But the hereditary background is not that relevant.

Even without the hereditary background, landscape preferences have some significance. The value of landscapes is felt to lie in their beauty rather than their awesomeness. Often landscapes include trees and plants, but these living things are important only as components of the total landscape. The philosopher George Santayana's definition of beauty (which follows Kant's) is that the beautiful is pleasingness regarded as a property of an object. (1896) Beauty is therefore subjectively objective. Being pleasing is subjective and being a property of an object is objective. Although beauty is different from usefulness, it is still relevant only to humans. But it may lead to the preservation of non-human things for their own sake and not just for their use or value. Small objects such as flowers and gemstones can also be preserved for this reason. But in terms of ameliorating conflicts with the ecosystem, beauty can play a role, just not a large one.

CONCLUSION

Modern technology and the industrial revolution have accelerated the conflict between human activity and the environment. Human superiority becomes not only a belief, it is a dangerous practical reality. The disturbing aspects of the exercise of human control are those which severely damage the ecosystem, sometimes to the extent of threatening the very existence of the ecosystem. As human systems increase in size, their scope becomes global. The alarming feature of these systems is that there are no automatic features of the ecosystem which can bring things back into balance. As Miriam McGillis once put it, we have taken the planet off automatic pilot.[26] We are probably the first species on this planet with the capability of destroying the entire ecosystem, not just in one but a multitude of ways. The examples above strongly suggest that the human propensity to manipulate the environment may not be compatible with long-term species survival. Will this animal shortly make itself extinct and maybe leave room for another attempt not so imbalanced? Humanity's threatening the entire ecosystem means that the point of view of humans as superior is *actually incorrect.* The rest of this book is devoted to finding a correct alternative point of view.

Modern technology dramatically accelerates our separation from the ecosystem. It is important to understand why this acceleration has occurred. The nature of technology and modern technology is the topic of the next chapter. That chapter also makes clear why modern technological processes tend to have toxic side effects and why genetic engineering is flawed.

REFERENCES

Amaral, K. (2003). Plastics in our oceans. *Woods Hole Institute.* Retrieved 10/31/11 from www.whoi.edu/science/B/people/kamaral/plasticsarticle.html

Asimov, I., & Pohl, F. (1991). *Our angry earth.* New York: Tom Dougherty Associates.

Baxter, C. (2011, May 6). Koch industries: Still fueling climate denial. *Greenpeace Blogpost.* Retrieved 11/2/11 from www.greenpeace.org/usa/en/news-and-blogs/campaign-blog/koch-industries-still-fueling-climate-denial/blog/34609/

Berkeley Plastics Task Force. (2009). Campaign against the plastic plague: Background info. *Earth Resource Foundation.* Retrieved 10/31/11 from www.earthresource.org/campaigns/capp/capp-background-info.html

Biology Online. (2009). *Pollution definition.* Retrieved 10/2/11 from www.biology-online.org/dictionary/Pollution

Co2gether. (2012) Understand climate change. *Co2gether.* Retrieved 1/31/12 from www.co2gether.org/my_co2/understand/

Carson, R. (1962). *Silent spring.* New York: Houghton Mifflin.

CIESIN (Center for International Earth Science Information Network). (n.d.). The Montreal protocol on substances that deplete the ozone layer. *CIESIN Thematic Guides.* Retrieved 4/03/09 from www.ciesin.org/TG/PI/POLICY/montpro.html

Colby, C. (1996). *Introduction to evolutionary biology.* Retrieved 9/15/11 from www.talkorigins.org/faqs/faq-intro-to-biology.html

Cummins, J. (2007). Requiem for the honeybee. *Institute of Science in Society Report 24/04/07.* Retrieved 10/3/11 from www.i-sis.org.uk/requiemForTheHoneybee.php

Ecology Center. (2011). *Plastics task force: Seven misconceptions about plastic and plastic recycling.* Retrieved 10/31/11 from www.ecologycenter.org/ptf/misconceptions.html

Emerson, C. (1999). *Aquaculture impacts on the environment.* Retrieved 12/12/11 from www.csa.com/discoveryguides/aquacult/overview.php

Fahrenthold, D. (2006, September 29). Wildlife waste is major water polluter, studies say. *Washington Post.* Retrieved 10/3/11 from www.washingtonpost.com/wp-dyn/content/article/2006/09/28/AR2006092801994.html

Flannery, T. (2010). *Here on earth.* London: Allen Lane.

Fountain, H. (2012, July 26). Storms threaten ozone layer over U.S., study says. *New York Times*. Retrieved 8/10/12 from nytimes.com/2012/07/27/science/earth/strong-storms-threaten-ozone-layer-over-us-study-says.html

Ganzel, B. (2003). *The dust bowl*. Retrieved 4/10/13 from www.livinghistoryfarm.org/farminginthe30s/water_02.html

Hansen, J. (2009). *Storms of my grandchildren*. New York: Bloomsbury USA.

Hardin, G. (1968). The tragedy of the commons. *Science*, *162*, 1243–1248. doi:10.1126/science.162.3859.1243 PMID:5699198.

Heidegger, M. (1955). The question concerning technology. In *The Question Concerning Technology and other Essays*. New York: Harper & Row.

Kant, I. (1790). *Kritik der urteilskraft*. New York: Hafner Publishing.

Katers, R. (2009). *The history of PCBs*. Retrieved 10/28/11 from www.foxriverwatch.com/site_index_fox_river_bay_pcbs.html

Knud-Hansen, C. (1994). Historical perspective of the phosphate detergent conflict. *Conflict Resolution Consortium, University of Colorado*. Retrieved 10/31/11 from www.colorado.edu/conflict/full_text_search/AllCRCDocs/94-54.htm

Lautenberg Press Office. (2010). *Lautenberg introduces safe chemicals act to protect Americans from toxic chemicals*. Retrieved 10/29/11 from lautenberg.senate.gov/newsroom/record.cfm?id=323863

Leakey, R., & Lewin, R. (1995). *The sixth extinction*. New York: Doubleday.

MacIntosh, H. S. (2007). *What's the dirt on phosphate-free soaps?* Retrieved 11/1/11 from www.treehugger.com/culture/ask-treehugger-whats-the-dirt-on-phosphate-free-soaps.html

Mote, V. (1975). Air resource management in Soviet Russia. *Proceedings of the American Philosophical Society, 119*(3). Retrieved 10/30/11 from www.jstor.org/pss/986670

Mountain Justice. (2012). *What is mountain top removal mining?*. Retrieved 1/31/12 from mountainjustice.org/facts/steps.php

Murchison, K. (2007). *The snail darter case: TVA versus the endangered species act*. Lawrence, KS: University Press of Kansas.

Norris, J. (2009, June 25). Sugar is a poison, says UCSF obesity expert. *UCSF News*. Retrieved 1/31/2012 from www.ucsf.edu/news/2009/06/8187/obesity-and-metabolic-syndrome-driven-fructose-sugar-diet

Northwest Fisheries Science Center. (2009). *Domoic acid poisoning*. Retrieved 10/2/11 from www.nwfsc.noaa.gov/hab/habs_toxins/marine_biotoxins/da/index.html

NRDC (National Resources Defense Council). (2005). *Pollution from giant livestock farms threatens public health*. Retrieved 12/12/11 from www.nrdc.org/water/pollution/nspills.asp

Relaxnews. (2011, October 28). Iceland to help France save trees from global warming. *Relaxnews*. Retrieved 11/1/11 from news.yahoo.com/iceland-help-france-save-trees-global-warming-101350209.html

Santayana, G. (1896). *The sense of beauty*. New York: Scribner's & sons.

Sapien, J. (2007, April 26). Superfund today: Massive undertaking to clean up hazardous waste sites has lost both momentum and funding. *The Center for Public Integrity*. Retrieved 10/30/11 from projects.publicintegrity.org/Superfund/report.aspx?aid=851

Stanford University. (2007). Harmful environmental effects of livestock production. *ScienceDaily*. Retrieved 11/3/11, from http://www.sciencedaily.com/releases/2007/02/070220145244.htm

UN Food and Agricultural Organization. (2005). *General situation of world fish stock*. Retrieved 11/2/11 from www.fao.org/newsroom/common/ecg/1000505/en/stocks.pdf

Walsh, B. (2011, October 4). Who's bankrolling the climate-change deniers? *Time*. Retrieved 11/4/11 from http://www.time.com/time/health/article/0,8599,2096055,00.html

Wikipedia. (2011). *Extinction*. Retrieved 10/17/11 from en.wikipedia.org/wiki/Extinction

ENDNOTES

1 Some plants do not grow by means of photosynthesis, notably the fungi, which recycle other plant matter.

2 Whether humans *should* be eating other animals will be discussed in Chapter 12, "Human Animal Relations."

3 Terms such as "scenery," "vista," and even "countryside" already include the idea of usefulness to humans.

4 "Instability" is not enough to make a change in the environment into pollution.

5 Insecticides have other problems. See below, "Harm to other Species."

6 Social insects do not just defecate and urinate anywhere in their nests or hives. For example, ants have special chambers reserved for their waste. My thanks to Jane Whitehill for raising the issue.

7 This is likely to have been the case when humans were hunter-gatherers.

8 Or, for that matter, domesticated animals. Before the advent of the automobile, horse manure pollution was a problem in major human cities.

9 There is surely a way to determine how many people (or animals) would be compatible with the ecosystem without extra waste disposal measures.

10 I was startled to see Monsanto involved in clear attempts at what I will term "knowledge pollution" in 1937. The significance of this tactic will be discussed in Chapter 3, "Science, Modern Technology, and Corporations." Monsanto is currently involved in disturbing uses of genetic engineering. Genetic engineering is considered later in this chapter.

11 The phenomenon is called *biomagnification*. If a persistent substance occurs at a certain percentage in an animal low on the food chain, the higher animal eating the lower one will accumulate the persistent substance present in a large number of lower ones. A tuna will accumulate all the mercury present in the hundreds of anchovies it eats over its lifetime.

12 Most references in this area are politically polarized. Mote 1975 appears to contain facts in addition to invective.

13 The effectiveness of recycling strategies will be dealt with in Chapter 9, "Environmentalism and Sustainability."

14 As long ago as the early 1970s, woodland rivers in the rural Yellow Springs, Ohio, area had become unsafe for swimming because of fertilizer pollution from farms.

15 Eutrophication occurs when a lake overdoses on nutrients such as nitrogen and phosphorus. Algae bloom in large numbers, making the lake cloudy and removing oxygen for fish. At the end process, the lake becomes a stagnant, smelly, and loaded with dead fish.

16 Selective breeding by humans still is checked against the environment.

17 U.S. Presidential candidate Michelle Bachman made in 2011 public statements indicating she does not grasp this point. As will be discussed later, much of the U.S. is hostile to the idea of global warming as a threat. In this, the U.S. is pretty much alone amongst developed nations and even developing nations such as China.

18 Humans have proposed helping to move the trees. See Relaxnews 2011. The proposal to recreate French trees in Iceland does not involve actually recreating a forest but rather preserving tree DNA.

[19] Incredibly, blogger Doug L. Hoffman uses an almost identical diagram to deny man-made influence in the carbon cycle. The name of his Website, *theresilientearth.com* is a tipoff: Humans can do whatever they please to the planet and the "resilient earth" will take care of it. A convenient but dangerous fantasy.

[20] The pragmatic theory of truth, *pragmatism,* was developed in the early 20th century by such American philosophers as William James and John Dewey.

[21] The phenomenon is called *biomagnification*. If a persistent substance occurs at a certain percentage in an animal low on the food chain, the higher animal eating the lower one will accumulate the persistent substance present in a large number of lower ones. A tuna will accumulate all the mercury present in the hundreds of anchovies it eats over its lifetime.

[22] See Murchison (2007).

[23] This situation is an instance of the Prisoners Dilemma, where higher-order self-interest requires reining in naïve self-interest. See note 15, Chapter 1, "Humans as Masters of the World."

[24] There is a strong tendency within modern technology to ignore the basis in nature for many of its products. This tendency leads to the overvaluation of the usefulness of modern technology. See further Chapter 3, "Science, Modern Technology, and Corporations."

[25] The preference for coffee over chocolate is not common. Yet both my daughters, who did not grow up with me, have this preference.

[26] http://www.du.edu/enviro/Events_and_Orgs.htm

Chapter 3
Science, Modern Technology, and Corporations

ABSTRACT

Why modern science is a superior source of truth. Modern science contains the essence of modern technology. The distinguishing feature of modern technology is that it treats everything as nothing but resource. This explains why modern technology must conflict with the ecosystem, because it does not recognize the ecosystem as anything other than a set of resources it can exploit. Corporations are the institution par excellence for embodying modern technology, since their overriding concern is to maximize profits, and treating everything as resource is the best way to maximize profits.

INTRODUCTION

As we saw in the preceding chapter, the human institutions of science, modern technology, and corporations bear much responsibility for a dramatic increase in human conflicts with the ecosystem. In this chapter, my aim will be to examine all three in more depth, solely in order to determine what features of these institutions and practices produce these conflicts with the ecosystem. Eventually ultimate causes need to be determined and alternatives suggested.

DOI: 10.4018/978-1-4666-4586-8.ch003

Science has become the gold standard for truth about the world. And modern technology has become a main driver of economic growth. Yet both modern science and modern technology are fairly recent arrivals in human history. Much of modern science has its roots in renaissance Italy in the late 1500s and early 1600s. Galileo Galilei (1564-1642) is known as "the father of modern science." According to Stephen Hawking, "Galileo, perhaps more than any other single person, was responsible for the birth of modern science" (2009). A dramatic change in our ability to make things emerged with the industrial revolution starting in the late 1700s. The phenomenon most characteristic of modern technology is the widespread use of electricity, which dates from the late 1800s. Finally, corporations, the legal and economic institutions most suited to exploit modern technology, assumed close to their current form in the early 1800s (Adams 2013).

These three human institutions—science, modern technology, and corporations—are related. All three embody recent and large changes in the way human beings do things: Modern science has been around for just over 400 years. Modern technology has been around for just over 200 years. And modern corporations have existed for just under 200 years. Compare these time spans of 200 to 400 years with our 7,000 years of civilization and our roughly 200,000 years of existence as a species. If serious problems are caused by these human institutions of science, modern technology, and corporations, we need to alter these institutions—perhaps drastically. They have just been around for too short a time to have proved their enduring value for the human species.

SCIENCE

What makes science so reliable? Why does it deserve to become the touchstone of truth? Can it be distinguished from earlier ways of understanding the world? A usual definition is that science uses systematic observation coupled with experiment as part of an ongoing tradition to obtain knowledge of the world. In fact, much of this definition also fits proto-science. The practice of improving one's relation to the world by observing it and seeing what works and what doesn't is at least as old as *Homo sapiens*. Indeed, it is hard to see how the ability to make tools could exist in the absence of accumulated empirical knowledge. If so, our ancestors *Homo erectus* and *Homo habilis* certainly had *proto-science,* which is the practice of learning about the world to improve one's relation to it. If *Australopithecus afarensis* made tools, then proto-science could go back 2 to 4 million years.

In fact, practical trial-and-error/learn-by-experience behavior, which we can call *proto-science*, is also practiced by a number of nonhuman animals:

- Chimpanzees eat various non-food plants which they have apparently discovered help relieve various ailments such as indigestion and worms. We call them medicines (Burne 2002).
- Three dolphins in concert create a wave at the water's edge which washes small fish onto the shore. Whereupon the dolphins eat the fish. This practice is called strand feeding (Weiss 2011).
- Killer whales have learned to turn sharks, even great white sharks, upside down for periods of time. Being upside down immobilizes the shark, which then suffocates and is eaten (Killer whales actually belong to the same genus as dolphins and hence are not whales at all.) (Heimbuch 2010).
- Gulls drop oysters on rocks to break open their shells and then swoop down to eat the oyster.

Non-civilized peoples practice proto-science extensively. The anthropologist Claude Levi-Strauss, in his *The Savage Mind,* describes the "thirst for objective knowledge" as one of the most neglected aspects of the thought of those people we call primitive (1962, p. 3). Levi-Strauss describes the detailed knowledge of plants and animals common amongst non-civilized peoples. He tells of the anthropologist E. Smith Bowen who attempted to learn the language of an African tribe she had just begun studying. Her informants collected a large amount of plants and proceeded to give her their names for them. This was useless for her because she was unable to tell one plant from another (1962, p. 6).

Levi-Strauss points out that it was in Neolithic times—beginning about 12,000 years ago—that the major techniques of civilization were perfected. These include pottery, weaving, agriculture and the domestication of animals. He observes:

. . .these enormous advances [cannot be attributed to] the fortuitous accumulation of a series of chance discoveries or. . .[cannot] have been revealed by . . . passive perception. Each of these techniques assumes centuries of active and methodical observation, of bold hypotheses tested by means of endlessly repeated experiments (1962, pp. 13-14).

So, much of what is attributed to modern science is present at the dawn of civilization and likely before. The practice of observation, hypothesis, and experiment seems to be the property of any intelligent being. Perhaps this is not a bad definition of 'intelligence.'

These considerations make it easier to articulate what is distinctive about modern science. Any accumulation of knowledge of the world was informal and ad hoc until the ancient Greeks gave us *scientific disciplines*. It was Aristotle (384-322 BCE) who essentially founded the disciplines of physics, biology, psychology, botany, and logic—in addition to political science, aesthetics and rhetoric. Although Aristotle was clear that empirical knowledge is based on observation, his works were taken as gospel (that is, not subject to verification or alteration) during the middle ages. Things changed with the rise of modern science in Italy in the late 1600s.

It is generally felt that modern science arose in Italy in the late 1500s and early 1600s. Galileo Galilei (1564-1642) was perhaps the most important figure. So what did Galileo and the early scientists add to proto-science? Five significant things:

1. Specialized apparatus.
2. A methodological commitment to abstract from purpose.
3. The requirement that results be repeatable and reproducible.
4. The continued availability of research results; and finally.
5. The commitment to continue testing results against reality.

I will consider each in turn.

1. The science museum in Florence, the Museo Galileo, until recently the Museo della Storia di Scienzia (Museum of the History of Science), contains much of the apparatus Galileo devised to conduct his experiments.[1] In addition to telescopes, there are various machines designed to test laws of physics, as well as navigational and medical instruments. (In the central well of the museum is a two-story-high armillarium, a model of the solar system with (complicated) paths for the sun, moon, and planets as they would have to be if the earth were the center of the solar system.) So instead of waiting for nature to provide a test of some hypothesis, or even arranging natural objects to provide such a test, scientists now designed and constructed apparatus to advance knowledge. In our day this has resulted in atom-smashers, dedicated telescopes, and a large variety of dedicated lab equipment. Clearly this is one respect in which technology and science are intimately interrelated, because often scientific advance depends on developments in the technology required to construct more sophisticated apparatus.
2. The commitment to abstract from purpose is sometimes expressed as a commitment to physicalism. Physicalism is the view that the universe is fundamentally composed of bodies interacting with each other. As Noam Chomsky has pointed out, this view has been untenable since Newton developed the concept of gravity. Gravity, after all, is action at a distance. So physicalism as the theory that reality is only bodies interacting is false. Physics now ad-

mits four fundamental interactions in addition to bodies directly interacting: Gravity, strong nuclear force, weak nuclear force, and electromagnetism.[2] Nevertheless, modern science does not acknowledge purpose as part of scientific knowledge. Sir Francis Bacon in 1620 published his *Novum Organum.* This work was specifically intended to update Aristotle. In addition to formulating experimental methods, Bacon asserted that explanations in terms of purpose have little value in science. Since value is fitness for a purpose, value is therefore not included as part of scientific knowledge. One way to think of the situation is that science considers the world as *present at hand.* In our everyday lives, we are conscious of the world as *ready to hand.*[3] Science, in constructing apparatus, has to consider whether one apparatus is better than another. Therefore scientific apparatus is part of the world as ready to hand, the world of purpose and value. Science sets up its apparatus in the world as ready to hand. But any resulting knowledge is of the world only as present at hand, minus any purposes or value.

3. The requirement that results be repeatable and reproducible was already part of the practice of Muslim scientists such as Alhazen (Ibn al-Haytham) around 1000AD.[4] These features of experimentation are critical for reducing the latitude for disagreement and dispute in science. The fact that scientific conclusions themselves are not always accepted is another issue, which will be dealt with later in this chapter.
4. The continued availability of research results is partially the result of the continuity of civilization since Galileo's time. Further, science may have been the first globalized institution. Even in Galileo's time, people came from some distances to spend time learning the new science. Robert Dudley, son of Queen Elizabeth's close friend Lord Robert Dudley, spent much of his life in Florence working on scientific experiments with Galileo. Except for relatively brief hiatuses, all contributors to science have considered themselves as adding to the same enterprise. The Nazis and the Soviets had their own versions of some sciences. It is important to remember these cases where political beliefs have taken precedence over science. In our day, economic corporate interests are attempting to take precedence over science on the matter of climate change.
   The continued availability of scientific results is important: It allows the acquisition of knowledge to be cumulative.[5] The professionalization of the acquisition of knowledge, with clear standards for who is able to contribute to the scientific enterprise, adds to the cumulative nature of knowledge gained through science. The term 'scientist' was coined by the philosopher of science William Whewell in 1834 (Markel 2010). Science became professionalized in the 19th century.

5. The commitment to continued testing results against reality is what distinguishes science from other traditions of putative cumulative knowledge. Religious traditions such as Catholicism are cumulative in some respects. But the commitment to test against reality is absent. Instead, there is continuity of authoritative pronouncements.[6]

Why should these characteristics and methods of modern science lead us to regard it as the touchstone of truth? As philosopher Hilary Putnam has observed, for the layman, the high prestige of science is due more to the spectacular results of science than the efficacy of its methods[7] (Putnam 1981, p. 185). Scientists, of course, do not regard the truth of science as dependent on the usefulness of its results. The question then becomes, what about scientific methods enables them to produce better results?

My answer to the question of why science deserves to become the touchstone of truth is this: For the non-scientist layman, it is the prodigious results of science. For the scientist, it is adherence to rigorous methods of producing reliable knowledge. The prodigious results the layman admires would not exist without the methods scientists adhere to. To recap, science uses systematic observation of the world coupled with experiment as part of an ongoing tradition to obtain knowledge of the world. Modern science adds: the use of specialized apparatus; a methodological commitment to abstract from purpose; repeatability of results; the continued availability of research results; and finally a commitment to continued testing of results against reality.

FAILURE TO ACCEPT SCIENTIFIC RESULTS

As we saw in the case of genetic engineering, organizations can accept the scientific basis of genetics and yet ignore its implications through the theory of natural selection.[8] Yet there is also outright rejection of such scientifically established results as natural selection and man-made climate change. Both these results are critical for correctly understanding our situation in the world as human beings: Natural selection (along with genetics and ecology) is critical for correctly placing humanity in the world. And the research supporting climate change is likely critical for the survival of civilization or possibly even the species *Homo sapiens*. Or, in the worst case scenario, all life on earth.

So why are these two results at all controversial? These rejected results have been established by the same methods that produce most of the technology everyone relies on. For example, everyone relies on technological results in the fields of health and transportation. In the case of natural selection, as the major mechanism of evolution, the major competing account is based on religion alone. The rejection of natural selection may be based on a complete rejection of any scientific account of the past and present of earth. Deniers usually feel the authority of a text—the Bible—written well before the advent of any science is superior. Since the Bible contains many contradictory passages, there are usually ulterior contemporary motives for the anti-scientific interpretation.[9] These motives are usually to protect human superiority and human exceptionalism. Also, in the case of climate change denial, the motive, at least on the part of the powerful corporations who fund this denial, is protection of their short-term economic interests.

This book does not assume the truth of human superiority and human exceptionalism. So in this book, neither of these doctrines can support the reliability of any kind of knowledge. And further, since both doctrines are implicated in genuine threats to continued human existence, any answer to those threats cannot assume either human superiority or human exceptionalism. We have to proceed from recognized scientific results.

Professor Stephan Lewandowsky has done some interesting research on why scientific results on climate change are rejected. He found that in many countries the left/right political divide was responsible for when climate change science was accepted and when not. He notes that many studies have found that "…there is a strong correlation between a person's endorsement of unregulated free markets as the solution to society's needs on the one hand, and rejection of climate science on the other" (Lewandowsky 2011). Lewandowsky feels that the intense emotions of climate change denial are caused by the fact that action on climate change must be done by regulation, and that is anathema to free market believers.[10] I do not believe this is enough to account for the intensity and the irrationality of climate change denial Rather, the idea of humans no longer being in control may be behind the denial. In other words, climate change implies that humans are not superior. So in order to maintain the belief in human superiority that they have based their lives on, they must deny climate change.

LIMITS OF SCIENCE

In spite of all of the good features of science as a way of gaining knowledge, it has important limitations. It is part of scientific method to abstract from purpose and value, and instead describe the world as present at hand, at arm's length from everyday human concerns. This is both a strength and a limitation. Even scientists must work in the everyday world of purpose and value when they set up their experiments. Whether one telescope is better than another, whether one location or time is better than another for performing an experiment—these issues of purpose and value must be their concern. Such concerns are extra-scientific in the sense of not being included in scientific theory. But they are still legitimate human concerns, and must concern scientists living in the world as ready to hand.[11]

Another kind of limitation for science was mentioned in a previous chapter. It has to do with the fact that important kinds of scientific explanation cannot be completed. In physics, one way we explain what something is, is by saying what it is composed of. Thus, we explain what matter is by saying it is composed of molecules of certain types. We explain what molecules are by saying they are composed of atoms. We explain what atoms are by saying they are composed of electrons, neutrons, and protons. We explain . . . (and so on). At this point, note that the Greek word 'atom' (ατομος) means 'without parts.' So atoms now have the wrong name because we know that they have parts. This error raises a question about this whole pattern of explanation. The philosopher Immanuel Kant noticed that at each stage of this explanatory pattern, there are two possibilities: Either the entities at that stage cannot be further decomposed. But that means they cannot be explained. Or the entities at that stage can be further decomposed and we never reach any entities that cannot be decomposed. In this case, the explanation is never completed (Kant 1781, pp. A434 B462).

In his *Critique of Pure Reason,* Kant sets out four of what he calls *antinomies.* An *antinomy* is a contradiction between two apparently equally valid principles. Two of the antinomies are about scientific principles. The one concerning simple parts was just discussed. The other is whether the world has a beginning in time or not. This is a recent issue concerning whether there was anything before the Big Bang. Kant's point is that these issues *cannot be settled.* They lie outside the scope of human reason. Mathematicians have had to live with undecidable issues since the 1930s, so there is no reason why physicists can't also do so.[12]

Perhaps the most important limitation of science concerns consciousness, which in many ways is closest to each of us. Consciousness is one of the main topics of the next chapter. There we will consider the relation of consciousness to the world. Here my concern is with the place of consciousness within science. The philosopher Mark Rowlands has presented a convincing argument that consciousness cannot be reduced to anything within science (2001). Rowlands points out that when things known through consciousness are explained scientifically ("reduced"), what is removed is the contribution of consciousness to that knowledge. For example, when the feeling of *heat* is scientifically explained as a certain level of mean molecular kinetic energy, the conscious description in terms of feeling disappears. Therefore, to explain consciousness scientifically what would have to be removed is the contribution of consciousness to consciousness! So there is nothing left to explain scientifically. Some scientifically-minded philosophers have for other reasons actually gone so far as to deny that consciousness exists.

A related consequence is that the ready-to-hand of everyday experience cannot be reduced to the present-at-hand of science. This consequence is important in understanding the true nature of consciousness. These issues will be pursued in the next chapter.

SCIENCE AND TECHNOLOGY

What is the relation between science and technology? Is science the basis for technology or is technology the basis for science? Chronologically, modern science came before modern technology, with modern science dating from about 1600 and modern technology dating from about 1800. The philosopher Martin Heidegger's outlook on the relation between science and technology is striking[13] (1955). Although he agrees that modern physics arose almost two centuries before modern technology, he thinks it incorrect to say that "surely technology got under way only when it could be supported by exact physical science" (pp. 21-22). Heidegger says this is chronologically correct but not historically true.

Heidegger's view is that modern technology embodies a new way of bringing things into being, and this fact is obscured by viewing it as merely applied science. Science implicitly contains this new way of bringing things into being, and modern technology embodies it. Science demands that "nature reports itself in some way or other that is identifiable through calculation and that [nature] remains orderable as a system of information" (1955, p. 23). The new way of bringing things into being is to treat everything as standing-reserve, on call for ordering as further standing-reserve. We will shortly have more to say about this notion, which is critical for the correct placement of technology in reality.[14]

I believe Heidegger is correct, that the relation of science to technology is not just a matter of the later application of science to practical issues. Rather, when science is applied to technology, the view of nature encapsulated in science becomes the way technology treats nature. As we will see, this way of bringing things into being is inherently disruptive and destructive.

So it is barely conceivable that modern science could have existed without modern technology ever developing. However, two constituents of science make this unlikely: The requirement for progress built into science; and the need for ever more complex equipment to wrest secrets from nature. Progress is going to require more elaborate and resource-intensive equipment, and without other payoffs than science, substantial support for this equipment will not be likely. These payoffs other than scientific knowledge are provided by modern technology. In addition, independent technological advances call for scientific advancements from more sophisticated experimental equipment (Computers provide myriad examples.). The advance of technology itself requires appropriate scientific research.

In other words, modern technology and modern science have a symbiotic relationship. Science's world as present-at-hand is the basis for modern technology's new ordering of everything as standing-reserve, on call for ordering as further standing-reserve. All purposes and values are stripped from the world, except for the purposes and values of modern technology itself. In order to better illuminate modern technology, we will now examine a brief history of technology.

A VERY BRIEF HISTORY OF TECHNOLOGY

Technology has four stages:

- Proto-technology, early tool development before civilization. One million, possibly 2 million years old.
- The classical technology of agriculture and cities that enabled the rise of civilization, roughly 10,000 years old.
- Modern technology enabled science, about 500 years old.
- Postmodern technology, about 100 years old.

The use of any technology enabling more than hunter-gatherer societies takes up a remarkably short portion of human existence on the planet. Using current estimates of the presence of human beings of our species for the last 200,000 years, civilization has been a possible human mode of existence for 7,000 years, no more than 4% of humanity's time on earth. Joseph Tainter argues in his *The Collapse of Complex Societies* that so far, civilizations other than our current one have been unable to avoid collapsing under their own weight. He thinks our current civilization is different from previous ones in having modern technology to increase resources and thus at least postpone collapse (Tainter 1988, p. 216.).[15]

Indeed, it is because of technology—at least technology in the service of science—that we can even begin to have any hope of overcoming the difficulties in our current position. For it is largely science and the technological apparatus necessary for science which is responsible knowledge of our situation. Science and technology in the service of science have increased our consciousness of our place on the planet. If our strengths have failed us in leading us in the direction of destroying our own ecosystem, science has not failed us in revealing this very situation to us.

Technology does not contain within itself a way out of our conflicts with nature. Indeed, proponents of technology tend to argue that for any problem, there is a further technological fix. But such fixes take us farther and farther away from any recognizably sustainable world. Not only do we end up existing for the sake of our technology, but the technological apparatus becomes increasingly susceptible to catastrophic failure, as Isaac Asimov illustrated in his science-fiction novel *The Caves of Steel.*[16] (Asimov 1954).

Modern technology takes us in directions not previously encountered in our environment. Our changes tend to have "side effects," unpredictable and dangerous consequences. As we saw in the previous chapter, it is common for changes introduced by technology to have harmful side effects. Chlorofluorocarbons and their effect on the ozone layer ought to be more than sufficient to put us on notice that we really do not know what we are doing with our technological changes. Further technological fixes to the ozone layer may well have unintended consequences for the rest of the ecosystem. There is simply no way of ruling out such possibilities.

THE POINT OF VIEW OF MODERN TECHNOLOGY

It is important to recognize fully the depth of problems with modern technology. Technologists tend to be intelligent people. It may seem obvious to me that manipulation of genetic material is extremely dangerous given the nature of how species relate to the environment. But bioengineers are similarly baffled by objections to their seemingly totally beneficial attempts to provide more food for people who are malnourished. The philosopher Martin Heidegger saw most clearly that *modern technology has its own point of view* which is completely separate from any other structure of human aims and purposes. The critical feature of modern technology is its willingness to treat anything as a resource to be reordered in the furtherance of other aims, including its own. Heidegger, in his essay "The Question Concerning Technology," concludes that modern technology is an independent force in human existence. It builds a new and incompatible order on top of what was there, primarily in order to extract and store energy for later uses (Heidegger 1955, pp. 14-17).

The point of view of modern technology regards everything as a potential resource, as "standing reserve" to be used or reused later in other related processes. A forest has status only as a timber resource. Land itself is only a resource for the building industry. Even human beings themselves become "human resources" from this point of view. Or they become "consumers." Or ill people become a "supply of patients for a clinic". Many distinctive modern technologies embody this notion of "standing reserve" in their very conception. Thus, electric power, whether in the form of available current or batteries, is always entirely standing reserve, on hand for potential use.

This way of looking at things, insofar as it ignores the previous pattern of processes, uses, and ends, is inherently "violent" in its effects on those processes.

Technologists very often view technology as a neutral means. An anonymous reviewer of this book put it thus: "What counts is how technology is being applied, and for which purposes." In other words, technology itself cannot be assessed. One way to see the disastrous consequences of this view is to apply it to an ecosystem-threatening application of technology. As we saw in Chapter 2, "Conflicts with the Ecosystem," chlorofluorocarbons were carefully tested before being released into the environment.[17] It was an unintended and unforeseeable side effect of the application that produced the ecosystem-threatening thinning of the ozone layer. There was nothing wrong with the application. So if applications are all that needs to be considered, there is nothing further to be considered in this case. In which case we are helpless before our own fate of likely destruction.

The alternative is Heidegger's realization, that modern technology is a new way of bringing things into being. Only then does it become possible to assess modern technology's relation to the ecosystem. As we have seen, it is characteristic of modern technology that it regards everything as potential resources on call for use in its own processes. From this point of view, we can clearly see the danger in bioengineering. Bioengineering takes DNA, the genetic material governing the continuance of life and makes changes to it in accordance with short-term human needs. Natural selection, which produced that DNA in the first place, is completely left out. Here the problem is not the technical faultiness or soundness of bioengineering applications but rather the very idea of treating DNA as a mere resource subject to the techniques of modern technology. Here too modern technology shows itself to be a distinctive way of bringing things into being and not just a neutral means.

Some authors actually take bioengineering further and embrace the possibility of leaving the current basis of life completely behind. Perhaps the most extreme version of this view is that of the roboticist Hans Moravec. Moravec suggests that it is time to leave DNA-based life behind in favor of the superior creations of the human mind. These "mind children" (the title of his book) will become our progeny through a process of "downloading" (Moravec 1988, pp. 115-117). To think that such a process is even possible requires belief in a definite metaphysical view of human beings. This view is that human beings are radically different from other animals in having minds as well as physical bodies. This version of human exceptionalism is a major topic in the next chapter.

It is thus a part of the point of view of modern technology that technology continues to present itself to us as a mere means, an enabler for our other ends. But it is impossible to understand modern technology correctly without realizing that this is wrong: That *modern technology has its own ends,* which are to reorder everything as standing reserve for yet-undisclosed uses. As Julian Young puts it in his lucid study of Heidegger on these issues,

> *...human beings [become] the most important [raw material] because as well as being available for manipulation and exploitation as productive units they are also the manipulators and exploiters, as it were the portals through which the global system of production and consumption—a system which, to repeat, is circular and so pointless—maintains itself (Young 2002, p. 46).*[18]

In the previous chapter, the question was raised of why it should be that technological advances so often had adverse side effects. The answer is that the likelihood of adverse side effects is built into the nature of modern technology, just because it conflicts with a previous order, that is, the ecosystem.

Can any use of technology be completely compatible with the ecosystem? Clearly modern technology has compatibility problems. But what about earlier technology? We think of agricultural technology as a paradigm of nature itself. However, the classical technology of cities has always collapsed, normally through adopting practices which were incompatible with sustaining their populations. So it seems that even this technology has compatibility problems.

Modern technology, though powerful and useful, also contains great dangers in unforeseen and unforeseeable side effects. Thus to minimize serious harm to the ecosystem, new uses of modern technology should be as carefully investigated as possible. Any potential for harm should be evaluated, if possible, in advance of deployment of the technology. However, our actual practice is the exact opposite. The initiation and deployment of new technology is almost entirely the responsibility of corporations. And corporations are motivated to downplay side effects on the ecosystem or even work toward denial of the very existence of side effects.

CORPORATIONS

A corporation is a legal entity capable of acting in some respects as an individual, mainly in terms of property rights, legal liability, and political rights. Theoretically, corporations are created to serve the public benefit,[19] and their trans-individual status allows then to function more efficiently, without constant shifting of property rights and responsibilities (The free dictionary.com 2004). A corporation, as a legal construct created for reasons of efficiency, clearly should not inherit all the rights of the individuals making it up.[20] When corporations have the rights of people, they should also be liable to the same punishments as people—if they kill, they should be executed or prohibited from doing business for some period. This is currently not the case.

When in the 1800s, corporations were relieved of any public interest concerns, what was left was the goal of maximizing profits. Corporations per se have no social or ethical goals. Some observers have pointed out that corporate behavior is similar to sociopathic behavior.[21] Many corporations claim that they practice 'corporate social responsibility.' But on inspection, corporate social responsibility always takes a back seat to the aim of maximizing profits.[22] And the people running corporations are currently constrained only by the directive to maximize profits. Social activist Michael Lerner states clearly that "Even the corporate executives with the highest level of spiritual sensitivity . . . have no choice but to accept corporate profits as the absolute bottom line" (Lerner 2000, p. 311). It is the overriding goal of maximizing profits which dovetails so neatly with the essence of modern technology.

Ultimately modern technology challenges nature to unlock energy and apply it to further uses, always with an eye toward getting the maximum yield at minimum expense (Heidegger 1955, pp. 14-15). So it is the modern for-profit corporation, free from any other aim, that can best serve the ends of modern technology.

Corporations and other human financial systems are good examples of what I will call *purely human systems. Purely human systems* are those which would become completely meaningless if all human beings were to vanish from the earth. If human beings were to vanish from the earth, all currency and all financial systems would become completely meaningless. So therefore would the corporate aim of maximizing profits. Corporations are also purely human institutions because their goal is profit seeking. Also as an ownership arrangement, they are purely human since property and ownership are also purely human.

Separation from the rest of the environment is built into the nature of purely human institutions. So they ignore or discount the environment in pursuing their prescribed goals. The environmental activist Jakob von Uexkull reports an encounter with Lawrence Summers, former Secretary of the Treasury, former President of Harvard and former chief economic adviser to President Obama. Before a lecture, von Uexkull drew a diagram with a large rectangle labeled 'environment' and a circle inside labeled 'economy.' Summers objected and relabeled the diagram, with the economy as the outside rectangle and the environment as the small interior circle (von Uexkull 2007).

Summers' perspective is exactly that of the for-profit corporation embodying the aims of modern technology in its striving to maximize profits. Of course Summers' perspective ignores the fact that "the economy" could not exist without the environment to support it, in the form of natural resources, air, water, and food. But, as Heidegger observed, the point of view of modern technology leads humanity to believe that everything humans encounter are their own constructs (1955, p. 27).

CORPORATIONS AND KNOWLEDGE POLLUTION

The ethical basis for the wrongness of lying is that lying cannot be a cooperative principle. In order for the lie to succeed, the person being lied to has to believe in the truth of what is being said. And if lying were made a universal principle, this could not happen. No one would believe what was said.[23] Even though a corporation is not an ethical individual, this universal principle consideration still applies. However, since a corporation exists to maximize profits, it often regards its profitability as taking precedence over the cooperative principle of telling the truth.

But because of the impact of corporations on the environment and on environmental policy, there is an additional consideration. Truth telling has another role besides coordinating cooperative behavior between human beings. Truth telling aligns us with the world we live in. Corporate behavior aligns itself only with whatever will maximize profits. Perhaps that would not be the worst if profit maximization were for the long term. But the normal timeframe for profit maximization is often the next calendar quarter or fiscal year.

Even if a corporation maximizes profits over a long time period, there is an underlying insoluble problem. Being ruled by profit maximization, a corporation is not using language to reveal the truth. Thus, it can have no relationship to the being of the world through its use of language. Whatever happens to the world as the result of its actions cannot be something it is concerned with. The environment is the being of the world. Thus corporate dissembling behavior shows that it has no relation to the environment. This can only have one result if persisted in over time—extinction of just about everything in sight, the ecosystem, plants, animals, humans, and finally the corporation itself. The continuing misalignment of corporate activity and what is actually taking place can lead nowhere else.

Exxon Mobil had the head of the Intergovernmental Panel on Climate Change (IPCC) replaced, not for scientific reasons but because they thought if fossil fuels were held responsible for climate change it would threaten their profits. If it turns out that we miss the window for avoiding drastic climate change effects because of the resistance of Exxon Mobil and its fellow corporations, I hope that will lead to stripping corporations of influence on public policy. If the missed window will mean we can no longer prevent the runaway global warming Stephen Hawking and James Hanson are concerned about, no action will be necessary because no action will be possible (Hanson 2009, p. 236). We—and our species—will shortly all be dead.

Previous cases of corporate knowledge pollution should be a warning: Dupont declared that the theory that its chlorofluorocarbons were causing the disappearance of the ozone layer was "rubbish." Tobacco companies for years disputed a causal link between smoking and lung cancer. Science is one of our most powerful and reliable tools for relating to reality. There is no question which should be followed if there is a direct dispute between science and corporations.

Thus, the real danger is that corporations distort reality to serve their own ends. They are a human legal construct which has acquired the powers of action of actual human beings, in fact greater powers. Not only are they not ethical individuals with all that that implies, they are not even sentient beings—with all that that implies. They can neither suffer nor be glad; they can neither be angry nor serene; they can neither be hungry or satisfied; they can neither wonder at a marvel of nature nor be appalled at nature's destructive force. They can be productive and efficient but dangerous in their ability to impact beings whose interests are intimately tied up with all the feelings just enumerated.

Steps should therefore be taken to prevent corporate knowledge pollution. Corporations should be prohibited from attempting to influence public policy by advertising or campaign contributions or by financing electoral initiatives. Given the nature of the danger, this prohibition may be a bare minimum. Corporations also fund biased pseudo-scientific studies to back their dissembling positions. So the prohibition has to be against any corporate influence on public policy in any form. Can we hope that the disconnect of corporations and the environment will ultimately be recognized by enough human beings so that we can deal with it appropriately?

REFERENCES

Achbar, M., Abbott, J., & Bakan, J. (2004). *The corporation.* [Motion Picture]. Zeitgeist Films Ltd. Retrieved from www.zeitgeistvideo.com

Adams, G. M. (2013). *Our hidden history of corporations in the United States.* Retrieved 4/11/2013 from reclaimdemocracy.org/corporate-accountability-history-corporations-us/

Asimov, I. (1954). *The caves of steel.* New York: HarperCollins.

Bacon, F. (1620). *Novum organum.* (B. Montague, Trans.). Retrieved 10/14/11 from history.hanover.edu/texts/Bacon/novorg.html

Burne, J. (2002, January 17). Animal instinct. *The Guardian.*

Dawkins, R. (2006). *The god delusion.* New York: Houghton Mifflin.

Hawking, S. (2009). Galileo and the birth of modern science. *American Heritage of Invention & Technology, 24*(1), 36.

Heidegger, M. (1927). *Sein und zeit.* New York: Harper & Row.

Heidegger, M. (1955). The question concerning technology. In *The Question Concerning Technology and other Essays*. New York: Harper & Row.

Heimbuch, J. (2010). *Ocean film fest 2010: A killer whale takes down a... great white shark??* Retrieved 11/8/11 from www.treehugger.com/natural-sciences/ocean-film-fest-2010-a-killer-whale-takes-down-a-great-white-shark-video.html

Kant, I. (1781). *Critique of pure reason*. London: Macmillan.

Kant, I. (1785). *Groundwork of the metaphysics of morals*. Retrieved on 5/10/04 from www.swan.ac.uk/poli/texts/kant/kantcon.htm

Klein, N. (2011, November 9). Capitalism vs. the climate. *Nation (New York, N.Y.)*.

Kuhn, T. (1962). *The structure of scientific revolutions*. Chicago: University of Chicago Press.

Lerner, M. (2000). *Spirit matters*. Charlottesville, VA: Hampton Roads Publishing.

Levi-Strauss, C. (1962). *The savage mind*. Chicago: University of Chicago Press.

Lewandowsky, S. (2011). Acceptance of science and ideology. *Shaping Tomorrow's World*. Retrieved 11/9/11 from www.shapingtomorrowsworld.org/ideologyScience.html

Markel, H. (2010, October 15). How the word 'scientist' came to be. *NPR*. Retrieved 10/15/11 from www.npr.org/templates/story/story.php?storyId=127037417

Nagel, E., & Newman, J. R. (1958). *Gödel's proof*. New York: New York University Press.

Putnam, H. (1981). *Reason, truth and history*. Cambridge, UK: Cambridge University Press. doi:10.1017/CBO9780511625398.

Rowlands, M. (2001). *The nature of consciousness*. Cambridge, UK: Cambridge University Press. doi:10.1017/CBO9780511487538.

Schultz, R. (2006). *Contemporary issues in ethics and information technology*. Hershey, PA: IGI Global.

Schultz, R. (2009). *Information technology and the ethics of globalization*. Hershey, PA: IGI Global Press. doi:10.4018/978-1-60566-922-9.

The free dictionary. (2011). *Definition of corporation*. Retrieved 11/1/04 from legal-dictionary.thefreedictionary.com/Corporation

Toedt, D. C. (2004). *Serious inconsistencies in the new testament writings: Six reasons for skepticism about the traditionalist account.* Retrieved 11/12/11 from www.questioningchristian.com/2004/10/troubling_incon.html

Tolkein, J. R. R. (1954). *The lord of the rings.* London: Allen & Unwin.

von Uexkull, J. (2007, February 21) *The post corporate world.* Paper presented at the 6th annual Frank Kelly Lecture. Santa Barbara, CA: University of California.

Weiss, J. (2011). *FAQ: Bottlenose dolphins.* Retrieved 11/8/11 from www.fieldtripearth.org/article.xml?id=1496&ordinal=3

ENDNOTES

1 Until recently, the Museum also contained Galileo's finger, preserved in a glass jar. The museum is well worth a visit if you happen to be in Florence. It is located on the river Arno next to the Uffizi.

2 Electromagnetism and weak nuclear force are ultimately the same force, called electroweak force.

3 These terms are due to the philosopher Martin Heidegger (1927).

4 Muslim contributions around the same time were important in other scientific areas. The mathematician Al-Khwarizmi gave his name to the *algorithm.*

5 Thomas Kuhn, in his *Structure of Scientific Revolutions* (1962), made popular the distinction between normal science (which is cumulative) and paradigm shifts (which are not straightforwardly cumulative).

6 The doctrine of the infallibility of the Pope in matters of faith and morals is recent, dating from 1870. It was vehemently opposed by many Catholics including John Henry Cardinal Newman. Richard Dawkins points out that doctrines such as the existence of Limbo have been affirmed by one pope and denied by another. So, as a matter of logic, at least one pope could not have been infallible (Dawkins 2006).

7 Putnam believes that science gains its respect by the layman because it is much more definitive than other means of gaining knowledge of the world such as religion. I don't see why this would be a concern to the non-scientist (1981, p. 185).

8 The uneducated view is that something is less believable because it is a "theory." This is not what "theory" means in science. The heliocentric theory is the theory that the earth goes around the sun. Not less believable because it is a theory.

9 The Christian Bible contains a purely Christian part (the New Testament) and an older part, containing Jewish scriptures. It is interesting that the Jews regard their scriptures as inconsistent and incomplete. There is even a tradition of the *midrach,* which is stories made up to remove inconsistency and incompleteness. Thus there is no basis whatever for the inerrancy of the Old Testament. And contradictions between the four Gospel books in the New Testament are well documented. See Toedt (2004).

10 Naomi Klein (2011) went to a conference of corporate climate change deniers and discovered that they believed, probably correctly, that dealing with climate change would disrupt their profitable economic system. So it is not just the free market ideology but rather the threat to the actual corporate economic system that causes climate change denial.

11 A substantial but incorrect philosophical tradition declares value not to be a rational issue at all. For a thorough account of why issues of purpose and value are indeed rational issues, see Putnam (1981), chapter 6.

9. Ancient philosopher Aristotle and modern philosopher Rawls agree that value is what it would be rational to choose. For example, in thing X with a function, a good X would be one that performs that function to a greater degree. Thus, a good knife would be one that cuts well. Value is clearly relative to point of view. A good bank for an investor is different from a good bank for a robber. But within each point of view, value is an objective issue. See Schultz (2009, Chapter 4, p. 55).

12 Kurt Goedel in 1933 published the first results about undecidability in mathematics—actually in arithmetic. See Nagel and Newman (1958).

13 Heidegger's views on modern technology are largely followed in this book.

14 Heidegger coins the term *Gestell* (translated *Enframing)* for what he calls the essence of modern technology. Some commentators such as Julian Young prefer to leave it untranslated. I prefer to use the English phrase, "the point of view of modern technology" to describe the concept.

15 Jared Diamond in his book *Collapse* is not as positive about the ability of technology to prevent collapse (Diamond 2005).

16 In Asimov's novel, set several thousand years in the future, human beings have colonized planets in other star systems. However, further colonization from earth is prevented by the evolution of a disease-free environment on the other planets. Meanwhile, earth itself has evolved into enclosed mega-cities with totally engineered environments and growing populations of 8 billion each. The environment on earth extrapolated in the novel is present-day New York City, and the criticism made of the environment is a cogent one, namely that the complexity required to sustain such an artificial and complex environment is fragile and that unexpected disruptions are likely to be catastrophic. The

solution in the novel, however, is merely to export people to new planets where there is more space. One must wonder why a species so clearly unable to live within the parameters of its environment will do better when given more space.

17 As reported in Chapter 2, the creator of chlorofluorocarbons drank a glassful of the chemical at a press conference to demonstrate their safety.

18 The most chilling image of the film *The Matrix* (Wachowski Brothers, 1999) is of banks of human beings become pure standing reserve.

19 The first versions of corporations in the 1800s included explicit requirements for corporations to serve the public interest. These requirements were dropped by the end of the 1800s (Adams 2013).

20 The 2010 Supreme Court decision Federal Elections Commission vs. Citizens United decided that corporations deserved the individual right of free speech, continuing a string of thoughtless Supreme Court decisions. As my student, Tim Duncan, observed, although corporations can buy and sell each other, and people can buy and sell corporations, corporations cannot buy and sell people, nor can people buy and sell each other. Therefore, corporations are not people and there is no reason to treat them as such.

21 See the documentary *The Corporation* (2004).

22 See Schultz (2009), Chapter 11, "Corporate Social Responsibility."

23 This is Kant's universal principle argument for the wrongness of lying (Kant 1785).

Chapter 4
Mind Body Consciousness

ABSTRACT

Humans also view themselves as not part of nature when they believe they are primarily nonphysical minds or souls or spirits, but we know that humans came to be on this planet through natural selection. Consciousness, although it is in some ways nonphysical, is still very much a part of nature. The work of two contemporary philosophers, Richard Taylor and John Rowlands, is used to explain the relation of consciousness to the natural world.

INTRODUCTION

How did we get into our false position of being in charge of the rest of the world?

Three partial answers are: Ego, consciousness, and intelligence. These are *necessary* conditions for the gap between us and the world to arise. But they are not *sufficient*. We could have egos, we could have consciousness, we could have intelligence without being estranged from the world we live in. Animals have egos and consciousness. Domestic cats have egos. If two cats are involved in an intense fight and you spray them with water, they will forget the fight and start furiously licking themselves to deal with the threat to their appearance. Elephants and primates such as chimpanzees, bonobos, gorillas, and orangutans are some animals that have both egos and consciousness. They have selves and can recognize themselves in a mirror. Elephants are upset at the deaths of fellow elephants.

DOI: 10.4018/978-1-4666-4586-8.ch004

Intelligence is closer to an answer to the question of how we got to think of ourselves as in charge of the world. Chimpanzees that make tools are intelligent. However, their ability to make tools does not divorce them from the ecosystem. Other species with similar or possibly greater intelligence do not appropriate the world, for example dolphins. The 64,000 dollar question is what do dolphins use their intelligence for if not to appropriate the world? Our intelligence used as though we are separate from the rest of nature could lead to our extinction.

One major source of the disconnect between human beings and the rest of nature is the belief that human beings are special in having *minds* or *souls* which are not physical or material. One prime motivation for these beliefs is that the nonphysical self survives death. There is even more of a disconnect when our world is felt to be defective or inferior to some better immaterial world (heaven, in many religions) which humans will go to when they die. Many religious people who believe in an afterlife also believe that this world is a place of trial: If one is good, one goes to heaven. If one is bad, one goes to hell. Most Christian sects believe in an afterlife and reward or punishment after death. Such beliefs go back as far as we have any record of religions. Ancient Egyptians mummified their rulers so that their immaterial spirits could continue to have sustenance. Both Egyptians and Mayans built elaborate tombs for their rulers.

These doctrines were not so dangerous when humans had less control over the workings of the world. But they can easily lead people to discount the necessity of preserving and protecting the world we live in. Some fundamentalist American Christians in positions of power over what happens in this world have the view that we need not care about this material world. Senator James Inhofe of Oklahoma, former chair of the Senate Environment Committee, said on national TV that those with environmental concerns were at risk of idolatry by worshipping "God's creation" rather than God. Inhofe belongs to a Christian sect described as End-Timers. End-Timers believe that Jesus will return in the near future and then the world will end. End-Time beliefs make such problems as global warming, the damaged ozone layer, and poisoning caused by industrial by-products inconsequential. Because the world will end soon, End-Timers are concerned only with short-term political-theological outcomes, not long-term solutions. Dealing with environmental issues such as the conservation of endangered species and curbing climate change require belief in an enduring earth[1] (Scherer 2004).

The enduring earth with humans on it should not be abandoned because of beliefs not based on any evidence. In any case, whatever the nature of any afterlife, we each begin here and live in this world. Until recently it was common for human beings to believe that the universe as a three-storied place, with the living in the middle, the dead in the underground, and the gods and spirits in the heavens. These views continue to be advocated by many religions and believed by many people.

Yet as soon as it turned out the earth was not a flat indefinite plane, such views can be at best figuratively true. It is a measure of the intelligence of most believers in Heaven and Hell and the afterlife that suicide is rarely committed to gain entry to a place that does not exist in physical reality. Yet the old beliefs persist in some form even when knowledge has advanced beyond them. The Heavens Gate group in San Diego combined both world views by committing mass suicide in 1975 to join aliens supposedly circling the earth in spaceships. Thus, they combined a nonphysical transition from life to "another plane" with a physical location their disembodied spirits would be moving to.[2]

Perhaps such beliefs continue to be held because of the experience of a benevolent force in the world whose favor ought to be sought, not only for the afterlife but also during this life. Sometimes called providence or God's providence, it is the belief that God intervenes for our benefit in the course of our affairs. We ask God for some special favor, such as passing an exam or having someone fall in love with us. When we get what we ask for, we believe God has provided the favor. Such experiences may seem valid, but it is hard to see why an all-powerful creator god would take his creation back, so to speak. It is more likely that natural selection leads to an astonishing but imperfect harmony in the biosphere in which we live. Living things are going to evolve in proximity to their food sources and other necessities of life. So these necessities will be 'providently' provided. However, predators and prey also evolved in the same ecosystem and it can hardly seem 'provident' to a gazelle to be eaten by a lion. Or, for that matter, a hiker to be eaten by a bear.

The complex interdependence characteristic of the ecosystem, although unplanned, does seem to exhibit intelligence. We see plants adapted in complex ways to fit animal behavior, such as trumpet vines adapted to be pollinated by hummingbirds and hummingbirds adapted to get nectar from those and similar flowers (Hilton 2012). Traditional theists see the action of an intelligent designer. But there is no reason why complex design could not simply be the result of natural processes.

The multiple interdependencies of environment, plant, and animal have led some to regard the entire planet as a living system called Gaia. The "Gaia Hypothesis" is that entire earth is a system with the goal of regulating conditions so that they are favorable for current life (Flannery, p. 34). There is some evidence for this hypothesis, but some scientists cannot accept the purposiveness required by the Gaia hypothesis and some theistic religious leaders regard the Gaia hypothesis as blasphemously usurping the powers of their preferred God. However, the episode of the stromatolites in earth's history shows that Gaia does not particularly care which life is preserved.[3]

Some climate change deniers attempt to use something like the Gaia hypothesis to bolster their case.[4] The idea is that human imbalances will be automatically corrected by the appearance of a correcting force in nature. In such circumstances, humans are not superior. Their missteps are taken care of by a superior power, which could just as well be the traditional monotheistic all-powerful God. But the existence of the stromatolites shows that Gaia does not care. If humans manage to wreck the environment to the extent that only very different life forms can survive, that is fine for Gaia.

The true overall situation is this: The universe is composed of different levels which are largely unaware of each other and whose interests conflict. I became aware of this situation when I was attempting to water some plants in my front yard. I severely disturbed a nest of ants, who began furiously running around and rescuing their larvae from this unforeseen threat. I had absolutely no intention of causing this crisis. So too with the relation of predators to prey. The lion has no concern whatever for the gazelle's concerns. That's how things have been organized by natural selection. So providence is sometimes, maybe be even usually, provident for individuals. But by no means always.

MINDS AS NON-PHYSICAL

Regardless of a religious basis for human beings possessing nonphysical souls, it can seem that consciousness can only be explained by our having non-physical minds. The clearest and most dramatic statement of this belief is due to the early modern philosopher Descartes. In his *Meditations* (1641), Descartes asserts that the universe consists of two different kinds of things, minds and bodies. Human beings (and God) are the only things that have minds. Everything else is bodies, which are more-or-less complicated arrangements of matter. Thus other living things, that is, animals and plants, are basically just complicated natural machines (Descartes 1637). Mind is completely non-physical, and its main property is *thought.* Bodies are physical and their main property is *extension* (being extended in space).

Descartes' view that mind and body are completely different kinds of things leads to an immediate problem, namely how they can interact with each other. Descartes' own solution is implausible: His solution is that mind and body interact in the pineal gland. Actually, it does not matter what choice of physical object we pick to do the interaction, even whether it is the brain or some part of it. There is still no possibility of a non-physical thing having a causal influence on a physical thing. Or vice versa.

One possible solution to this dilemma is just to deny the existence of mind altogether and say that there are only physical things in the universe. This is the doctrine of *physicalism,* which is felt by some to be required by science. It seems to me to be a pointless doctrine in the face of the subjective experience of *consciousness.*[5] But some believers in physicalism actually deny that there is such a thing as consciousness![6] Science does not require physicalism, for reasons given in Chapter 3, "Science, Modern Technology, and Corporations." Physicalism, the view that the universe is composed of bodies interacting with each other, has been untenable since Newton. Newton's concept of gravity is action at a distance. Physicalism has become even more untenable in contemporary physical theory. In addition to bodies directly interacting and gravity, physics now admits three other fundamental interactions: Strong nuclear force, weak nuclear force, and electromagnetism. Electromagnetism also acts at a distance.

A coherent alternative to mind/body is that of the 20th century German philosopher Martin Heidegger. Heidegger finds that a human being is actually *Dasein (presence)*. The German *Dasein* literally means 'being here' or 'being present.' For Heidegger, human beings are not disembodied minds trying to find some impossible connection to the material world. Rather, human beings begin as beings that are present in a world. This presence in the world is manifested in consciousness (Heidegger 1926, p. 211). Although Heidegger's account may give us the proper placement of consciousness in reality, it leaves untouched the same problem presented by Descartes: We know that we actually are in some respects natural machines, and that the phenomenon of consciousness must somehow be linked to neural processes in the brain. But once again, the problem is how can this connection be made? Also, we are not, as Heidegger has it, presence in a *world.* Rather, we are presence as a point of view of a world. This small alteration makes a big difference in understanding consciousness, a topic we will return to shortly.

HUMAN BEINGS AS NON-PHYSICAL ENTITIES

Descartes' view that human beings are primarily non-physical beings echoes a widespread religious doctrine: That human beings possess, or even are, *souls* or *spirits.*

These souls are often said to survive death (they are *immortal* souls). Like Descartes' minds, they are not material (Indeed, Descartes thought his concept of mind was a clearer version of the concept of immortal soul.) (1641, Synopsis of Meditation 2).

Usually, believers in souls confine souls to human beings, reinforcing the chasm between human beings and other animals. These beliefs can be used as the basis for the treatment of animals as mere objects. The belief that the important part of our nature is non-material can also lead to the view that the material world (that is, nature or the environment) is not of great importance to human beings.

Occasionally people speculate that humans were introduced to this planet from elsewhere by alien beings. This would make humans a completely different form of life from everything else on the planet. If so, humans would not be directly responsible for relating to the rest of life on this planet. However, all DNA on this planet, including human DNA, is written in an "alphabet" of exactly four amino acids: adenine, guanine, cytosine, and thymine. However, there were an additional 16 amino acids with the same chemical properties which could have been used. So chances are high that all DNA-based life on this planet had the same origin.

Some authors actually embrace the possibility of leaving the current basis of life completely behind. Perhaps the most extreme version of this view is that of the roboticist Hans Moravec. Moravec suggests that it is time to leave DNA-based life behind in favor of the superior creations of the human mind. These "Mind Children" (the title of his book) will become our progeny through a process of "downloading" (Moravec 1988, pp. 115-117). To think that such a process is even possible requires that human beings be radically different from other animals in having minds as well as physical bodies. The same problems about the relation of mind and body which faced Descartes are still no closer to a solution. But no account of how a relation is possible between a physical and a nonphysical thing has ever made much sense. Moravec merely makes the mind physical: He is a good scientific physicalist. We can "download" patterns of information, as if that is all there were to human consciousness and intelligence. A (very) large collection of leather-bound volumes would be interchangeable with any human being.

CONSCIOUSNESS

It is the necessity to explain the existence of consciousness which gives rise to the idea that consciousness belongs to a different, nonphysical realm. The 17th century philosopher Spinoza is reported to have observed that if a thrown stone had consciousness, it would think it had determined its own trajectory.[7] Here is a conscious stone's likely internal monologue: "I'm going up, up, and up. I think I want to just keep going up. Wait a minute, maybe level would be good, yes, I'll try level. Wait a minute, maybe down would be even better. Yes, let's try down. Wait a minute. Here's the ground. I don't feel like moving now." Of course, we know the stone's path, a parabola, is determined by the initial force of the throw and by gravity. But the conscious stone thinks otherwise. The conscious human's more complicated trajectory is ultimately determined by the same physical laws, the causes and conditions to which that human is connected. However, consciousness itself cannot be completely reduced to those laws. This insight is due to the philosopher Mark Rowlands (2001).

His reasoning, mentioned in the preceding chapter, is this: When things known through consciousness are explained scientifically ("reduced"), what is removed is the contribution of consciousness to that knowledge. For example, when the feeling of *heat* is scientifically explained as a certain level of mean molecular kinetic energy, the conscious phenomenal description disappears from a scientific point of view. Therefore, to explain consciousness scientifically what would have to be removed, is the contribution of consciousness to consciousness! So there is nothing left to explain scientifically.

If consciousness cannot be reduced to neural states, the problem remains to characterize the interaction of consciousness and physical nature. That is, how can conscious states take part in the physical causal order? Rowlands' proposed solution is this: The sensation of pain cannot take part in the physical order, but the consciousness of bodily damage brought about the pain, can take part in the physical order (2001, p. 234). But the sensation of pain *does* take part in the physical causal order. I jerk my hand away from the top of the hot stove not because I am conscious of bodily damage, but to remove the pain. Thus there can be *two* causes for the same action differently viewed, the pain causing me to jerk my hand away, and the complex sequence of neural events which results in the contraction of muscle fibers in my arm. Rowlands explicitly rejects this account because it involves what he calls "the . . . untenable doctrine of causal overdetermination" (2001, p. 227) Causal overdetermination is untenable because if there is more than one cause sufficient for an effect, neither can be strictly speaking the cause.

Another possibility is implicit in Rowland's non-reduction argument. The distinction between the ready-to-hand and the present-at-hand from Chapter 3 "Science, Modern Technology, and Corporations," is relevant here. Science considers the world as *present-at-hand.* In our everyday lives, we consider the world as *ready-to-hand.*[8] The world as ready-to-hand is the world as we interact with and alter it, with our equipment and constructions. The (scientific) world as present-at-hand is the world we measure and describe at a distance, a world without purposes or values. Conscious experience is of the ready-to-hand, the everyday world of purposes and values. Conscious experience is not of the world as present-at-hand to science. Thus to "reduce" the jerk of my hand to neural states and muscle contractions, we remove the pain and my purpose in getting rid of it. The world as ready-to-hand is redolent with purposes which vanish when reduced to the world of physical science. There are not multiple causes, but rather two entirely different points of view.

Scientists become part of the world as ready-to-hand when setting up experiments, for example considering whether some equipment is better than others. But any resulting knowledge from the experiment is of the world only as present-at-hand, minus any purposes or value.

Scientists and those sympathetic to science very often get themselves into deep trouble by supposing there is a world present-at-hand independent of the world as ready-to-hand. Two accomplished scientists, Israel Rosenfield and Edward Ziff, began a review of a book on neurophysiology with the following statement:

The worlds we see, hear, feel, smell, and taste exist independently, but we know them only through the fabrications of our brains. The colors we see do not exist apart from our perception of them. The words and sentences we believe we are hearing are a jumble of sounds, whistles, grunts, and silences. From a variety of external signals our brains create something that is not there (Rosenfield & Ziff 2012).

The first sentence slams headlong into a major philosophical problem. If we know the world only through *fabrications* of our brains, we do not know the world at all. We can picture Rosenfield and Ziff in white lab coats with clipboards taking notes on their experimental subjects, who we can suppose are reclining in chairs and hooked up by wires to various monitors. Rosenfield and Ziff are recording what their subjects' brains create "from the variety of external signals." And what these brains create is "something that is not there," according to them. However, they are in the same boat as their experimental subjects. Although they think they are recording the brain activity of their subjects, the procedures of this scientific experiment are really just creations of their brains from external signals, and are "something that is not there."

The problem here is none other than that of the relation of consciousness to the world. This is not a trivial problem, and one that so far does not have a completely satisfactory solution. It was one of the main problems of Kant's *Critique of Pure Reason*. Once brain activity is divorced from knowledge of the world, there is no way to get the world back. In the next two sections, I will try to show how brain activity gives us knowledge of the world.

CONSCIOUSNESS AND POINT OF VIEW

We know that consciousness is not reducible to physical properties, which does not tell us much about how consciousness relates to the world. In the next two sections, two topics will be discussed important to an account of consciousness which is faithful to our experience of it: First, the role of point of view in consciousness; and second, philosopher Richard Taylor's argument about the relation of consciousness to the world.

We have to remember that human beings are animals, and their initial presence is as a point of view in a world.[9] Our idea of space, as the place where we are present together with others and the rest of reality, is a construct. Animals exhibiting extensive coordinated purposive cooperative behavior may for that reason also have this concept of space.[10] The construct is an unfilled space of indefinite extension, available for entities to appear in it. No one has ever *experienced* this space directly. You never directly see the space behind you. Yet space is conceptualized as indefinite in extent and three-dimensional. It contains the world as ready-to-hand. Conversely, even the most theoretical physics does not and cannot completely abstract from a human presence. Both quantum physics and relativity theory reinstate an *observer,* a point of consciousness which reflects its environment. Tucked into every representation of the world, whether scientific present-at-hand or everyday ready-to-hand, is a conscious point of view.

However, because of science's abstraction from purpose and value, its observers are only points of view and nothing more. The set of elaborate interdependencies which make human life possible are missing. For science, evolution does not have a purpose. Thus, the amazing increased complexity of the ecosystem and the appropriateness of its parts to each other are reduced to the operation of natural selection. Just as with consciousness and its phenomenal properties, value and purpose properties are left out.

The belief that human beings are the pinnacle of evolution is merely another form of cheerleader human superiority.[11] But the greater power of human beings over the environment manifests itself in human actions for purposes and toward goals. So that power cannot be dealt with except on those terms, the terms of the world as ready-to-hand.

THE RELATION OF CONSCIOUSNESS TO THE WORLD

From these observations about point of view, we now turn to an argument by philosopher Richard Taylor about the relation of consciousness to the world. Taylor's version of an argument from design shows that intelligence and consciousness cannot exist in the universe only within human beings. It presents an unusual perspective on the relation of consciousness to the world (Taylor 1974).[12]

Although Taylor introduced the argument as a version of an argument intended to prove the existence of God, he notes at the end of his discussion that whatever is proved lacks important traditional properties of God such as benevolence. After stating the argument, I will state what Taylor says he believes it proves and then give my own somewhat different account of its relevance for consciousness.

Taylor begins as follows:

Suppose, then, that you are riding in a railway coach and glancing from the window at one of the stops, you see numerous white stones scattered about on a small hillside near the train in a pattern resembling these letters:

THE BRITISH RAILWAYS WELCOMES YOU TO WALES

Now you could scarcely doubt that these stones do not just accidentally happen to exhibit that pattern. You would, in fact, feel quite certain that they were purposefully arranged that way to convey an intelligible message. At the same time, however, you could not prove, just from a consideration of their arrangement alone, that they were arranged by a purposeful being (Taylor, p. 114).

Taylor continues by noting that *apparently* purposeful arrangements often have such nonpurposeful explanations such as chance variation and natural selection.[13] But he notes:

Here, however, is the important point . . . that if, upon seeing from the train window a group of stones arranged as described, you were to conclude that you were entering Wales, and if your sole reason for thinking this . . . was that the stones were so arranged, then you could not, consistently . . . suppose that the arrangement of the stones was accidental. You would, in fact, be presupposing that they were arranged that way by an intelligent and purposeful being or beings for the purpose of conveying a certain message having nothing to do with the stones themselves (Taylor, p. 115).

Taylor then gives a similar example of marks which apparently state a message in some ancient language, and concludes:

It would . . . be irrational for anyone to suppose both that what is seemingly expressed by the marks is true, and also that they appeared as the result of nonpurposeful forces, provided the marks are his sole evidence for believing that what they seem to say is (Taylor, p. 116).

The *coup de grace* then follows. Taylor asks us to consider our sense organs, brains, and nervous system which enable us to have knowledge of the world. Eyes in particular have greater functionality and complexity than any camera designed by men. Although it is natural to think of eyes as having been constructed *in order to* allow people to see, most biologists hold that eyes are the result of long years of chance variation and natural selection, without any "in order to" or purposiveness. Taylor continues:

The important point . . . is that we do not simply marvel at these structures, and wonder how they came to be that way. We do not simply view them as amazing and striking things, and speculate upon their origins. We, in fact, whether justifiably or not, rely on them for the discovery of things that we suppose to be true and that we suppose to exist quite independently of those organs themselves. We suppose, without even thinking about it, that they reveal to us things that have nothing to do with themselves, their structures, or their origins. Just as we supposed that the stones on the hill told us that we were entering Wales—a fact having nothing to do with the stones themselves—so also we suppose that our senses in some manner "tell us" what is true, at least sometimes. The stones on the hill could, to be sure, have been an accident, in which case we cannot suppose that they really tell us anything at all. So also, our senses and all our faculties could be accidental in their origins, and in that case they do not really tell us anything either. But the fact remains that we do trust them, without the slightest reflection on the matter (Taylor, pp. 117-118).

Taylor concludes that if we trust our sense organs to tell us things about a world other than themselves, we must believe they have been created with that purpose by some purposive being. He concludes that the argument has very little connection with any of the usual properties attributed to God, and therefore that it has only philosophical interest (Taylor, pp. 119-120). The lesson I draw from his argument is different. My belief is that consciousness is of a world and not merely a succession of mental states. By Taylor's argument, this belief requires me also to believe that the universe is ordered *in order to* make this possible. That is, purposiveness does not reside solely within human beings. And, as Taylor's argument makes clear, it could also be true that consciousness arose as the result of chance variation together with natural selection. That in no way removes the necessity of my having a belief about purposiveness in the universe. That belief is still necessary for me to trust that my consciousness is of a world.

God is often felt to be a big human being with more power. This may be because human beings are only familiar with themselves as agents with purposes. But what Taylor's argument establishes is that without purposiveness other than human purposiveness, we cannot regard ourselves as having knowledge of anything outside ourselves. We can call this "Taylor's Trust Principle" (TTP):

Anything taken as informative other than about itself must be taken as having the purpose of being informative.

Using Taylor's Trust Principle, we conclude the following: Consciousness is not brain activity, nor is it located anywhere. Consciousness is of reality, and what we (for the most part) experience is reality itself. By Taylor's Trust Principle, we are not "brains in a vat," like the storehoused humans in the film *The Matrix* (Wachowsky, 1999). Consciousness does not tell us what is going on in our brains. Consciousness has the purpose of telling us about the world around us. Either it can be trusted to do that, or there is no world around us. Therefore, Taylor's Trust Principle implies that consciousness cannot be another world in itself. There is no separate spirit world and thus no ground for discounting the importance of our world, the real world.

The relation of our consciousness to the world is in some important ways like our relation to a documentary film. Suppose we are watching Robert Flaherty's astonishing 1922 documentary *Nanook of the North.* This film is about Inuit life in the Arctic, filmed with actual Inuit people performing their everyday tasks. Consider the scenes showing the building of an igloo. We have had enough experience with movies to believe what we are seeing on the screen is some Inuits building an igloo. We do not believe the film is not a record of what happened because it is only light projected on a reflective screen. Even more so with our conscious experience. We don't need to know the mechanism behind our contact with the world. Of course these days, just about anything in a film can be digitally simulated. But a successful CGI simulation is one that fools us into thinking something real was photographed. So there is a standard for seeming real which successful CGI attains. The real world, fortunately, cannot be simulated. The distortions of hallucinogens and mental illnesses are just that, distortions.[14]

The most important lesson to be learned from this account of consciousness is that it does not make human beings different from all other beings. We also belong to this world and any belief that makes us essentially different from the world, whether nonmaterial or spiritual, is simply mistaken. Any solution to the problem of our conflict with nature has to keep in mind our presence in this world as an evolved animal. We know enough from science about our place in the environment to make any view that says we are not part of this world and its ecosystem, dangerous nonsense.

CONCLUSION OF SECTION 1: WHERE WE ARE NOW

It is clear that human beings have *physical* superiority enhanced by their technology over other living things and to some extent over the environment. It is also clear that there is no objective justification for any kind of *practical* or *ethical* human superiority. Quite the contrary, the greater power of human beings seems to be often used in ways which threaten the destruction of not only much of the rest of the environment, but even themselves. In the past few hundred years, practices embodying modern technology have frequently been in conflict with the rest of the ecosystem and have damaged it. Perhaps this behavior is not surprising, given that many humans believe they are essentially different from any other beings on the planet. Human beings, in the words of anthropologist Claude Levi-Strauss, "appropriat[e] the world as if it were a thing and behav[e] on it with neither decency nor discretion" (1968).

But human beings did not always see themselves apart from and above the environment. At some point, their ancestors were animals among other animals. Always, to be sure, intelligent and capable animals, but during their lifetimes poor candidates for masters of the universe. The next part of this book, Section 2, "The Origins of Technology and Human Superiority," will explore how the transition took place from animal in the ecosystem to (imperfect) master of the ecosystem.

REFERENCES

Descartes, R. (1637). *Discourse on method.* Retrieved 10/10/11 from records.viu.ca/~johnstoi/descartes/descartes1.htm

Descartes, R. (1641). *Meditations concerning first philosophy.* Retrieved 9/14/11 from www.wright.edu/cola/descartes/mede.html

Flaherty, R. (1922). *Nanook of the north.* [Motion Picture]. New York: The Criterion Collection.

Flannery, T. (2010). *Here on earth.* London: Penguin Books.

Heidegger, M. (1926). *Sein und zeit.* New York: Harper & Row.

Hilton, B. (2012, July 8). Hummingbirds and flowers: A study of co-adaptive relationships. *Howard Hughes Medical Institute.* Retrieved 7/8/12 from www.hhmi.org/coolscience/resources/SPT—FullRecord.php?ResourceId=11

Jaynes, J. (1976). *The origin of consciousness in the breakdown of the bicameral mind.* Boston: Houghton-Mifflin.

Kirby, P. (2011, February 15). Religion the ultimate tyranny. *Washington Post.* Retrieved 4/27/12 from onfaith.washingtonpost.com/onfaith/panelists/paula_kirby/2011/02/religion_the_ultimate_tyranny.html

Kuijsten, M. (2006). *Reflections on the dawn of consciousness.* Henderson, NV: Julian Jaynes Society.

Levi-Strauss, C. (1968). *The origin of table manners.* New York: Harper & Row.

Nolan, C. (2010). *Inception.* [Motion picture]. Burbank, CA: Warner Bros.

Rosenfield, I., & Ziff, E. (2012). Review of Shepherd's Neurogastronomy. *The New York Review of Books*, *59*(11).

Rowlands, M. (2001). *The nature of consciousness.* Cambridge, UK: Cambridge University Press. doi:10.1017/CBO9780511487538.

Scherer, G. (2004, October 27). Christian-right views are swaying politicians and threatening the environment. *Grist*. Retrieved 11/18/11 from www.grist.org/article/scherer-christian

Taylor, R. (1974). *Metaphysics*. Englewood Cliffs, NJ: Prentice-Hall.

Wachowski, A., & Wachowski, L. (1999). *The matrix*. [Motion picture]. Burbank, CA: Warner Bros.

ENDNOTES

1 Recently, some influential Christians have become environmentalists. Pat Robertson, long opposed to environmentalism in any form, now believes climate change is real and that action should be taken to deal with it.

2 Perhaps one of the reasons these beliefs persist is the ability of religions to enforce them as part of a package of beliefs which must be accepted on faith. See Kirby (2011) on the baneful effects of religion on freedom of thought.

3 See Chapter 2, "Conflicts with the Ecosystem," Species Extinction.

4 See Doug L. Hoffman 's Website, *theresilientearth.com*

5 This book is primarily concerned with consciousness as awareness of the world. Other facets of consciousness such as the experience of agency and interior monologue have drawn the attention of psychologists such as Julian Jaynes (1976). Jaynes claims that consciousness as the experience of agency together with interior monologue did not occur until about 3000 years ago. This type of consciousness is a "cultural construct." But so is modern technology. And, unlike modern technology, it is difficult to work out exactly what the impact of this internal change has been on the relation of humanity to the world. See also Kuijsten (2006).

6 Philosopher Daniel Dennett is perhaps the best known and most prolific.

7 Stanley Martens conveyed this statement to me in 1971. He said it was due to the philosopher Benedict Spinoza, and it is certainly consistent with Spinoza's views. I have not been able to locate the saying in Spinoza, so it may be Prof. Martens' interpretation.

8 These terms are due to the philosopher Martin Heidegger (1927).

9 This is *not* the view that human beings start with unorganized sensations.

10 Predators hunting in packs whose behavior must be consciously tailored to the behavior of others may be the best example: Lions, wild dogs, wolves, even dolphins herding schools of fish. Defensive behavior such as the schooling of

fish may not require awareness of the individual animal's place in the cooperative activity. Similarly with social insects such as ants and bees.

11 See Chapter 1, "Humans as Masters of the World," "Cheerleader Superiority"

12 References are to the second edition of Taylor's book *Metaphysics* (1974). Two later editions exist, but the present argument was not altered appreciably in those editions.

13 It is worth reading Taylor's (1974) full argument on pp. 114-119 for the elegance and convincingness of his detailed presentation. The summarized version here is sufficient for my purposes.

14 Dreams are not a good candidate for simulations of the real world. Dreams are notably fragmentary, often missing such details as any identity characteristics for a supposed person. Or sometimes the characteristics of a person change drastically and for no reason. For me, it is decisive that I can sometimes tell within a dream that I am dreaming. The film *Inception* (2010) has diminished plausibility for me because it presents dreams as full-featured worlds.

Section 2
The Origins of Technology and Human Superiority

Chapter 5
Humans as Hunter-Gatherers

ABSTRACT

The development of human separation from the environment is outlined beginning with our immediate ancestor Homo erectus. This ancestor made tools complex enough to save and store, thus creating a human space distinct from the rest of nature. Homo erectus later tamed fire and built temporary shelters, again adding to a human environment distinct from nature. When Homo sapiens, our species, appeared, they adopted additional practices that separated them from the rest of nature. They made and wore clothing, carried out ceremonial burials, began making representations, and exterminated nearly all animals larger than themselves.

INTRODUCTION

In the past four chapters, a number of factors emerged which contribute to humanity's conflicts with nature: Attitudes of human superiority and human exceptionalism; modern technology and its regarding nature as nothing but resource; and humans as beings that do not really belong to this world. We could try to return to the time before some of these factors came into play and call the previous state, the solution. There are problems with this approach. In some ways, for-profit corporations are currently the most serious threat to life on this planet. Yet, without the power and aims of modern technology they would probably not be that much of a threat. So one might consider eliminating the modern technology that powers them. This may

DOI: 10.4018/978-1-4666-4586-8.ch005

not be a good idea for a number of reasons. It is not just a matter of eliminating a few gadgets. Our entire economy is technology-dependent and as a result has also become globalized. Our food supply is dependent on the workings of globalized commerce. Modern technology was an addition to previously existing civilization. But even the people of pre-modern civilizations depended for their survival on complex institutionalized food production. And when those institutions collapse, most people die. When the classic Maya civilization in what is now Honduras collapsed, population went from a conservative estimate of 3 million people down to about 30,000 (Diamond 2005, p. 175).

Also, an earlier state which in itself does not cause ecological problems, may contain the basis for a later state which does cause problems. For example, as Heidegger (1955) saw, modern science contained within itself the point of view of modern technology. Thus to eliminate modern technology we would have to eliminate science. And eliminating science would eliminate almost all ability to deal with ecological problems. (Prescientific civilization had already produced ecological problems.) Reverting to civilization before science and modern technology is not an answer either. As Joseph Tainter has documented, all previous civilizations have collapsed. Tainter believes that technology has prevented the collapse of the current civilization (1988). But we already have sufficient evidence that technology both helps and hinders. So technology may not be able to prevent the collapse of our civilization.

Some look further back to pastoral life or to a hunter-gatherer existence. People living these lives may seem to have been more in harmony with nature than us civilized types. There is certainly some truth to this, because hunter-gatherers have to think of themselves as part of the ecosystem. If they fail to understand their relation to the ecosystem, they cease to exist. Yet we also know that wherever human hunter-gatherers have appeared, all large animals (megafauna) usually go extinct quickly. So much for a more spiritual relation to the environment. There is abundant evidence for human-caused megafauna extinctions going back at least 60,000 years (Leakey 1995).

Yet human beings (or at least their ancestors) were not always apart from and above the environment. At some point, our ancestors were animals among other animals. Always, to be sure, intelligent and capable animals, but during their lifetimes poor candidates for masters of the universe or even the ecosystem. This part of this book, Section 2, "The Origins of Technology and Human Superiority," will explore how the transition took place from animal in the ecosystem to imperfect but overconfident master of the ecosystem. Hopefully, this procedure will lead us to a usable diagnosis of what may be leading humanity to destroy itself. And without an adequate account of how we came to be as powerful and as destructive as we are, proposed remedies for our situation are likely not to help much.

Trying to picture what it was like to be a human in the various earlier periods is not easy. As soon as we leave written history, evidence of attitudes and inner life is inferential. Yet although there is no direct evidence of attitudes of human superiority in prehistory, there are ample physical traces demonstrating the growing separation of humans from the rest of nature. For example, when humans began wearing clothing, they were no longer an animal amongst animals. Various kinds of separation from the ecosystem together with humanity's greater physical intelligence and power would surely be enough eventually to engender attitudes of human superiority to the rest of nature.

But even if my details and dates are incorrect—especially in prehistory—I do not believe my conclusions will be affected to any great degree. For example, I accept the current consensus that human tool use began with our ancestor *Homo habilis* about 2.5 million years ago. But there may turn out to be additional convincing evidence that our earlier ancestor *australopithecus afarensis* used tools much earlier, 3.5 million years ago. If so, that will not require me to revise the sequence of events humans and their ancestors experienced in getting to our present circumstances.

However, it is easy to fail to appreciate that we humans have lived very differently. Just within the past century, it is not that easy to remember the changes. I can personally remember back to mid-20th-century before television, when there was just radio. I can remember when there were few automobiles and lots of streetcars (now slowly returning); I can remember when there were no jet planes (For the record, I was born in 1942.). There were phones, but when you picked them up, a human being would ask who you wanted to call and make the connection. And my grandparents lived in a world without any automobiles and planes (The Wright brothers first 59-second flight took place in 1903.). There were lots of horses. But I am not so much concerned with the development of the products of technology but rather with the changes in human attitudes toward the world brought about by the widespread use of those artifacts.

A contemporary example would be the changes in attitude brought about by the automobile in a city like Los Angeles. With few exceptions, people drive to disparate destinations for all major life functions. Because people no longer have much contact with each other in shared communal spaces, alienation and distrust of one's fellow Angelenos has become widespread. Gated communities are the preferred new housing development, providing the illusion of security and further alienating people from each other. Ironically statistics show that gated communities only marginally improve protection from crime. The most effective protection is Neighborhood Watch, where a neighborhood communally undertakes its own surveillance[1] (Drew 2008).

Table 1. Human timeline

Period of	Beginning Date
Hunter-gatherers	4,000,000 years (earliest human ancestor)? 200,000 years (homo sapiens)
Farmers	11,000 years
Civilization	7,000 years
Modern Technology	250 years

The lesson is that technology is never a neutral tool. Technology always has consequences for how people can live and how they view their world. This is true now, and it was true millions of years ago.

THE HUMAN TIMELINE

For my purposes, the human timeline can be divided into four periods, each of which embodies a dramatically different way of life, including different ways of relating to the rest of nature (see Table 1).

I will now discuss each period in turn, beginning with a list of the main events for each period, especially concerning human control of the environment and separation from the environment.

PREHISTORIC HUNTER-GATHERERS

Not only did early modern humans (*Homo sapiens)* practice a hunter-gatherer lifestyle. So did our ancestor species *Homo habilis and Homo erectus.* Only members of our species, *Homo sapiens,* were subsequently farmers or were civilized. Our more remote ancestors also employed proto-technology (my term from Chapter 3). That is, they used tools, sometimes more persistent and advanced tools, and some had control of fire. Table 2 reflects our main ancestors from the period of hunter-gatherers in the context of the major species figuring in human evolution. (O'Neil 2012).

It can be difficult to pin down features of a hunter-gatherer lifestyle. The assumption is that features of the prehistoric lifestyle are in many respects the same as features of current hunter-gatherer populations. But unfortunately some claimed features are not found in either current or earlier hunter-gatherer populations. Or the features are not unique to humans. One disputed difference is a division of labor between men and women, with men usually doing the hunting for fish and game

Table 2. Human evolution

Name	First Appearance (years ago)	Extinct as of (years ago)
australopithecus afarensis	4,000,000	2,900,000
australopithecus africanus	3,000,000	2,000,000
homo habilis (see Figure 1)	2,500,000	1,400,000
homo erectus (see Figure 2)	1,800,000	70,000
homo sapiens neanderthalensis	600,000	30,000
homo sapiens sapiens	200,000	?

Figure 1. Scientific reconstruction of Homo habilis (Picture from Wikipedia Commons taken by Lillyundfreya)

and women foraging for plant food. Division of labor between women and men is actually rare in current hunter-gatherer societies and was also likely rare in Paleolithic societies (Kramarae & Spender, 2000).

Another major distinguishing difference between humans and apes, according to Richard Leakey, is sharing food. Leakey quotes an associate as remarking "[A chimpanzee] might say, 'You humans are very odd; when you get food, instead of eating it promptly like any sensible ape, you haul off and share it with others'" (Leakey 1992). However, further observation of chimps has found that they do share food, both meat and plant, as well as hunting tools (Pruetz 2011).

Another once-popular theory of the difference between humans and apes was that humans were hunters and apes were not. Thus, Robert Ardrey in his *The Hunting Hypothesis* claimed that "Man is man and not a chimpanzee, because for millions of years we killed for a living" (Ardrey 1977). Richard Leakey notes that the idea of man as a hunter played into notions of human superiority, of humans as "somehow triumphant in evolutionary ascendancy over the apes, right from the beginning" (1992, p. 183). But the archaeological record is not clear on whether animal remains

Figure 2. Scientific reconstruction of Homo erectus (Picture from Wikipedia Commons taken by Lillyundfreya)

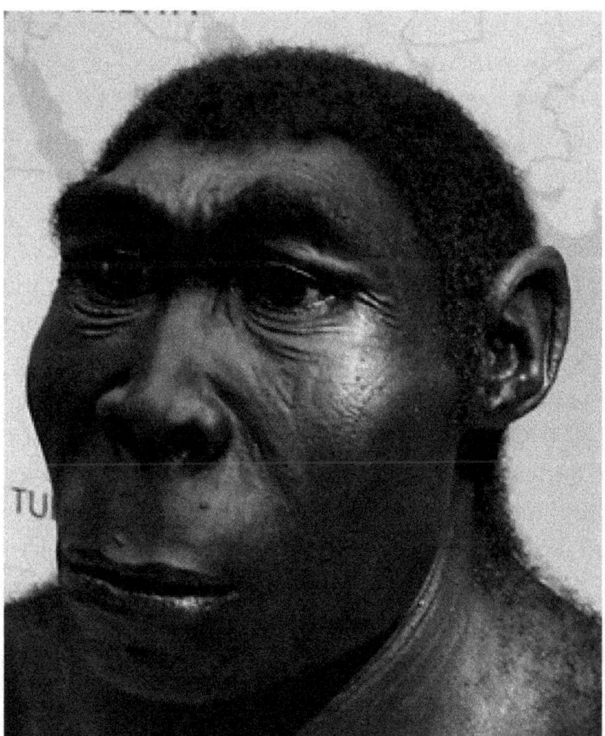

found in proximity to humans are due to kills by human males or due to scavenging (Almost all carnivores shamelessly do both.). Some archaeologists now feel the important behavior was the sharing of food, rather than the nature of that food. The behavior of current hunter-gatherers makes it clear both plant and animal food are essential. Hunting tends to be hit-or-miss and plant food tends to be steadily available. One anthropologist argued that food sharing itself was responsible for increased brain capacity in humans. Given that chimpanzees also share food without increasing their brain size dramatically, this is unlikely. It is more likely that eating more meat provided the additional nutrition that made an increase in brain size possible[2] (McBroom 1999). But as we have seen, the meat could have come from scavenging as well as from hunting. So in any case, there is no evidence for the triumphalist view of man-as-hunter.

Possibly the most important feature of the hunter-gatherer lifestyle is that it normally requires that any settlement not be permanent. The resources in a given area run out, and the group must move to an area where they are more plentiful. This applies both to game and to plant food. Some hunter-gatherer tribes are currently still nomadic for the same reason. So claims like those of Helmut Ziegert of permanent *Homo erectus* settlements 400,000 years ago may be difficult to sustain. Ziegert also claims that *Homo erectus* practiced, or could have practiced, agriculture at that time (Jones 2007). These claims go together because without persistent food resources, any permanent settlement could not survive. Since agriculture was such a major advance, it is difficult to see how ample traces of its practice would not have survived in the archaeological record between 300,000 years ago and 11,000 years ago, when the first evidence of agriculture was found. Even if no plant species were bred, selection and cultivation would have been necessary and should have left some traces.

However, some observers detect a bias toward *Homo sapiens* in these matters. Robert Bednarik (2001) sees a bias toward biblical teachings in the acceptance of the “Out of Africa” view. This view is that *Homo sapiens* arose once in Africa at about 200-300,000 years ago, and then about 60,000 years ago migrated to the other continents and then replaced earlier species. Bednarik points out that this view is based entirely on molecular genetic evidence. The time frame for the most recent common female human ancestor is about 200,000 years. The timeframe for this ancestor was determined by examination of DNA in the mitochondria, cell structures passed from female to female only. So this common female ancestor is called mitochondrial Eve. A similar molecular genetic search establishes a most recent common male ancestor, Y-chromosome Adam, dating from 140,000 years ago or possibly as recently as 60,000 years ago[3] (Bettinger 2007).

Bednarik argues that the evidence from molecular genetics does not establish anything like these firm dates, and that there is no archaeological evidence for the dispersion of *Homo sapiens* with more advanced culture at 60,000 years or so. He sees the "Out of Africa" view as an attempt to diminish the humanity of our predecessor species *Homo habilis* and *Homo erectus*. This view paves the way for seeing *Homo sapiens* as dramatically superior to its predecessors.

Bednarik's technical claims against the "Out of Africa" view are reputable but not accepted by the majority of paleoarchaeologists. But I hope they are correct. Richard Leakey notes a tendency for paleoarchaeologists to prefer theories of human origins that make the origin of *Homo sapiens* a relatively sudden event including large changes. Leakey surmises that those who believe such theories have the unconscious belief that "humans are so different from the rest of the animal world, they cannot accept the idea that we are a product of evolution, just like other species" (Leakey 1992, p. 199). The further consequence, which Leakey does not draw, is that if humans were a sudden and relatively recent arrival on the planet, our current conflict with the environment is much more likely an indication that we are a 'biological mistake.'

Indeed the very name of our species reflects not only prejudice toward our ancestors but encapsulates a fairly blatant version of human superiority. *Homo habilis* was named for his use of tools. *Homo erectus* was named for his upright posture. *Homo neanderthalensis* was named for the valley where his remains were first found. *Homo sapiens* was named not only for his cleverness, but his judicious wisdom. *Sapiens* is clearly an honorific. The name *Homo sapiens* was bestowed on us by the inventor of biological nomenclature, Linnaeus, in 1758. This was before the industrial revolution and modern technology, that is, before we began making unwise choices about our relation to the environment.

An Australian science writer, Julian Cribb, proposed in the journal *Nature* that the name of the species be changed. His reason for this proposal was that "an animal that imperils its own future and that of most other life forms and ecosystems does not merit a single '*sapiens*', let alone the two we now bear" (Cribb 2011). The scientific response was disturbing to me. Supposedly responsible scientists thought the proposal was not worth discussing and even thought it should not have appeared in *Nature*. Climate scientist Ken Caldeira opined that names do not matter. So he presumably wouldn't mind being called Asshole Caldeira (or to preserve ethnicity, Caldeira del Culo). Another baffling comment was that taxonomic names cannot be changed. Anyone examining field guides to fungi over the span of twenty years will see continuing extensive changes in fungal taxonomy, in both genus and species names. And the final coup de grace, by Professor Sander van der Leeuw, is good for avoiding dealing with most any issue: It's in the past and we need to deal with the future[4] (Parry 2011). This excuse was most recently deployed by President Obama as a way of dealing with the possible war crimes and violations of international law of the previous administration. As a result, violations of international law are now entrenched US policy.

These obtuse and dismissive rejoinders indicate that these scientists are still in the grip of human superiority. Further, that they believe the very real human conflicts with nature are not worth considering. They imply that there is nothing that could cause us to doubt the description of our species as 'wise and judicious.' More fitting would be *Homo urbanus* (clever) or *Homo rapaces* (rapacious). Once one does not take human superiority as a given, a recent, sudden origin for *Homo sapiens* is some reason to think our departure from the scene may also be recent and sudden.

DEVELOPMENTS IN THE HUNTER-GATHERER PERIOD

Here is the timeline for relevant human and human ancestor developments during the hunter-gatherer period:

- **3.5 million years ago:** Human ancestor *australopithecus afarensis* uses tools (uncertain)
- **2.5 million years ago:** Human ancestor *Homo habilis* uses tools
- **1.5 million years ago:** Human ancestor *Homo erectus* with cached tools
- **800,000 years ago:** Human ancestor *Homo erectus* uses controlled fire
- **400,000 years ago:** Human ancestor *Homo erectus* constructs shelters
- **200,000 years ago:** Humans (*Homo sapiens)* appear
- **190,000 years ago:** Humans (*sapiens and Neanderthal*) wear clothing
- **100,000 years ago:** Humans use pigment and body ornaments
- **90,000 years ago:** Humans carry out ceremonial burials
- **50,000-100,000 years ago:** Humans domesticate dogs
- **50,000-100,000 years ago:** Humans cause megafauna extinctions
- **35,000 years ago:** Humans make representations—figures and paintings

Later significant changes in how humans relate to the world are:

- **11,000 years ago:** Farming
- **7,000 years ago:** Civilization
- **250 years ago:** Modern technology

These later changes will be dealt with in the next three chapters.

Many other animals besides man use tools. Cases where other animals *store* tools are rare. So man early on treats the environment as not only ready to hand but as *standing reserve* or *resource*. So this distinctive feature of modern technology (treating everything as standing reserve) is actually present (to a limited extent) early in human development. A chimpanzee may pick up a stick for one time use to probe for termites. And the most primitive, chipped stone tools are usually not cached for further use.[5] When they are standing reserve, even at this early stage stored tools exhibit the human ability to appropriate the environment for its own purposes and to mark out a distinct part of the environment as human-rather-than-environment. Thus we eventually create a human world different from the natural world.[6]

The historical development of stone tools has been intensively studied by archaeologists. Their findings are directly relevant to the current topic. Archaeologists agree on seven periods (called *modes*) of tool use. Table 3 follows the common practice in archaeology, to name each period or mode for the location where the first tools of that kind were found.[7]

Recently, the ability to make tools may have been found in an ancestor to *Homo sapiens, Australopithecus afarensis*. There is disputed evidence for tool use at 3.5 million years ago. The earliest well-attested use of stone tools was by an earlier member of our genus, *Homo habilis,* 2.5 million years ago (Farivar 2011). However, these earliest tools belong to period 1 of toolmaking, the Oldowan period. The major difference between the Oldowan period and the next Acheulean period is that an Oldowan tool is the result of a fortuitous and probably ex tempore operation to obtain one sharp edge on a stone, whereas an Acheulean tool is a planned result of a manufacturing process. (Johnson 2013)

So the Oldowan toolmaking of (possibly) *australopithecus afarensis* and *Homo habilis* is basically no different in kind from what chimps now do: Making ad hoc

Table 3. Evolution of tools

Mode or Period	Name	Year of Origin	Ending Year	Species
1	Oldowan	2,600,000	1,600,000	habilis+erectus
2	Acheulean	1,700,000	100,000	Erectus
3	Mousterian	300,000	30,000	Neanderthal
4	Aurignacian	47,000	41,000	Sapiens
5	Microlithic	40,000	6,000	Sapiens
6	Neolithic	11,000	3,000	Sapiens
7	Modern	3,000		Sapiens

changes in found objects for current use. It is not until we get to the Acheulean tools of *Homo erectus* that we encounter specifically human tool use.[8] These tools were likely cached, creating for the first time a dedicated human space distinct from the rest of nature.

CACHED TOOLS (1.5 MILLION YEARS AGO)

So it is likely that *Homo erectus* stored his tools, since they were more than ad hoc improvisations on material at hand. What kind of storage was available to a nomadic hunter-gatherer constantly on the move? Clearly, no fixed location is available for this purpose. So he must have some way of carrying the tools with him, some sort of naturally occurring bag. Perhaps the dried skin of a small animal. Note that *Homo erectus* has not yet discovered clothing, so the function of storage is all the function that such a bag would have at this point.

Perhaps this storage does not happen at the beginning of the production of Acheulean tools. Maybe it does not happen until fairly late in the career of *Homo erectus*. But the sequence from more complex tools to some kind of storage must happen. As tools get more complex and therefore require more of an investment of time, it ceases to make sense to create them, use them, and discard them. Thus, storage is born, and thus is born a dedicated human space distinct from the rest of nature, whose purpose is to keep available instruments for altering aspects of the environment to suit human needs.

Can we say more about our *Homo erectus* ancestors? Richard Leakey, probably the world's foremost paleoarchaeologist specializing in early human ancestors, reconstructs a *Homo erectus* encampment by the side of a river in east Africa, 1.5 million years ago. A group of about a dozen *Homo erectus* individuals are staying for a few days by the side of a river. There are acacia trees for shelter and fruit, and the sandy river bank makes a soft place to sleep. Two defined areas of the 30-by-60-foot site were centers of bone processing and tool making. Animal bones found included hippo, a zebralike animal, giraffe, small antelope, and catfish. Leakey surmises that these individuals had simple snares constructed from bark and sharpened sticks for hunting. He also surmises that the women had bags of animal skin for collecting the results of foraging and for carrying children (Leakey 1992, pp. 175-181). Of course, anything except the stone tools and bones would have left no traces after 1.5 million years. But the sequence from tools to storage doesn't depend on suppositions about vanished artifacts.

Storage might even have begun as a way of collecting plant foods and later repurposed for storing tools. We know that at least temporary storage of gathered plant foods was likely because food was cooperatively obtained and then shared.

Then it would be natural for non-perishable plant foods such as nuts and seeds to be stored when the group moved and there was some reason to be doubtful about future food. These types of storage separate humans from the environment less than the storage of tools. Local collection of plant food is merely a convenience in getting the food to the consumption site for sharing. And carrying along nuts and seeds is just an extension of this.

Handling food gathering and consumption cooperatively does not in itself contribute to distancing humans from the environment. Even though food sharing is a distinctive human trait, which would not have been at all apparent to our *Homo erectus* group.[9] But beyond the hunter-gathering lifestyle, food sharing combined with agriculture enables civilization. And in civilization, humans have already taken giant steps to distance themselves from nature and the ecosystem.

Some archaeologists argue that *Homo erectus* had language, possibly as early as 1.5 million years ago. This conclusion is on the basis of at least two considerations: First, *Homo erectus* had physiological and brain capabilities for speech similar to modern humans (Leakey 1992, pp. 257–258). Second, *Homo erectus* crossed open ocean to get to some of the Pacific sites where its remains were found. This ability is generally regarded as requiring a level of cooperation to build rafts that would not be possible without language.

However, the question of whether language adds to distance from the environment is complex. This issue will be discussed below, in the section on Symbolic Representation.

FIRE (800,000 YEARS AGO)

Homo erectus had control of fire 800,000 years ago. Control of fire means that humans no longer accept the environment as given. The direct use of fire is for heating and protection from predators. Then, later, fire is used for cooking, possibly about 250,000 years ago.[10] And much later, about 10,000 years ago, humans use enhanced fire to produce tools out of metal, a material better for many uses than material supplied by nature. The control of fire 800,000 years ago also means that for the first time, humans directly control energy.

SHELTERS (500,000 YEARS AGO)

In spite of the popular characterization of early humans as "cavemen," not that many early humans actually used caves as shelters. The earliest evidence of cave dwelling, by *Homo erectus,* is from 770,000 years ago (Rafferty 2009). But even back then, suitable caves were relatively scarce. Evidence for the first persistent human-made shelter is from about 500,000 years ago, in Japan. The shelter would have been built by *Homo erectus.* The remains consist of 10 post holes forming two irregular pentagons which may have been two huts. A number of stone tools were also found at the site. Japanese dating is considered to be reliable (see Figure 3).

Dr. Chris Stringer, head of the human origins group at London's Natural History Museum, commented "It's evidence that they [*Homo erectus*] built structures but how permanent this was we don't know. Who knows whether this was a shelter they stayed in for one week, or one month?" John Rick, an anthropologist at Stanford University, observed, "They [*Homo erectus*] had the idea of actually making a structure, a place where you might sleep. It represents a conceptual division between inside and outside" (BBC News 2000). Japanese archaeologists studying the site observed that the buildings could have been used as a place to rest, a lookout for hunting, or a place to store hunting tools (Japan Times 2000).

These commentators are careful to place this discovery of shelter within the context of a hunter-gatherer lifestyle. As discussed previously, it is difficult to credit the claims of Prof. Helmut Ziegert of permanent *Homo erectus* settlements 400,000 years ago. Prof. Ziegert also claimed that *Homo erectus* practiced, or could have practiced, agriculture at that time (Jones 2007). Of course without persistent food resources, any permanent settlement could not survive. But how could there be no traces of the practice of agriculture in the fossil record between 300,000 years ago and 11,000 years ago, when the first evidence of agriculture was found? It is much more likely that *Homo erectus* had constructed shelters but not permanent settlements.

Figure 3. Oldest human structure (From blog.ounodesign.com by permission)

The development of an inhabitable purely human space separate from the environment is a huge conceptual breakthrough. But it is not followed by suburban tract houses for *Homo erectus*. It is not until the advent of farming that permanent dwellings become sustainable. The development of human-constructed structures does increase the separation of humans from the environment by conceptually dividing inside human space from outside nature. But this development does not lead to changes in lifestyle until nearly 500,000 years later. And those changes—farming and civilization—were initiated by another species, *Homo sapiens*, which itself does not appear for another 300,000 years.

MODERN HUMANS APPEAR (200,000 YEARS AGO)

There are currently two competing views about the origin of *Homo sapiens*: The multiregional hypothesis holds that *Homo sapiens* gradually replaced earlier humans in several areas: Africa, Asia, Europe, and Australia. Some also believe that the earlier humans and *Homo sapiens* all belonged to the same species. The 'Out of Africa' hypothesis (discussed above) holds that *Homo sapiens* arose once, in Africa, about 200,000 years ago. *Homo sapiens* then spread out to other areas about 60,000 years ago and completely replaced earlier humans. Some hold that *Homo sapiens* was a superior new species, with superior technology. The main evidence for 'Out of Africa' comes from molecular genetics: The DNA in the cell structures mitochondria and in the Y-chromosome of different current humans was examined, and it was possible to calculate the time of the most recent common ancestor of those people.

The 'Out of Africa' hypothesis is currently by far the most popular with paleontologists. As indicated above, this hypothesis may have some possible disturbing implications. Here are a few more implications and their consequences for the project of this book.

One major question for the 'Out of Africa' hypothesis is how the replacement of older humans took place. Some claim there is no archaeological evidence for the claimed superior technology, but we can waive this objection. For at this point in human development, humans are still hunter-gatherers. Humans are pre-agriculture and pre-civilization. So what could this superior technology have been other than more effective weapons? It is most likely on this hypothesis that *Homo sapiens* simply exterminated the earlier humans. The megafauna shortly to become extinct would not have been surprised. And certainly the subsequent history of *Homo sapiens* has shown a striking tendency to exterminate not only related species, but also large groups of its own species. Genocide has been a regular feature of human history, from Genghis Khan through the unfortunate Midianites, whose genocide was approved by the Bible (Numbers 31:18). More recently there were 12 million

Africans killed in the course of implementing the slave trade in the 17th to 19th centuries. And then recently there was the holocaust of the middle 20th century, killing 6 million.[11] On a somewhat smaller scale, American wars have produced civilian collateral deaths of 1 to 2 million in Vietnam and perhaps 1 million in Iraq.[12]

So what are the consequences for the separation of humans from the ecosystem? If we have a separate species *Homo sapiens* whose main claim to fame is its superior ability to kill, the ability of that species to coexist with the ecosystem must be an acquired trait. If, however, the multiregional hypothesis is correct, the source of our disconnect with nature may not be so clear cut. In the end, the two hypothesis make a difference only in the emphasis we place on various factors leading to our disconnect with nature.

CLOTHING (170,000 YEARS AGO)

Clothing is unique to humans. Evidence has been found for the possession of clothing by both *Homo neanderthalensis* and *Homo sapiens*. The most interesting evidence is the date human body lice diverged from human head lice. This is likely 170,000 years ago. This divergence dates the origin of clothing because human body lice are adapted to live on clothing (Joyce 2011).

Clothing separates our bodies from the rest of the world. It means we encounter the world mediated by artifacts of our creation. Clothing also cannot be ad hoc: It needs to be stored, in a semi permanent location or a portable location. Even animal skins require a lot of effort to prepare, and would not be tossed aside. Although it is possible to conceptualize the acquisition of many dramatically new human practices, the wearing of clothing is not so easy to do. First of all, *wearing* implies the persistent carrying of something considered as a covering. The only suitable objects around initially would have been unprocessed animal skins.[13] Processing the animal skin would itself be difficult and hard to justify if one had not already formed the intention of carrying it around for some time. So there is a sort of chicken-and-egg problem.[14]

But perhaps it might have gone like this: They have just skinned an animal and the skin has come off cleanly and in one sizable piece. Deer hides may have been easiest. (Boren 2004). Then some bad weather came up, a sudden storm possibly, and someone drew the deerskin around them. If the storm went on for some time, it would have dawned on the first-wearer that it was worth hanging on to the skin. They probably already had sacks of some kind to carry their tools, because those tools (Acheulean or even Mousterian) had required some effort to produce and were therefore not just created, produced and discarded. So it would have been natural to make the sacks larger to carry the clothing[15], or maybe also they found some way of attaching the sacks to the body. They do construct shelters, so that is a place to store both tools and clothing. Since all social animals are concerned with status, it is likely that clothing was used to mark status early on.

PIGMENT AND BODY ORNAMENTS (100,000 YEARS AGO)

To some paleontologists, the presence of body pigments implies the presence of a socially accepted symbol system. There is some evidence for use of pigments as early as 800,000 years ago, but the evidence for body pigments is equivocal, because the pigment substance also occurs naturally. The evidence for human-made pigments and ornaments dating to 100,000 years ago is firmer (Bower 2011).

As discussed earlier, commentators such as Robert Bednarik believe that a large part of the motivation to believe the "Out of Africa" hypothesis is a belief in the superiority of *Homo sapiens*. That is, *Homo sapiens* arose about 200,000 years ago in Africa and migrated out of Africa about 60,000-100,000 years ago. Further, this new, superior species was responsible for nearly all major cultural developments including the use of symbols. Bednarik observes:

If there had been a mass migration out of Africa, by a technologically, cognitively and intellectually superior human species, one would expect to find their arrival marked by a new technology, new tools, new types of subsistence extraction methods and so forth. There is not one iota of evidence, anywhere in the world, that would suggest the arrival of any innovation coinciding with the arrival of Eve's supposed [progeny][16]. On the contrary, there is ample evidence that, wherever [modern Homo sapiens] appeared and co-existed, often for long time spans, with archaic Homo sapiens (such as neanderthaloids), they invariably adopted the life style and technology of the resident archaic populations (Bednarik 2001, pp. 4-5).

The artificial perforation of small objects used as beads or pendants appears about 300,000 to 200,000 years ago (2001, p. 10). Bednarik believes that such artificially produced objects have "symbolic meaning." Because they are non-utilitarian, they clearly have some social significance beyond themselves. But determining what significance they might have requires much more of the social context than is available. It is not plausible to conclude, as Bednarik does, that ". . . the use of such sophisticated objects as beads and pendants [200,000 or more years ago] demonstrates, beyond reasonable doubt, that... [*Homo erectus*] possessed well-established semiotic systems of various types" (2001, p. 23).

It is not until somewhat later, 75,000 years ago for figurines, and 35,000 years ago with cave art, that we actually have representational symbols. Their significance for the separation of humans from the environment will be discussed in the section below which deals with such symbols.

CEREMONIAL BURIALS (90,000 YEARS AGO)

The earliest undisputed ceremonial burial dates from 90,000 years ago. A skeleton was stained and there were grave goods accompanying the skeletons. There is some evidence for possible much earlier burial rituals. There is some evidence that the Neanderthals buried their dead along with animal bones and stone tools (Solecki 1975). To interpret evidence of ceremonial burial as indicating religious belief in an afterlife may be a bit of a stretch. It is true that all historic civilized cultures as well as many non-civilized cultures believe in an afterlife. But elephants have burial rituals and mourn their dead. The evidence for elephant religion is just about as good as that for Paleolithic religion, namely both elephants and Paleolithic humans regarded the bones of the deceased as an object of care.

Testimony about the content of supernatural belief is essential in order to attribute it to any group. Nevertheless, as we will see with representation, it is unlikely that such beliefs arose just at the point there were records of them. A being would need to possess two concepts in order to begin thinking about an afterlife: First, consciousness of oneself; and second, awareness that the self one is conscious of will die. Many animals have self consciousness, notably chimpanzees and elephants. The acute awareness of elephants of the passing of their fellows together with their own self-consciousness, makes it extremely likely that elephants are aware of their own deaths. Yet there is no evidence elephants believe in an afterlife.

So something in addition to self-consciousness and awareness of one's own death produces the human belief in an afterlife. It may be that humans are hardwired to create imaginary worlds other than the one they inhabit.[17] Preliterate cultures almost universally have a belief in an afterlife. This belief, in the context of their stage of development, does not contribute to a separation from the ecosystem. However, in the context of the later development of civilization, a belief in another, better world can be a potent contributor to a greatly diminished regard for this world. The example of Senator James Inhofe was mentioned in Chapter 4 "Mind Body Consciousness." Inhofe said on national TV said those with environmental concerns were at risk of idolatry by worshipping "God's creation" rather than God. Inhofe's statement very clearly exhibits how belief in God can lead to discounting the environment in which we live.

DOMESTICATED DOG (50,000-100,000 YEARS AGO)

A recent experiment showed that domesticated dogs are genetically different from wolves. Wolf cubs and puppies were given identical treatment from birth, but the wolves never accepted humans as the puppies did (Brockmann 2009). Dogs are the only domestication which takes place before agriculture began about 11,000 years ago. Most domestications would by definition require a *domus* (Latin for *home*). A nomadic hunter-gatherer would have difficulty in maintaining animals kept for food or other products. Bison, cattle, sheep, goats—livestock—usually require a persistent area but some livestock are sometimes herded by nomads.[18] Once corralled, livestock could be subject to artificial (that is, human) selection for traits humans wanted in the animals. Cats probably were kept because of their skill at eliminating rodents, but again they are not easily transported and not that useful to a hunter-gatherer.

Historian of science Edmund Russell (2011) states the difficulty well: "Wolves do not obey human commands, and it is hard to imagine that people persisted in raising dangerous animals for uncertain benefits far in the future." Artificial selection definitely was not operating in the initial domestication of dogs. Hunter-gatherers of that time would never have seen it done before. And the end product—a domesticated animal—would have been entirely outside their experience.

An alternative view has the wolves doing the selection. Calmer wolves scavenging garbage mate with other calm wolves on the outskirts of human settlements, eventually producing "village wolves" like current village dogs (Coppinger 2001). This view seems to require persistent settlements for the wolves to self-select. And they don't exist yet. Also village dogs are not useful. They are tolerated. Hunter-gatherers would probably not keep them around unless they were useful. And it is hard to see how such usefulness could have arisen on this scenario.

These explanations are instances of what I will call *reverse temporalization*. To reverse temporalize is to explain an entirely new human activity by requiring humans originating it with no related experience of the activity, to know in advance why it will be beneficial. This is simply not possible. Of course, it is possible for people to create previously nonexisting things. But in reverse temporalization, one requires people who could have had absolutely no concept of the item under discussion, to have such a concept.

The two accounts of the domestication of wolves just given are examples of reverse temporalization. There will be more examples when we come to the origins of farming and civilization.

Here is an account, which does not use reverse temporalization. For people to keep wolves around, the wolves must have provided some immediate benefit to hunter-gatherers. Bamber Gascoigne explains:

Wolves are good hunters. Humans and wolves are competing for the same prey, but there are advantages for both in teaming up. For the wolf, human ingenuity and the use of weapons mean a share in a greater number of kills - and perhaps even an occasional taste of larger victims, such as mammoth. For humans, the wolf's speed and ferocity is equivalent to a new weapon (Gascoigne 2001).

So there is mutually beneficial cooperation in hunting between wolf and human hunter. Through repeated success, the hunter would want to continue the association with the wolf and its pack, with wolves being selected (at first) more-or-less naturally for friendliness to humans, and then later artificially for other traits humans want. We might speculate whether there was also natural selection of humans friendly to wolves. Supposing that wolf plus human was more successful in hunting, there could have been selection pressure of this kind. One dog scholar, Mark Derr, believes that it was not until much later, with agriculture and civilization with large-scale artificial selection, that our proliferation of more-or-less decorative dog breeds occurred (2011).

The significance of later artificial selection associated with agriculture will be discussed in the next chapter. But the domestication of dogs in a hunter-gatherer context seems not to contribute to human separation from the ecosystem. In fact, in establishing a mutually beneficial relationship with a non-human species, it tends to have the opposite effect. Humans actually have a greater involvement in the ecosystem, even if not in a large or dramatic way.

MEGAFAUNA EXTINCTIONS (50,000-100,000 YEARS AGO)

The megafauna extinctions caused by humans from 50,000 to 100,000 years ago and subsequently are perhaps the most disturbing feature of human history before agriculture and civilization. The name 'overkill hypothesis' is given to the theory that humans are responsible for the extinctions. As mentioned earlier, it is natural to suppose that hunter-gatherers do not produce environmental devastation in a big way. After all, as hunter-gatherers, they need to respect and care for the environment that is sustaining them. And indeed, later documented human-caused extinctions were by aboriginal peoples previously thought to have been in harmony with nature. These are Australian extinctions about 60,000 years ago; extinctions by native Americans about 12,000 years ago; and extinctions by the Maori in New Zealand (Leakey 1995, ch. 10).

Robert Feranec states the situation concisely:

Between 50,000 and 10,000 years ago, . . . roughly 100 genera of megafauna (animals weighing more than 100 pounds) became extinct worldwide. Among them are such well-known creatures as mammoths and saber-toothed tigers and the more obscure, though no less significant, Diprotodon (an Australian marsupial the size of a hippopotamus) and Coelodonta (a woolly rhinoceros found in Europe). Whether their disappearance was caused by changes in climate or by "overkill" (being hunted to extinction by humans) has been hotly debated for the past 40 years (Feranec 2011).

But it is hard to avoid the feeling that objections to the overkill hypothesis are attempts to avoid the conclusion that humans could behave so irresponsibly. Objections normally take the form: The extinctions *might* have been caused by other factors, so humans are not responsible. But since evidence is incontrovertible for overkill by hunter-gatherers in more recent cases, there is really no reason not to accept the overkill hypothesis for the earlier extinctions.

The consequences for the separation of humans from the ecosystem are also disturbing: Within a hunter-gatherer lifestyle, humans were able to amass sufficient technological knowledge and cooperative skills to dramatically change the ecosystem from what it had been in their absence. These losses of biodiversity clearly were negative, since they did not materially improve the survival chances of human beings, any more than an urban riot improves the survival chances of inner city dwellers.

But what could the hunter-gatherers who committed these extinctions have been thinking? The answer is they may not have been thinking about much beyond how easy it was to kill the megafauna. 20th century Americans were thinking the same when they drove the passenger pigeon to extinction. And the Maori were thinking the same when they drove the moa to extinction. Moa were ostrichlike flightless birds in New Zealand, represented by several species ranging from 3 to 10 feet tall. Over 500,000 moa skeletons have been recovered, indicating the killing went on for many generations (Leakey 1995, p. 185).

Of course some use was made of the catch, but it was literally overkill that drove these modern species to extinction. It is possible that early hunter-gatherers may have driven megafauna to extinction just by killing what they needed. But their thinking was short term: These are good to get, now. In any case, neither modern nor ancient hunters had the concept of sustainability. Whatever sustainability was achieved, was achieved by a balance between human population and animal prey population. The megafauna extinction shows that this natural balance was not sufficient in the case of humans to prevent extinctions.

The important question is when and how they were able to obtain the power to commit megafauna extinctions. Most megafauna extinctions occurred as a result of humans colonizing new territory, where the local animals had no hereditary protection against humans (Leakey 1995). If this was the major cause, then the humans causing the megafauna extinctions were behaving exactly as invasive species. Then, when humans are able to be present everywhere on the globe, they are invasive species everywhere. The consequences for the ecosystem are not good.

But one wonders how early humans were able to kill powerful predators (such as saber tooth tigers) and massive mammoths. Even if these animals were unfamiliar with humans, it would not have been easy to kill them in large numbers. Perhaps some new hunting techniques would have increased their hunting skills. One possibility was wolves, also known as dogs, whose services would have become available at about this time.

So either humans are by nature an invasive species upsetting the ecosystem, or human techniques are powerful enough to overwhelm any natural equilibrium.

PICTORIAL REPRESENTATIONS (35,000 YEARS AGO)

So far the caves found at Chauvet, France contain the earliest human pictorial representations known. The paintings in the Chauvet cave date from about 30,000-32,000 years ago. Fortunately, they are available to a large audience through the permission of the cave custodians and the efforts of the filmmaker Werner Herzog. His *Cave of Forgotten Dreams* (2010) makes it clear that these drawings were not the work of beginner artists.[19] How long would it take in hunter-gatherer days to establish such a representational tradition? One possible yardstick is the development of European representation from the dark ages to medieval pictures, roughly 800 years. But given that there were no supporting institutions such as permanent settlements, a guess anywhere from 2000 to 5000 years earlier would not be out of line.

The development of representation marks an entirely different human relation to the world. The British author Virginia Stephens once remarked, "[Representation] is not a copy of the real world. One of the damn things is enough."[20] But although representation may not be a copy of the real world, it *is* a surrogate for the real world. Being able to deal with surrogates of the real world instead of the world itself makes a huge difference. However, practical uses are limited when the representations are immobile on the wall of a cave.

Yet even in the days when representations were only on the walls of caves, we need to ask why humans of those days would have gone to the trouble of making them. Hypotheses about religious rites are rather fanciful. The Chauvet Cave paintings are clearly representational and stand on their own. We have to regard the ability to produce representations as a fundamental human skill. This means we are already wired to create another world in lieu of the one we inhabit. And, as we will see later, representation allows us to deal with the world we inhabit, the ecosystem, through our representation of it. The danger is that of replacing the real thing with the representation.

There are naturally occurring representations such as reflections on the surface of water. But then the separate idea of marking on a surface to create such a reflection already contains the idea of representation. But perhaps some naturally occurring formations came close to representing people or animals, and attempts to copy these led to the production of representations. However, the motive to create representations still has to pre-exist. It is for this reason that the ability to produce representations has to be a fundamental human skill.[21] And, as we saw in Chapter 4, "Mind Body Consciousness," our presence in a world is always the presence of a point of view on that world.[22] Therefore, any world created by humans is with and from a point of view. And our ability to produce representations includes our ability to take points of view different from our actual point of view.

Language, like representation, is a human tool which can be used to create worlds different from the real world. We call them fictions. And although it is possible to lie (convey an untruth) without language, language makes it much easier. However, unlike tools or representations, we have no record of language until writing with phonetic alphabets arises around 5,000 years ago.[23] Thus our first information about spoken language is not available until well after the rise of agriculture and civilization. Some think language arose about 40,000 years ago, at about the same time as figures and cave paintings. Professor Francisco D'Errico argues that body pigments and body ornaments are material "proxies" for symbolic communication. Body markings are symbolic representations and require shared understanding. So too with body ornaments, whose meaning requires social learning. The use of pigments dates back 300,000 years, and the use of body ornaments 100,000 years. Both are present in Neanderthal sites (Bolles 2008). Also Neanderthals have the same "speech" gene as humans and also have the same bone structure in the larynx. This makes it plausible that Neanderthals had speech, possibly as early as 300,000 years ago.

However, even if Neanderthals (and maybe even earlier *Homo erectus*) had speech, we can tell almost nothing about its content and structure. Professor D'Errico's reasoning is sound that the presence of body pigments and body ornaments show that these species had the capacity for communicated symbolic representation. But by his own admission, all we can say on the basis of the archaeological record is that these proxies, as he calls them, are correlated with some form of spoken language.

But there may be other clues. Chimpanzees and gorillas have been taught to communicate using physical signs. Nevertheless, their communications have the structure of complete thoughts, that is, what we call sentences. Therefore, the ability to structure communication as complete thoughts most likely was present in all our ancestors right back to even before we diverged from the apes.

A main function of language is to coordinate complex cooperative activity. So the impact of language on our separation from the environment is not clear; language could just as easily lead to cooperative activity to preserve the environment as to neglect it.[24] It is difficult enough to speculate on attitudes toward the environment when we have clear physical clues. But speculating on the impact of any highly uncertain linguistic structures is speculation squared. And speculation squared could be indistinguishable from making things up.

CONSEQUENCES FOR SEPARATION FROM THE ECOSYSTEM

What can we say about humans as hunter-gatherers and the development of separation from the ecosystem? We have some idea about when humans acquired various skills and created various kinds of objects. But it is important not to assume that each historical development is inevitable. It is all too common to present history as a preordained march to our current glory. Of course, this makes it impossible to diagnose any wrong moves or negative developments. Instead, we have to ask, when would it have been difficult for humans not to proceed to the next stage? Only then can we regard that transition as inevitable for all practical purposes.

Here is the sequence:

- About 2.5 million years ago our hunter-gatherer ancestors first created ad hoc tools.
- Then about 1.5 million years ago our ancestor *Homo erectus* invested time in making more complex tools which need to be stored. Stored tools create a human space separate from nature, and these stored tools are used to change rest of nature.
- Roughly half a million years later, *Homo erectus* learned to control fire. For the first time, humans control energy and can modify the environment to suit human needs.
- After another half a million years, *Homo erectus* constructs shelters, creating the distinction between inside and outside. 'Inside' is a human space to live in.

The innovations of *Homo erectus* create a separate human space, both for storage and shelter. They also allow modifying the environment through fire to serve human needs. Nevertheless, these innovations were deployed over a period of 500,000 to 1.5 million years without any significant evidence of ecological damage. These innovations are also relatively independent. The invention and elaboration of purely human space does not in itself seem to lead to any harmful separation from the ecosystem. Later elaborations are not as benign.

Modern humans (*Homo sapiens)* appear after another 200,000 years. They (and the Neanderthals *Homo sapiens neanderthalensis)* accomplish the following:

- About the time *Homo sapiens* appears, they and the Neanderthals begin wearing clothing. Clothing separates us from the environment, which is now mediated by artifacts we created.
- After another 100,000 years, both modern humans and Neanderthals begin carrying out ceremonial burials. Such burials are also carried out by animals, for example elephants. No evidence for belief in afterlife one way or the other.
- Within the next 50,000 years, humans domesticate dogs. This first domestication, unlike later ones, had to be mutually beneficial for both wolves and humans. Separation from the ecosystem is actually decreased.
- Also within this 50,000 years, humans cause megafauna extinctions. Human power is great enough to alter ecological balance significantly.
- Perhaps 20,000 years later, 35,000 years ago, humans make representations—figures and paintings. Representation allows humans to create another world in lieu of the one they inhabit. Making representations seems to be a fundamental human skill. Since worlds come to us from a point of view, the ability to assume multiple points of view is also a fundamental human skill.

Clothing distances humans from the environment and also for the first time creates highly visible differentiation between humans and other animals. This differentiation easily leads to an attitude of the superiority of humans to other animals. Further, the demand for clothing requires the hunting of more animals. Conversely, the domestication of dogs increases the connection of humans with the ecosystem, at least with wolves. This first domestication had to be mutually beneficial for humans and wolves.

Both burial rituals and the making of representations concern the creation of another world other than the one we live in. I conclude that there is an innate human propensity to imagine worlds different from the actual one which we consider as good as real. This propensity obviously can play a substantial role in distancing us from the actual world and its ecosystem. But since for us worlds come with and from a point of view, the ability to take multiple points of view can actually be important in drawing us closer to the actual world and its ecosystem. We are not limited to our own point of view and thus can come to appreciate other beings from their own point of view.

Finally, megafauna extinctions reveal that as long ago as 60,000 years, human were able to produce significant damage to the ecosystem. Either humans are by nature an invasive species upsetting the ecosystem, or human techniques are powerful enough to overwhelm any natural equilibrium.

So the innovations of *Homo sapiens*, unlike those of *Homo erectus*, are either already conflicts with the ecosystem or lend themselves to creating conflicts. The only possible exception is the domestication of dogs. But even that innovation might have led to the enhancement of hunting techniques to the point of mass extinctions.

In the next three chapters, the rapidly increasing conflicts of humans with the ecosystem over the past 11,000 years are outlined. In this chapter, we examined 2.5 million years of history, and potential conflicts of humans with the ecosystem appeared only in the last 100,000 years or so. 11,000 years is a short enough period so that major evolutionary changes in humanity are not possible. Consequently, any relevant innate behavior would likely have expressed itself long before then.

REFERENCES

Ardrey, R. (1977). *The hunting hypothesis*. New York: Bantam Books.

Barton, M., Riel-Salvatore, J., Anderies, J., & Popescu, G. (2011). Modeling human ecodynamics and biocultural interactions in the Late Pleistocene of Western Eurasia. *Human Ecology*. Retrieved 11/22/11 from www.openabm.org/node/2734

Bednarik, R. (2001). *Beads and the origins of symbolism*. Retrieved 11/27/11 from http://www.semioticon.com/frontline/bednarik.htm

Bettinger, B. (2007, July 20). Mitochondrial Eve and y-chromosomal Adam. *The Genetic Genealogist*. Retrieved 11/28/11 from www.thegeneticgenealogist.com/2007/07/20/mitochondrial-eve-and-y-chromosomal-adam/

Bolles, E. B. (2008). Neanderthals had speech. *Babel's Dawn*. Retrieved 11/23/11 from www.babelsdawn.com/babels_dawn/2008/03/neanderthals—1.html

Boren, J., Baker, T., Hurd, B., & Mason, G. (2004). Tanning deer hides and small fur skins. *New Mexico State University*. Retrieved 11/27/11 from aces.nmsu.edu/pubs/_l/l-103.pdf

Bower, B. (2011). Stone age paint shop unearthed. *Science News, 180*(11). Retrieved 12/11/11 from www.sciencenews.org/view/generic/id/335172/title/Stone_Age_paint_shop_unearthed

Brockmann, H. J. et al. (2009). Interspecific behavioral comparisons of hand-reared wolves and dogs. *Advances in the Study of Behavior*, *39*, 87.

Coppinger, R., & Lorna. (2001). *Dogs*. New York: Scribner.

Cribb, J. (2011). Taxonomy: New name needed for unwise Homo? *Nature*, *476*(282). PMID:21850091.

Derr, M. (2011). *How the dog became the dog*. New York: Overlook Press.

Diamond, J. (1997). *Guns, germs, and steel*. New York: Norton.

Diamond, J. (2005). *Collapse*. New York: Viking.

Drew, E., & McGuigan, J. (2008). Prevention of crime: An overview of gated communities and neighborhood watch. *International Foundation for Prevention Officers*. Retrieved 12/09/11 from www.ifpo.org/articlebank/gatedcommunity.html

Farivar, C. (2011). *Ancient humans used stone tools earlier than previously thought*. Retrieved 1/07/11 from www.dw-world.de/dw/article/0,5891406,00.html

Feranec, R. (2011). What became of the megafauna? *American Scientist, 99*(6). Retrieved 12/8/11 from www.americanscientist.org/bookshelf/pub/what-became-of-the-megafauna

Gascoigne, B. (2001). History of the domestication of animals. *History World*. Retrieved 12/13/11 from www.historyworld.net/wrldhis/PlainTextHistories.asp?historyid=ab57

Goldenberg, D. (n.d.). *Not-so-stupid animal tricks*. Retrieved 1/2/11 from mentalfloss.com

Gombrich, E. H. (1996). *Art and illusion*. Princeton, NJ: Princeton University Press.

Goodman, N. (1976). *Languages of art*. Indianapolis, IN: Hackett.

Heidegger, M. (1955). The question concerning technology. In *The Question Concerning Technology and other Essays*. New York: Harper & Row.

Herzog, W. (2010). *Cave of forgotten dreams*. [Motion Picture]. New York: IFC Films.

Humphrey, N. (1998). Cave art, autism, and the evolution of the human mind. *Cambridge Archaeological Journal*, *8*(2), 165–191. doi:10.1017/S0959774300001827.

Japan Times. (2000, October 26). Archaeologists find 'oldest' habitation. *Japan Times*. Retrieved 12/4/11 from www.japantimes.co.jp/text/nn20001024a1.html

Johnson, S. (2013). *Acheulean stone tools*. Retrieved 4/23/13 from www.ehow.com/info_8242594_acheulean-stone-tools.html#ixzz2RJbSn6dE

Jones, T. (2007). *Did homo erectus settle down 400,000 years bp?*. Retrieved 11/19/11 from anthropology.net/2007/06/25/did-homo-erectus-settle-down-400000-years-bp/

Joyce, R. (2011). Dressing for the Pleistocene. *Ancient Bodies, Ancient Lives*. Retrieved 12/6/11 from ancientbodies.wordpress.com/2011/01/07/dressing-for-the-pleistocene/

Kramarae, C., & Spender, D. (Eds.). (2000). *Routledge international encyclopedia of women*. Florence, KY: Taylor & Francis.

Leakey, R., & Lewin, R. (1992). *Origins reconsidered*. New York: Anchor.

Leakey, R., & Lewin, R. (1995). *The sixth extinction*. New York: Doubleday.

Lindemans, M. F. (1998). Xipe totec. *Encyclopedia Mythica*. Retrieved 11/27/11 from www.pantheon.org/articles/x/xipe_totec.html

News, B. B. C. (2000, March 1). World's oldest building discovered. *BBC News Sci/Tech*. Retrieved 12/4/11 from news.bbc.co.uk/2/hi/science/nature/662794.stm

McBroom, P. (1999). Meat-eating was essential for human evolution, says UC Berkeley anthropologist specializing in diet. University of California, Berkeley. Retrieved 11/27/11 from berkeley.edu/news/media/releases/99legacy/6-14-1999a.html.

McKibben, B. (2007). *Deep economy*. New York: St Martin's.

Norman, J. (2013). The earliest hearths. *From Cave Paintings to the Internet*. Retrieved 4/11/13 from www.historyofinformation.com/index.php

O'Neil, D. (2012). *Early transitional humans*. Retrieved 4/11/13 from anthro.palomar.edu/homo/homo_1.htm

Parry, W. (2011). Bid to rename Homo sapiens called unwise: Most scoff as science writer says Latin term for 'wise man' no longer deserved. *MSNBC on Science*. Retrieved 11/27/11 from www.msnbc.msn.com/id/44182081/ns/technology_and_science-science/t/bid-rename-homo-sapiens-called-unwise/#.TtQZpLIr1dQ

Pennisi, E. (1999). Did cooked tubers spur the evolution of big brains? *Science*, *283*(5410), 2004–2005. doi:10.1126/science.283.5410.2004 PMID:10206901.

Pruetz, J., & Lindshield, S. (2011). Plant-food and tool transfer among savanna chimpanzees at Fongoli, Senegal. *Primates, Online First*™. Retrieved 12/7/11 from www.springerlink.com/content/?k=(au%3a(pruetz)+OR+ed%3a(pruetz))

Rafferty, J. (2009). Zoudoudian. *Brittanica Online*. Retrieved 4/11/13 from www.britannica.com/EBchecked/topic/114721/Zhoukoudian

Russell, E. (2011). *Evolutionary history*. Cambridge, UK: Cambridge University Press.

Solecki, R. S. (1975). Shanidar IV, a Neanderthal flower burial in northern Iraq. *Science*, *190*(28), 880. doi:10.1126/science.190.4217.880.

Tainter, J. (1988). *The collapse of complex societies*. Cambridge, UK: Cambridge University Press.

Wieland, C. (2006, July 6). Mitochondrial Eve and biblical Eve are looking good: Criticism of young age is premature. *Creation Ministries International*. Retrieved 11/27/11 from creation.com/mitochondrial-eve-and-biblical-eve-are-looking-good-criticism-of-young-age-is-premature

ENDNOTES

1 Bill McKibben has a thorough analysis of how technology can destroy the values of community in his *Deep Economy* (2007), chapter 4, The Wealth of Communities.

2 Many paleontologists think other species are involved in the evolution of *Homo sapiens* from *Homo erectus*. Others think the supposed other species are variants of *Homo erectus*. All paleontologists agree that Neanderthals and human beings evolved separately.

3 It is not surprising that believers in the Bible as the correct and possibly only correct storehouse of human wisdom jumped to accept this conclusion. See for example Wieland (2006).

4 Ethicist Randy Cohen began a TV discussion with the comment that he had just murdered someone on the way over to the show, but of course, that was in the past and did not matter.

5 One band of chimpanzees (in the Tai forest) has been caching stone tools for cracking nuts for some time. See Goldenberg 2009. It may be that a million years or so might see chimpanzees on the same road we humans have followed.

6 Other animals such as ants, termites, bees, and beavers dramatically alter their environments and could be thought of as creating distinctive parts of the environment. There doesn't seem to be any evidence in the past 200,000 years of their dramatically increasing their appropriation of the environment, either quantitatively or qualitatively. Thanks to Catherine Schultz for reminding me

about beavers. Also none of these animals accumulates "stuff" to the extent that human beings do. Thanks to John Dittmeier for this point.

7 This was also true of Neanderthals, who were named for the Neander valley in Germany where their remains were first found. This can make for confusing Google searches. I searched for Neanderthal clothing and was directed to present day boutiques in the Neander valley.

8 Richard Leakey was reported to have maintained that all *habilis* tools were Oldowan and all *erectus* tools were Acheulean. He is also reported to allow that early *Homo erectus* probably still had Oldowan tools and developed the more advanced tools later.

9 Attributing consciousness that food-sharing is a distinctive human trait to *Homo erectus* is an example of what later in the chapter is described as *reverse temporalization.*

10 Dates for the origin of cooking vary more wildly than most prehistoric dates, from about 2 million years ago, to as late as 10,000 years ago. Both extremes seem implausible. The appearance of hearths at 250,000 years ago is good evidence that cooking had begun by that time. See Norman (2013) and Pennisi (1999).

11 Other genocides include the Armenian genocide by Ottoman Turks at the beginning of the 20th century. Nazis later used this genocide as a model for their own genocide of the Jews. Other later genocides include those in Rwanda by Rwandans, in Bosnia by Serbs, and Darfur by Sudanese.

12 In retrospect, it is hard to give any sensible reason for these wars. They may be wars for the sake of wars, to maintain what Eisenhower named "the military-industrial complex." The former Costa Rican President Oscar Arias may be responsible for the observation that if you have an army you will use it.

13 Priests of the Aztec god Xipe Totec wore the flayed skin of a captive as a coat in the Spring. Priests wore it until it rotted off. So perhaps a chilly early human might not have minded the odor. See Lindemans (1998).

14 The temptation here would be to employ "reverse temporalization" and attribute to individuals with no concept of clothing the ability to aim to make clothing. Reverse temporalization is characterized when we come to "Domesticated Dog."

15 This would then be the first clothes closet.

16 The reference is to the common ancestor Eve determined by genetic evidence, not the biblical Eve.

17 Making representations may be another example. See the section on Pictorial Representation.

18 Nomadic herders usually herd sheep.

19 This is the view of many commentators including art historian E. H. Gombrich (1996). However, Humphrey (1998) argues that the cave paintings resemble the line drawings of a 3-year-old autistic girl and therefore were not the work of a sophisticated "modern" mind. Although there is a resemblance in the pattern of lines, Humphrey ignores the elaboration present in the cave drawings such as the exploitation of irregularities in the cave surface. This elaboration is clearly visible in Herzog's film (2010). For my purposes, it is enough that they were deliberate representations.

20 Quoted in Goodman, *Languages of Art* (1976).

21 Thus any other account of representation seems to be basically reverse temporalization, an unhelpful explanatory mode.

22 Chapter 4, "Mind Body Consciousness," "Consciousness and Point of View."

23 Sumerians were using writing around 5,300 years ago. The Egyptians were using writing around 5,000 years ago, the Chinese around 3,300 years ago. The Mayans independently developed writing about 2,600 years ago (Diamond 1997).

24 The possession of language is frequently used as a criterion for being human. Both the 16th century philosopher Descartes and the 20th century linguist Noam Chomsky maintain that the ability to speak is a necessary condition for being human.

Chapter 6
Humans as Farmers

ABSTRACT

The developments separating humans the most from the ecosystem were farming and civilization. Farming dates from about 11,000 years ago, and civilization dates from about 7,000 years ago. Farming separates us from the ecosystem by replacing the ecosystem with plants and animals of our own choice. However, farmers must still respect the environment, which allows those plants and animals to thrive. Once humans begin farming, there are forces that make it difficult to return to a hunter-gatherer lifestyle, even when that lifestyle is preferable in many ways.

INTRODUCTION

The transition from hunter-gatherer lifestyle to settled farmer lifestyle made a huge difference in how humans relate to the ecosystem. The later transitions to civilization and modern technology arguably had greater impact. But farming for the first time involves deliberately altering the ecosystem for human benefit. In fact, many centuries later, the pastoral farming landscape has become the paradigm of nature. Nature has become a garden. According to Christian origin myth, a garden—the Garden of Eden—was the original dwelling place for humans.

Here is a timeline for the development of farming and agriculture:

- **20,000 years ago:** Humans cultivate wild plants
- **12,000 years ago:** Humans plant domesticated crops

DOI: 10.4018/978-1-4666-4586-8.ch006

- **12,000 years ago:** Humans domesticate sheep
- **11,000 years ago:** Humans domesticate pigs
- **10,000 years ago:** Widespread farming in Fertile Crescent, Egypt, China
- **10,000 years ago:** Humans domesticate cats and diversify dog breeds
- **10,000 years ago:** Humans domesticate cows and goats
- **9,500 years ago:** Villages (Çatal Höyük, Turkey)
- **8,000 years ago:** Humans domesticate chickens
- **7,000 years ago:** Elements of civilization
- **7,000 years ago:** Humans domesticate donkeys
- **6,000 years ago:** Humans domesticate water buffalo, horses, camels, llamas
- **6,000 years ago:** Humans domesticate honeybee, silkworm
- **4,000 years ago:** Humans domesticate elephants

Before civilization arises about 7,000 years ago, animals are domesticated for two purposes: Either for food—animal products like milk and egg or meat; or for performing specific tasks such as helping with hunting or catching rodents. Beasts of burden are not domesticated or used until civilization arises. Although beasts of burden help tremendously in transporting goods and people, they are not essential for civilization. The classic Maya developed an elaborate civilization without the use of beasts of burden of any kind. In any case, it appears that pre-civilization human farmers did not employ beasts of burden.

Farming had become well-established by 12,000 years ago. The following crops appear at the same time in the near east: wheat, barley, peas, lentils, bitter vetch, chick peas, and flax. By 9,000 years ago, a farm economy exists in Greece consisting of wheat, barley, sheep, pigs, and goats.

ORIGIN OF FARMING

So how did the transition between hunter/gathering and farming come about? And what are the implications of this development for the way humans relate to the ecosystem? The first step appears to have been cultivation, that is, the deliberate planting and care of wild plant varieties. This may have taken place as long ago as 20,000 years ago. As previously noted, a sedentary lifestyle is important for agriculture. Crops must usually be cared for. Sheep, the first domesticated species, can be raised by nomads. Cattle, to some extent, and pigs not so much.

Perhaps at first wild plant cultivation was a way of enhancing the yield of gathering activities. Seeds of more productive plants could have been planted with the expectation of returning to the site later.[1] When there was an accumulation of such plants producing sufficient food, it would become possible to reverse priorities and make gathering a secondary activity for providing food. The other big step would have been domesticating animals for food. Sheep were the earliest domesticated animals, descended from wild mouflon. Hans-Peter Uerpmann has proposed a plausible domestication scenario: Hunters might have brought orphaned baby mouflon back to their homes. Because mouflon are docile and cute as babies, they might have been kept at households long enough to reach sexual maturity and begin breeding (Hirst 2011). For a species to become domesticated, it must be willing to breed in human company. This domestication did not occur just once, but rather at least three times, in the Fertile Crescent.[2]

Although Uerpmann's scenario is quite plausible, one wonders why it did not happen earlier. Hunting dogs-descended-from-wolves have accompanied humans for at least 40,000 years. So the idea of humans keeping animals had existed. But keeping sheep for food and clothing is somewhat different. The multidisciplinary scientist Jared Diamond considers and answers this question, that is, the question of why the transition from hunter-gathering to farming did not take place earlier. Chapter 6 of his *Guns, Germs, and Steel* (1997), titled "To Farm or not to Farm," deals with exactly this question. Diamond's term for farming, "food production," is not appropriate. It is a mild form of reverse temporalization. It looks forward to agriculture as a support for civilization and even echoes the point of view of modern technology—plants and animals are standing reserve, food resources to be employed for other purposes. It is important to determine the extent to which agriculture can be practiced without civilization. This will be impossible if we define agriculture as food production, that is, in terms of its role in civilization. So I will continue to use the more neutral term 'farming.' Farming can be characterized as the human lifestyle which modifies land to make it suitable for raising plants and raises animals beneficial to humans, both mainly for food. These plants and animals are often modified by humans through artificial selection to increase their value to humans.

Diamond's answer in his *Guns, Germs, and Steel* as to why farming did not begin 10,000 or 20,000 years before it did is this: Wild mammals were still abundant and high-quality wild plants such as grain were not abundant. This part of his answer is somewhat plausible. Increased human hunting skills and climate change resulted in mammals becoming scarcer and wild grain becoming more abundant over this particular time span. This answer by itself makes the development of farming a somewhat contingent matter. But since farming arose independently at several locations, there must have been a similar trigger at each location. It is implausible that farming arose accidentally.

In any case, changes in the availability of food for hunter-gatherers is only one part of Diamond's answer about why farming arose when it did. The next part of his answer is that agriculture arose when the tools to practice it became available. But this reason is circular. The first farmers could not have had advance knowledge of how to construct the tools necessary for agriculture. Just before the appearance of farming, people would have to say, 'although we don't currently have any use for these inventions, we will, as soon as we switch from hunter-gathering to farming.' This part of his explanation is a full-blown case of reverse temporalization. Diamond himself calls attention to a similar case of reverse temporalization just before discussing agricultural tools:

The first people who adopted [farming] could obviously not have been making a conscious choice or consciously striving toward farming as a goal, because they had never seen farming and had no way of knowing what it would be like (1997, p. 105).

The final part of Diamond's answer is the most intriguing: Human population growth required greater productivity from the land than was possible under hunter-gathering. However, in an earlier paper, Diamond notes that farmers are markedly less healthy and less well nourished than hunter-gatherers. Skeletons from Greece and Turkey show that the average height of hunter-gatherers about 12,000 years ago was 5'9" for men, 5'5" for women. With the adoption of agriculture, by 5000 years ago height had decreased to 5'3" for men, 5' for women. Study of a transition from hunter-gathering to farming by Native Americans in 1150 A.D. revealed that

[c]ompared to the hunter-gatherers who preceded them, the farmers had a nearly fifty percent increase in enamel defects indicative of malnutrition, a fourfold increase in iron-deficiency anemia (evidenced by a bone condition called porotic hyperostosis), a threefold rise in bone lesions reflecting infectious disease in general, and an increase in degenerative conditions of the spine, probably reflecting a lot of hard physical labor (Diamond 1987, p. 3).

Further, hunter-gatherers typically spend twelve to nineteen hours a week collecting food, while farmers typically work from dawn to dusk.

So why would a way of live so seemingly disadvantageous be adopted? Farming could support many more people than hunting, but with a poorer quality of life. Hunter-gatherer population density is usually under one person per ten square miles, whereas farmers average 10 persons per square mile. A planted field produces much more food than a forest with scattered edible plants. Nomadic hunter-gatherer women have children at four-year intervals because babies must be carried until old enough to keep up with the adults.[3] Farm mothers, on the other hand, can have children every two years or less (1987). The result will be that the population of farmers grows faster than the population of hunter-gatherers. And more mouths to feed will require more farming and less hunter-gathering.

Diamond continues:

As population densities of hunter-gatherers slowly rose at the end of the ice ages [about 12,000 years ago], bands had to choose between feeding more mouths by taking the first steps toward agriculture, or else finding [other] ways to limit growth. Some bands chose [farming], unable to anticipate the evils of farming, and seduced by the transient abundance they enjoyed until population growth caught up with increased food production. Such bands outbred and then drove off or killed the bands that chose to remain hunter-gatherers, because a hundred malnourished farmers can still outfight one healthy hunter. It's not that hunter-gatherers abandoned their life style, but that those sensible enough not to abandon it were forced out of all areas except the ones farmers didn't want (1987).

Diamond concludes that agriculture was "the worst mistake of the human race." Not only was it disadvantageous to the health and well-being of the farmers, agriculture leads, according to Diamond, directly to gross social and sexual inequality, starvation, warfare, and tyranny. These are more the fruits of civilization, with its division of labor. So the real question is whether civilization is inevitable once agriculture is practiced. This question is addressed in the next chapter.

Diamond believes the move from hunter-gathering to agriculture was the wrong response to increased population density. The role of increased population density in triggering major changes in humanity's relation to the environment will be discussed in Chapter 9, "Causes of Conflict with the Ecosystem." One major issue is whether more people is in itself a good thing.

Diamond in a much later article notes that

All humans were originally hunter/gatherers who independently became farmers in only about nine small areas scattered around the world. A century of research . . . has shown that what made those areas exceptional was their wealth of wild plant and animal species suitable for domestication (2012).

We can now articulate a coherent view of the origin of farming. In a number of small areas there was a concurrence of wild animals and plants sufficient to make a sedentary lifestyle more appealing than hunter-gathering. Perhaps some steps toward domestication had already been taken by humans who still remained primarily hunter-gatherers. We can also conclude that previous attempts at a sedentary lifestyle had not been successful. We can also conclude that the necessary conditions obtained at somewhat different times—from about 11,000 years ago in the fertile crescent of the Middle East to 6,500 years ago in North America. Some sources note that the climatic changes of the last glacial period may have produced the climate change and resulting flora and fauna changes making farming possible (This would help to explain why farming was not initiated earlier.). However, once sedentary farming is successful for a number of years, hunter-gatherer population controls no longer obtain.[4] And then the increase in population beyond what hunter-gathering can sustain, means that farming has to continue. It is indeed wired into human beings that they want to provide for their progeny. So farming continues even when, as Diamond notes, it may not be advantageous.[5]

THE BOUNDARY BETWEEN HUNTER-GATHERING AND ITS SUCCESSORS

Diamond also attempts to show that there is not a sharp distinction between hunter-gatherers and farmers. He produces examples of hunter-gatherers who are sedentary and farmers who are nomadic. Neither type of counterexample is a reason to abandon the distinction, because both depend on exceptionally favorable circumstances.

Coastal Pacific hunter-gatherers became sedentary—but still hunter-gatherers—when there were ample resources in the environment so that moving wasn't required. If this happened a lot, there would have been no need to develop farming. The second type of counterexample is the nomadic farmer. Diamond instances New Guinea nomads who plant their crops and then go away, coming back periodically to check. Again, this has to be exceptional and growing conditions very favorable. Normally farmers have to pay close attention to their crops. In the Yucatan in the 19th century, the Mayans actually had defeated the Mexican army in what are called the Caste Wars. But they had to harvest their corn crop at a specific time, so they left their positions and the Mexican army resumed control. It is much more common for farming to require close attention to one's crops.

These counterexamples exhibit no significant change from the distinctive features of the hunter-gatherer lifestyle. Coastal hunter-gatherers had not discovered a new lifestyle and would most likely have started moving if something happened to the abundant supply of seafood. The New Guinea nomads who do not keep tabs on their crops would have been faced with two choices when they came back to crop failures: If the yield from the crops was enough to sustain them, switch to a sedentary lifestyle of farming. In this case the crop failure would be due to inattention. Or if the yield was not that great, continue planting and again leaving to pursue a hunter-gatherer lifestyle. In this case, they also would cease to depend on the planted crop.

Diamond's last example is hunter-gatherers in New Guinea who manage the landscape but not with domesticated plants. He notes that they almost meet the definition of farmer (1997, pp. 106-107). These people have clearly adopted the farming lifestyle. So none of Diamond's examples establish the existence of significant hybrid lifestyles between hunter-gathering and farming. This is what one would have expected, given the radical difference between the way each treats the environment.[6]

AN EARLIER START FOR FARMING?

There are carved stone monuments at Göbekli Tepe, Turkey, dating from 11,000 years ago. The existence of these monuments again raises the question of what hunter-gatherers were capable of. The paleontologist Klaus Schmidt holds that hunter-gatherers built the monuments, because there is no evidence of any nearby settlements (Batuman 2011). However, there are other sites from about the same time within 30 miles.[7] Several of these sites are settlements. Some such as Nevali Çori are extensive, almost too extensive for the very beginning of settled existence. In any case, the widespread presence of settlements in the area argues against even the presence of hunter-gatherers in the Göbekli Tepe area.

Schmidt holds that these hunter-gatherers somehow stumbled on the idea of division of labor in the course of building a religious monument and were thereby propelled into agriculture (Batuman 2011). While Prof. Schmidt deserves praise for his extensive site work at Göbekli Tepe, these ideas are extremely implausible at best. There is also the defect that there is no evidence for them, nor can there be. We have no access to the beliefs of the builders of Göbekli Tepe other than what we can discern from the site and our knowledge of the area at that time. There is absolutely nothing in the behavior of hunter-gatherers that would lead us to believe they would or could construct this monument for unknown religious reasons.

Even so, the date for Göbekli Tepe—11,000 years ago—is still anomalous. Division of labor is not a feature of early farmer society. Yet the construction of the monument of Göbekli Tepe clearly required division of labor (Batuman 2011). This suggests that some features of civilization were present way before the official starting date for civilization of 7,000 years. Further, it suggests that farming and civilization may have begun around the same time in this area.

IMPACT OF FARMING ON THE ECOSYSTEM

Domesticated plants and animals are still plants and animals and thus are part of the ecosystem. In the history of farming, it happens with some frequency that human creations substituted for the preexisting ecosystem have proven destructive both to the human creation and the underlying ecosystem. Large parts of the Fertile Crescent, the area where farming was first practiced, are now desert. Most of this transformation took place in the last 40 years and is mainly due to extensive damming of the two major rivers in the area, the Tigris and the Euphrates (National Geographic News 2001). Other formerly fertile areas such as North Africa became desert almost entirely due to human practices. Over-farming of wheat by the Romans is often blamed for the desertification of North Africa.

Some common farming practices often lead either to desertification or the loss of fertility of the land. In these cases, the original productive wild ecosystem is also lost, usually permanently. Clearing forest and grassland for agriculture has been done extensively in many parts of the world. Unless compensating measures are taken, clearing reduces the fertility of the land. If the area cleared is subject to winds, topsoil is lost to wind. This happened on a large scale in the US dustbowl of the 1930s. Tropical rainforest has thin topsoil, so when the forest is cleared, it is usually possible to grow crops for only a few years. When farming stops, the area becomes a wasteland. It does not return to rain forest. Irrigation to supply water for crops when natural rainfall is lacking, results in a buildup of salt in the soil. Eventually the land becomes unusable. This is currently happening in Brazil in the rainforest and in areas of Central America.

One relatively recent cause of trouble is the efforts of European colonizers to impose their own farming practices in areas ill-suited to those methods. In Australia, European sheep tear up the surface of land more fragile than England with their hooves. The soil blows away (Diamond 2005, pp. 390-391). In Europe, clearing cropland of trees to maximize crop exposure to sun is sensible. In central Africa, clearing cropland of trees usually results in complete drying of the soil, which then blows away. Instead, trees and crops have to be grown side by side: In tropical Africa, full sun is the enemy of crops (Bilger 2011).

As in the last example, there exist measures to combat these negative effects. And it is possible to farm sustainably. Farming in Egypt has continued without interruption from 7000 years ago. Most farming in Egypt was done within the area periodically enriched by the flooding of the Nile. Similarly, farming in China has been continuous for 7500 years. Techniques used to permit reuse of the land include crop rotation, fertilizers, and irrigation. However, the U. S. Department of Agriculture estimates that 25 million acres of land are lost per year because of increased salinity in irrigated land. Some measures to increase the fertility of the land have had other social effects, such as freeing farm workers for other types of labor.

Even when land is preserved for continued use, withdrawing land from the "wild" ecosystem and replacing wild animals with domestic ones, can have serious ecological effects. Loss of habitat impacts biodiversity. And why should biodiversity concern humans, who after all control the environment and maintain species they themselves have developed? Here we see again the connection of separation from the ecosystem and human superiority.

FARMING AND SEPARATION FROM THE ECOSYSTEM

When we see what agriculture does to the human relation to the ecosystem, it becomes clear that separation from the ecosystem and human superiority are actually the same thing. Hunter-gatherers were able to develop and use over a period of more than one million years what was called proto-technology in Chapter 3, "Science, Modern Technology, and Corporations." The only significant disturbance of the ecosystem came with overkill of megafauna, due to enhanced hunting skills. But the aim of the hunter-gatherers was not to alter the ecosystem. It is never the aim of hunters, prehistoric or otherwise, to exterminate the source of their prey. Certainly, it was not the aim of the Maori of New Zealand to drive the flightless birds called moas to extinction. But they did (Diamond 2007). It is not the aim of current international fishing fleets to destroy fisheries by overfishing. But they do.

Farming intentionally displaces preexistent plant species and replaces them with other plants selected for human benefit, usually regardless of the impact of this replacement on the ecosystem. The same is true with domesticated animals. "Wild" animals are replaced with domestic animals more suited to provide benefit to humans. Again, any impact to the ecosystem is a secondary concern.

In making these changes, humans implicitly believe they know better than the ecosystem what is better for the ecosystem. Of course, when farming was first developed, the ecosystem and its demands were unrecognized. So any limitation on the exploitation of the environment would only be to allow the exploitation to continue and not be thwarted by nature. It would not be for the sake of the ecosystem. Thus in farming human belief in human superiority first manifested itself. This is human superiority as formulated in Chapter 1, "Humans as Masters of the World":

(Human superiority). We humans are in charge of what happens to the environment, and our agency is the only agency humans need to consider.

Also, anything not under the aegis of human agency is deemed 'wild' and is a candidate for being brought under human control, often by elimination.

However, farmers clearly cannot accept human exceptionalism as formulated in Chapter 1:

(Human exceptionalism) We humans are not part of nature.

Plants and animals, whether wild or domesticated, grow and flourish only in accordance with the dictates of nature in providing water, sunlight, and food[8] for those plants and animals. No moderately sane farmer could entertain the idea that he or she was not part of nature.

If this explanation of the origin of human superiority in farming is correct, the account of the origin of farming takes on great importance. For farming embodies a radical change in how humans relate to the environment. It is not hard to see how domesticated plants and animals came to be. Humans first had to have two ideas: One is cultivating wild plants, which humans began doing about 20,000 years ago. Second, to raise the young of docile species orphaned during hunting. So far, no separation from the ecosystem nor any basis for human superiority. Ironically enough, the potential for domestication came about because of the localized bounty of the ecosystem in providing plants and animals exceptionally suited for human cultivation.

The potential for both separation from the ecosystem and human superiority occurs when humans begin to practice artificial selection to produce distinctively human domesticated animals and plants. The potential becomes actual when farming supplants hunter-gathering as a lifestyle. As science contains the potential for modern technology, artificial selection contains the potential for human superiority.

I previously explained Diamond's account of how the last transition, from hunter-gathering to farming, occurred. It is plausible that once the mechanisms of domestication are available, a jump in population may have resulted in the necessity to adopt farming.

Jared Diamond may have been right to call farming the worst mistake in the history of the human species. But not for his reasons. There is not an immediate connection between farming and the evils he lists, namely, gross social and sexual inequality, starvation, warfare, and tyranny. Rather, farming gives these clever animals called humans the idea that they are in charge of the ecosystem. And farming provides the basis for the next major development of humanity, civilization. In the next chapter, whether civilization must generate the evils Diamond cites is considered. But whether civilization generates these evils or not, the prior question is whether farming must lead to civilization.

FARMING AND CIVILIZATION

Can there be a stable agricultural society which does not generate civilization? Clearly once civilization arises, farming must still continue for food production. As far as civilization is concerned, farming is food production. It doesn't matter how the food is produced, it only matters that it *is* produced. So obviously we begin to see a much greater level of separation from the ecosystem in civilization.

There are two possible views about the relation of farming to the origin of civilization: (1) Once humans begin farming, the rise of civilization is inevitable; (2) Civilization will be a temptation, but is not inevitable.

Some have argued that population density by itself is enough to trigger major lifestyle changes. Three anthropologists—Marcus Hamilton, Bruce Milne, Robert Walke, and ecologist Jim Brown—claim to have discovered a fundamental law involving human population density. The greater the population, the greater its density on the land it lives on. They conclude that the greater density itself shows that this population is making more effective use of resources and that therefore more people are freed up to perform tasks other than producing food. So, according to these authors, density played an important part in causing the development of complex societies (de Chant 2011).

Unfortunately, it is hard to take these claims seriously. First of all, greater population densities could be caused by limited land for the population to live on. In such a case, there is no causal connection between population and density. More importantly, it is hard to see how density could be a cause at all. We have just seen that moving from hunter-gathering to farming may not make more effective use of resources, even though it permits higher densities. Similarly, density does not cause freeing up of individuals. In the transition from hunter-gatherer to farmer, the amount of work that farmers must perform *increases*. When we come to civilization, it is clear that people *are* freed up to do other things than produce food. But density does not seem to be the *cause*. How people are freed from food production is an important issue in the next chapter.

In order to decide the issue of whether the transition from farming to civilization is inevitable, we need to know a lot more about civilization and how it acquired its various properties. These are issues for the next chapter. We have seen that the transition from hunter-gathering to farming may have been almost inevitable, in the circumstances when it took place. But those circumstances were somewhat particular and contingent. So the transition from farming to civilization may or may not have been inevitable, all things considered.

REFERENCES

Batuman, E. (2011, December 19). The sanctuary: The world's oldest temple and the dawn of civilization. *The New Yorke*, 72-83.

Bilger, B. (2011, December 19). The great oasis. *New Yorker (New York, N.Y.)*, 110–121.

De Chant, T. (2011, August 17). Hunter gatherer populations show humans are hardwired for density. *Scientific American's Guest Blog*. Retrieved 11/29/11 from persquaremile.com/2011/08/17/hunter-gatherer-populations-show-humans-are-hardwired-for-density/

Diamond, J. (1987, May). The worst mistake in the history of the human race. *Discover*, 64–66.

Diamond, J. (1997). *Guns, germs, and steel.* New York: Norton.

Diamond, J. (2005). *Collapse*. New York: Viking.

Diamond, J. (2012). What makes countries rich or poor? *The New York Review of Books*, *59*(10).

Hirst, K. K. (2011). Domestication history of sheep. *About.com.Archaeology.* Retrieved 12/13/11 from archaeology.about.com/od/shthroughsiterms/qt/Sheep-History.htm

Kelly, R. (2005). The evolution of lethal intergroup violence. *Proceedings of the National Academy of Science, 102*(43), 15294–15298. Retrieved 12/22/11 from www.ncbi.nlm.nih.gov/pmc/articles/PMC1266108/?tool=pmcentrez

Lee, R. (1980). *Lactation, ovulation, infanticide, and women's work: A study of hunter-gatherer population regulation*. Retrieved 5/22/12 from tspace.library.utoronto.ca/bitstream/1807/18007/1/TSpace0104.pdf

National Geographic News. (2011, May 18). Ancient fertile crescent almost gone, satellite images show. *National Geographic News*.

ENDNOTES

1 This is the practice of some present day New Guinea hunter-gatherers. See Diamond (2002, p. 106).

2 The Fertile Crescent stretches from present day Syria through Iraq (Mesopotamia) up to western Iran. It was the site of many early developments in agriculture and, later, civilization.

3 Since Planned Parenthood was not available to hunter-gatherers, other means of keeping birth frequency to about four years had to be operative. Nursing had a role in postponing fertility, as did hormonal changes in hunter-gatherer women. Since abortion would have been quite dangerous in aseptic conditions, infanticide had to be an option. The behavior of current hunter-gatherers provides some clues.

4 See Lee (1980).

5 Indeed, current subsistence farmers often overpopulate to insure sufficient progeny and sufficient workers.

6 Diamond repeats these arguments in his 2012 article. My objections still apply.

7 This was established using Google Earth.

8 To speak of nutrients rather than food is to adopt the point of view of science. This may not be a problem unless one also adopts the point of view of modern technology.

Chapter 7
Civilization

ABSTRACT

Farming makes civilization possible by producing a food surplus, so that a number of people are free from work connected with producing food. This separation from the environment enables civilized humans to develop a wide range of activities, pursuits, and occupations, accompanied by many technological innovations. These new activities and innovations are often regarded as demonstrating the superiority of humans. However, since almost all these pursuits and innovations benefit humans only, it is only physical superiority that is demonstrated.

INTRODUCTION

The two previous periods of human existence we have examined, those of hunter-gatherers and farmers, are distinctive ways of making a living. There is no corresponding distinctive way of making a living for civilized human beings. In fact, civilization is often praised for allowing extensive division of labor. Division of labor is an important defining characteristic of civilization. For the first time, social organization allows for a diversity of occupational roles. Most people's occupations are also more distant from the supporting ecosystem.

What are the distinctive features of civilization? Most dictionary definitions include such evaluative language as 'advanced state of development' with a 'high level of culture, science, and social development.' The Website Dictionary.reference.com offers the following gloss on 'civilization':

DOI: 10.4018/978-1-4666-4586-8.ch007

Once a nation, culture, or group of people has been brought out of the 'savage' darkness into an enlightened and advanced state, it becomes a civilization (2011).

Building this evaluative language into the very characterization of civilization makes it close to impossible to give a neutral account of the features of civilization. It also makes it difficult to evaluate its flaws critically, or to compare civilized lifestyles with previous so-called 'savage' or 'primitive' lifestyles (a.k.a. hunter-gathering) or 'rustic' lifestyles (a.k.a. farming).[1] The question is, what makes civilized people not-rustic, not-savage, not-primitive?

A more neutral account of the characteristics of civilization begins with division of labor, that is, occupational specialization. Such occupational specialization usually comes accompanied by *hierarchy.* Hierarchy involves layers of authority, with the higher layer issuing orders and the lower layer obeying. Also, civilization is primarily seated in cities. A *city* is a large, permanent settlement, usually with common utilities and a governing class. Farming and hunter-gathering are minor activities so that the city is dependent on outside areas for its food. There may also be a legal system, record-keeping, currency, and the practice of arts and sciences. The core features seem to be division of labor, cities, government, and food dependence, and also a certain extension. A single city might be its own *society,* but it would not be thought of as a *civilization.* A civilization would normally involve a number of communicating cities with a common culture[2], also with political and (usually) religious ties.

The question is, how did social structures of this type develop out of farming communities? My aim is to uncover the changes in human relations to the ecosystem and in human attitudes of superiority brought about by civilization. Civilization is primarily a change of social organization with consequences for its relation to the environment.

Timeline:

- **9,500 years ago:** Villages (Çatal Höyük, Turkey)
- **7,000 years ago:** Division of labor
- **7,000 years ago:** Use of metal (ambient copper)[3]
- **6,000 years ago:** Domestication of horse, camel [4]
- **5,000 years ago:** Numerous city-states
- **5,000 years ago:** Use of concrete (Mesopotamia, Egypt)
- **5,000 years ago:** Writing (Sumer, Egypt);
    - (China 3,300 years ago);
    - (Mesoamerica 2,600 years ago)[5]
- **5,000 years ago:** Commodity money
- **3,700 years ago:** Codified laws (Hammurabi)

Civilization is a distinctive way for humans to live and relate to the environment. There are radical differences between civilization and even farming modes of life. The first thing we should notice about civilization is that it has not been around for all that long. The earliest currently known instance of civilization was at Sumer, in the fertile crescent of Mesopotamia, between 5,000 and 6,000 years ago (Gill 2000). Compared to the roughly 200,000 years that modern humans (*Homo sapiens)* has been around, this means that slightly less than 3% of humanity's time on earth has been spent living in civilization.

The second thing we should notice about civilization is that within that relatively short time, all civilizations with the exception of our own have collapsed. If we consider our own civilization to be Western or European civilization, we can date its origin to be roughly 1000-1200 years ago, when Europeans emerged from the dark ages following the collapse of the Roman Empire. We have lived in this, the only extant civilization, for about 1% of humanity's time on earth. Both of these figures suggest that it would be premature to argue that civilization is necessary for or distinctive of human life.

There is, instead, considerable evidence to the contrary. Two recent researchers, Joseph Tainter (1988) and Jared Diamond (2005), trace in some detail the collapse of previous civilizations. Tainter thinks that developed technology may prevent the collapse of the current civilization. Diamond thinks the same factors that caused the collapse of other civilizations could easily cause the collapse of ours. The factors involved in those collapses are: Environmental damage; climate change; hostile neighbors; decreased support by friendly trade partners; and the civilization's attitude toward its environmental problems (2005, pp. 11-15). Diamond is careful to note that not all factors are present in every collapse. But the last factor—attitude toward environmental problems—is always significant.

Although Diamond thinks that *societies* can collapse for reasons not having to do with environmental problems—the Soviet Union is his example—it is difficult to believe that anything short of environmental problems could cause the collapse of an entire *civilization.* Because civilizations are almost entirely divorced from the ecosystem, it is likely that they will have environmental problems. One question is whether the built-in environmental problems of civilizations are severe enough to make their eventual collapse either likely or inevitable.

ORIGIN OF THE DISTINCTIVE COMPONENTS OF CIVILIZATION

Sometimes the origin of agriculture is conflated with the origin of civilization. Thus, Paul Hartman writes:

We typically define "civilization" as that point at which a primitive people learn that the food supply can be made more reliable by domestication of crops and animals, at which point two other things happen. The first is the realization that the people need no longer be nomads—constantly moving from place to place, hunting and gathering—but can remain in one place. The second is to discover that with domestication, not all of the people are necessary to provide food. Some can be allowed other tasks, and thus, an artisan class can emerge. . . A division of labor in a fixed location is the evidence for "civilization," and it begins where crops grow most abundantly—in the fertile valleys of rivers such as the Tigris and Euphrates (1996).

Hartman's account explains the transition to civilization in terms of a sequence that would not have been possible for people making the transition. It is thus an instance of *reverse temporalization.* As we have seen, domestication generally *requires* settlement.[6] Also, assignment of people to fixed occupations (division of labor) generally requires some authority and therefore food surpluses cannot cause division of labor in the absence of an authority.

There is also a suggestion that areas of abundant crops will likely produce civilization. We have seen that areas of abundant crops were the cause of farming. A farming lifestyle clearly contains the potential for developing civilization, because civilization could not have arisen without food surpluses.

As discussed in the preceding chapter, some anthropologists believe that civilization is automatically triggered by density. Marcus Hamilton, Bruce Milne, Robert Walke, and ecologist Jim Brown claim to have discovered a fundamental law involving human population density: The greater the population, the greater its density on the land it lives on. They conclude that the greater density itself shows that this population is making more effective use of resources and that therefore more people are freed up to perform tasks other than producing food. So, according to these authors, density played an important part in causing the development of complex societies (de Chant 2011).

But it is hard to see how density could be a cause at all. We saw in the preceding chapter that moving from hunter-gathering to farming does not make more effective use of resources, even though it permits higher densities. Similarly, density does not cause freeing up of individuals. In the transition from hunter-gatherer to farmer, the amount of work that farmers must perform *increases.* When we come to civilization, it is clear that people *are* freed up to do other things than produce food. But density does not seem to be the *cause.* Density may be a necessary condition for food surplus, but it does not cause the institutions of civilization which make use of the food surplus. Density may also play an important role in providing the groundwork for technological and cultural innovation. This issue will be discussed in the following section.

THE DISTINCTIVE COMPONENTS OF CIVILIZATION

Human settlements can be villages, towns, or cities. A civilization is a collection of interdependent cities with significant common cultural aspects. There seem to be no generally accepted ways of distinguishing villages, towns and cities, except for size: Towns are bigger than villages, and cities are bigger than towns. However, there are important differences between human settlements of different sizes, and they can be marked as follows:

- A village is a small settlement all of whose inhabitants are directly involved in farming or some other primary resource-gathering activity (e.g., fishing or mining). A village has no or minimal government activity.
- A town is usually larger than a village. In addition to primary farming or resource-gathering individuals, it also has some dedicated government administration. (This administration most likely appeared *after* the formation of a state exercising its authority from a center.)
- A city is a substantial settlement whose inhabitants do not engage in primary production activities such as farming or resource gathering. Therefore, they are engaged in other activities such as government administration or the production of manufactured objects or the distribution of food and other products. As civilization develops, the number and variety of occupations proliferates.

City dwellers neither farm nor get food from hunter-gatherers. So how could this arrangement have gotten started? What could farmers have gotten out of it? Let us go back to early farmers and work out developments from there.

Early farmers will probably tend to live in villages, as protection from those who think taking already grown food is easier than growing it themselves, and from predators who could find irresistible large concentrations of edible prey. One sees this pattern of settlement in rural Europe, especially France or Italy: Farmers cluster in villages and farm the surrounding land. Farming thus leads to villages for protection. There are also potential cooperative benefits in knowing what one's neighbors are up to, for example being able to trade surpluses to make up for deficiencies.

In establishing the origin of civilization, it is important to remember that humans (and their ancestors) always lived in groups. So the transition to civilization is not a transition from no social organization to civilization, but rather a transition from an informal social organization to a much more elaborate formal social organization (Service 1975, ch. 3). Anthropologist Elman Service (1975) claims that hunter-gatherer societies and early farm societies were egalitarian. Service claims no one had authority; rather, leaders arose for particular ad hoc purposes without their holding an established office of any kind. Social sanctions are informal. However, the absence of defined job roles does not mean the absence of authority. All primates living in social groups have dominance relations (Welker 1985). So what needs to be explained in the transition to civilization is the rise of the authority structures distinctive of civilization.

A fully developed civilization has a governing authority embodied in a *state*. Many anthropologists believe that all states were preceded by *chiefdoms*, political organizations where one individual had highest authority but without the administrative structure of a state (Beliaev et al. 2001). Essentially an individual is recognized in the village or town as being better at activities requiring direction by an authority. His authority is made permanent and he becomes a chief. Such a chief can become hereditary and can be invested with supernatural powers. All this elaboration of authority can take place within hunter-gatherer or early farming societies. In any case, none of this by itself leads to separation from the ecosystem. But when the authority extends to economic matters as well as political, we begin to see patterns of authority more characteristic of civilization.

When political authorities gain control over distribution of goods, we begin to add the social structures distinctive of civilization. The key to this happening is the accumulation of a significant surplus of goods. Initially this will mainly be less perishable stored farm products. It is when individuals or groups (classes) gain control of goods they do not produce—especially agricultural goods—that civilization begins its divorce from the ecosystem.[7]

If goods are redistributed to the benefit of the villagers, they have a motive for continuing the arrangement. Also if public works are undertaken through the economic authority, that is more reason to continue the arrangement. As the social structure becomes more complex, eventually a class arises whose job it is to administer the various elements of the more complex society. One has a true government (a *state*) when this bureaucracy has authority which it can back up with repressive force.[8] There is a hint of reverse temporalization in this sequence. One would like to know more about the mechanism of several steps, especially how an administrative class arises or how it attains the ability to employ repressive force.

Jared Diamond, in his the *World until Yesterday* (2012), asks, "But how does the chief or king get the peasants to tolerate what is basically the theft of their food by classes of social parasites?" (p. 356). The answer, he claims, is that the chief or king proclaimed an organized religion with these tenets: The chief or king is related to the gods or is a god; the ruler can obtain important favors from the gods such as rain. Additionally, the ruler organizes public works to everyone's benefit and raises an army which can conquer neighboring lands to everyone's benefit including the peasant. Because of these benefits, the peasants supply food to the ruler and his entourage.

We have evidence for these uses of religion in all early civilizations such as those of Egypt, Mesopotamia, Polynesia, Mesoamerican, and Inca. But even so, the prior question is how religion acquired so much power. Before civilization, we will never have any firm evidence for the religious structures of precivilized hunter-gatherers. So although Diamond's explanation is quite plausible, it is also incomplete and incompletable.

These developments leading to civilization take place in single villages, towns, or cities. Other developments are important in transforming a collection of cities into a civilization. For example, when a collection of villages or chiefdoms accept domination by one center. This is called *synoecism,* from the Greek 'to colonize together.' There can be political synoecism, where several independent units accept the domination of a political center. Or there can be physical synoecism, where several independent units relocate to a center (Blanton et al. 1999). Again the question is, why would independent villages do this? We can suppose that at first the independent villages benefited from so doing, either from redistribution by the political center or for defensive reasons if there were external enemies.

Of course, once cities with the power of states arise, the answer may be that villages are subjugated by force. But the first city-states could not have arisen this way, with fully articulated armed force. They would have (impossibly) had to have foreknowledge of states and their powers and potentialities, before there were any states. This is another case of reverse temporalization.

Historically, military threat from external enemies was the reason for synoecism of several communities with Athens in classical Greece. But where such a threat is minimal, there can still be advantage in closer coordination between a number of separate villages. This can be achieved by choosing a center which is *disembedded*: It is neutral with respect to preexistent parts. In ancient Oaxaca in Mesoamerica, the neutral area chosen for the disembedded capital Monte Alban was an under-populated area of poor farmland which was also centrally located with respect to existing population centers.[9]

At this point in accounting for the development of civilization, anthropologists tend to wave their hands. Thus, Elman Service says:

[This center], now more like that of an empire than a simple local primitive homogeneous ethnic unit, may quite quickly acquire the specialized appurtenances usually attributed to the archaic civilizations: writing and mathematical notations with associated tax and census scribes, great specialized art, metallurgy, elaborate ceremonial religion, and grand public monuments and other works. They are the final benefits of . . . a centralized and expanding political organization (1975, p. 308).

It is hard to believe that all these things just *happen* as the result of the rise of a political organization. But, on the other hand, all these things did come about quickly—within two thousand years or less. So we do have to conclude that civilization did supply the conditions for these developments. These developments are technological. So it follows that civilization supplies the conditions for technological innovation.

And when extremely dramatic changes to human lifestyles occur independently at several locations, it is difficult not to suppose that major elements in the process are more-or-less inevitable. Certainly, civilization arose completely independently in the Americas and Eurasia. This strongly suggests that the sequence leading from food surplus to a ruling class of a city with a diversified economy is perhaps inevitable.

CIVILIZATION, HUMAN SUPERIORITY, AND SEPARATION FROM THE ECOSYSTEM

I am primarily interested in how the rise of civilization affected attitudes of human superiority and separation from the ecosystem. Anthropologists and paleontologists can trace the various steps from hunter-gatherers to civilization. But it remains somewhat of a mystery how the civilized lifestyle with its nearly complete divorce from the ecosystem could have arisen so quickly and independently at numerous sites. Hunter-gatherers initially have only ad hoc job specialization. When the position of chief arises, the person fulfilling that position is felt to have superior capabilities. But when the existence of food surpluses makes it possible for that person and those around him to have persistent authority, classes are created. And especially when the class with authority is hereditary, what is created is a class which is superior for no other reason than that it is superior. There could be something hard wired in this arrangement, because it persists into current times. Some hereditary rulers have skills useful to their countries. Some do not. But otherwise enlightened countries such as England, the Netherlands, and the Scandinavian countries still have hereditary rulers.

Of course there is always the danger that such arrangements may become kleptocracies. But the European monarchies mentioned above are all constitutional monarchies; the monarch has little political power and elected officials have oversight over the monarch. Elected representative governments are also not immune to kleptocracy. Over recent years, ten heads of state have siphoned off amounts ranging from an estimated $25 billion dollars (Suharto, Indonesia), to Alberto Fujimori's $600 million (Peru). (Infoplease 2012) But what seems to be wired in is the ability of human beings to believe they are superior for absolutely no reason, and for others to go along with it.

Writing also seems to be to some extent innate because it too arose independently at least three times, in Sumer, China, and in the Maya civilization. As stated in Chapter 5, "Humans as Hunter-Gatherers," it seemed that making representations was an innate human capability because there is no plausible pathway to representation. Pictorial representations are the basis for pictograms, which also turn out to be the basis for writing systems. The earliest pictograms and ideograms date from about 11,000 years ago. Writing that records language dates from about 5,000 years ago. Pictograms were gradually standardized to become alphabets and syllabaries (Partch 2004).

Once civilization produces the conditions for the creation and accumulation of technology, the actual progress is not that predictable. This is true both for 'classic' technology and modern technology.[10] Consider the invention of the wheel. In Eurasia the wheel appears 6500 years ago, at about the same time in Mesopotamia, the Caucasus, and central Europe (Ament 2007). The potter's wheel was invented at the same time, leading to speculation that the potter's wheel was the basis for the regular wheel (or vice versa). But in Mesoamerica, the wheel appears only on Olmec and Aztec children's toys. Although most pre-colonial American cultures had elaborate pottery, none used the potter's wheel (Adas 1992). However, other important technologies such as writing were invented independently in Eurasia and Mesoamerica (Diamond 1997). So it is clear that there is a substantial amount of contingency in the appearance of particular items of technology.

But civilization with the same main features also arose independently several times—including the propensity for technological innovation. Anthropologist Henri Claessen found by statistical analysis that a number of features were found in virtually all early civilizations (99%). These features are: A definite territory which they were able to defend; definite social classes; lower classes produce food and there is an agricultural surplus; a centralized authority able to enforce edicts (Claessen 1978). But again these features don't provide in themselves an explanation for the efflorescence of technology in early civilization. So, for now, note only that civilization itself seems to produce technological innovations for reasons we as yet do not understand.

An important value of civilization is as the infrastructure for innovation, mainly technological. But these innovations are always implemented at a distance from the ecosystem. There is nothing in the processes of technological innovation within a civilization that prevents those innovations from being detrimental to the ecosystem and even to that very civilization. But such innovations often are detrimental. As mentioned earlier, all previous civilizations have collapsed in large part because of conflicts with the environment. Both pre-classic and classic Mayan civilizations collapsed because of the taste of rulers for plaster coating on their monumental buildings. An enormous amount of wood was required to produce the plaster, which led to deforestation and serious damage to cropland. Eventually the populations of the cities could not be fed. This same scenario happened not once but twice, in the earlier city of el Mirador, about 100 CE (Brown 2011) and the later city of Copan, about 800 CE (Diamond 2004, p. 169).

CIVILIZATION AND ITS DISCONTENTS

In some ways, civilization may not be an improvement over pre-civilization. First, civilized individuals work a lot harder than hunter-gatherers. And, as Henry David Thoreau argued, the fruits of those labors may be of dubious value. Second, according to Sigmund Freud, civilization must to some extent frustrate its members in order to endure.

The anthropologist Marshall Sahlins (1968) found that ethnographic data showed that hunter-gatherers worked far fewer hours and enjoyed more leisure than typical members of industrial society, and they still ate well. Hunter-gatherers met the same requirements as their sedentary neighbors through much less complex means. As we saw earlier, it was also true that hunter-gatherers worked a lot less than farmers. The extra complexity in our way of life costs us extra effort.

This theme had already been put forward in the mid-19th-century by Henry David Thoreau in his *Walden* (1854). Thoreau believes that civilized humans have no time for anything of importance: They have become machines. Thoreau is writing at the time of the development of modern technology and of corporations, so his criticisms of civilization to some extent apply to civilization enhanced by modern technology.

Thoreau considers how civilization supplies the basic necessities of life. He observes that civilized clothing does not have as its object that humans be "well and honestly clad," but that clothing be fashionable, with the ultimate aim of enriching corporations. (23) Civilized shelter, according to Thoreau, usually requires a person to spend more than half his life paying for it, and sometimes more. The question is whether the more elaborate shelter is worth this effort. And above and beyond this, Thoreau asks what is the value of division of labor itself. He calls for college students to *live* life rather than *play* life or *study* it:

Even the poor student studies and is taught only political economy, while that economy of living which is synonymous with philosophy is not even sincerely professed in our colleges. The consequence is, that while he is reading Adam Smith, Ricardo, and Say, he runs his father in debt irretrievably (p. 40).

Thoreau finds that many "modern improvements" are "improved means to unimproved ends."

Thoreau's account of how he lived at Walden Pond shows that he did not completely live apart from civilization, but rather that he minimized his dependence on it. His main objection to civilization is that it stunts humans by redirecting their energies toward worthless or frivolous pursuits. His own simple lifestyle at Walden Pond is meant not so much as an example to be followed, as an example of how it is possible to live well without (much) civilization. His lifestyle points the way beyond civilization. *Walden* ends with the following striking passage:

I do not say that John or Jonathan will realize all this; but such is the character of that morrow which mere lapse of time can never make to dawn. The light which puts out our eyes is darkness to us. Only that day dawns to which we are awake. There is more day to dawn. The sun is but a morning star (p. 221).

Thoreau is clear that that lifestyle is best which respects and treasures non-human nature ("the wild"). Thoreau's specific lifestyle recommendations are quite ascetic. Over and above the specific evils of civilization, there are the evils of being too animal or too beastlike. It is a higher realization of one's humanity not to eat meat. Additionally, Thoreau recommends abstinence from sex, alcohol and gluttony: All sensuality is the same, unclean and shameful (pp. 149-150). It is somewhat surprising to see this asceticism combined with reverence for non-human nature. But the net result is the same as when religious doctrines about human beings make them separate from and "above" brutish nature. The respect we pay to the rest of nature does not include respecting ourselves as part of nature.

Thoreau's critique of civilization is worth keeping in mind, especially as a counterbalance to exaggerated claims of the value of civilization. But the recommendations for individual action make sense only if human beings are not part of nature.

Ironically, Sigmund Freud's views echo Thoreau's in a curious way. Freud, in his classic essay, *Civilization and its Discontents* (1930), claims that civilization must create a certain level of frustration in its members in order to exist. For civilization to exist, according to Freud, pre-civilized instincts, especially those having to do with aggression, must be repressed. These instincts are the "animal" ones Thoreau wanted us to overcome. For Freud, this repression—which is actually turned inward as conscience—must necessarily lead to a certain amount of unhappiness. But for Freud, this is the price we must pay for civilization. And, for Freud, unlike Thoreau, civilization is more-or-less a good thing. Freud says:

. . . civilization is a process in the service of Eros [the life instinct], whose purpose is to combine single human individuals, and after that families, then races, peoples, and nation, into one great unity, the unity of mankind. Why this has to happen, we do not know; the work of Eros is precisely this (1930, p. 69).

The aggressive instincts are in the service of Thanatos [the death instinct], which, when moderated, provide the ego ". . .with the satisfaction of its vital needs and with control over nature" (p. 68).

Ironically, Freud calls on civilization to tame those instincts. Thoreau wanted more advanced individuals to tame these instincts on their own. But equally ironically, civilization in the service of Eros, must tame the destructive instincts of control of nature. If we step outside Freud's realm of inner drives, we have to recognize that civilization is a major step toward more control of nature. So how can this added control of nature—in the service of the death instinct—also be in the service of Eros?

Of course Freud does not deal at all with the relation of civilization to nature. Nor does he deal with the historical origins of civilization. The question we need to ask is, when and how did Eros hold sway in pre-civilized lifestyles such as farming and hunter-gathering? Unfortunately, Freud's answer will be in terms of the action of a mysterious force striving to produce "one great human unity." So Freud's answer is ultimately no help in determining how humans came to regard themselves in charge of the environment. He has nothing to say about the impact of this force on the environment. Further, this force embodies aspects of both the life instinct and the death instinct. So it cannot be regarded as necessarily positive. And finally, invoking this force toward unity does nothing to explain the transitions from hunter-gatherer to farmer and farmer to civilized human being.

CITY VS. COUNTRY

Civilized individuals have a definite view of their own superiority. This is embodied in the sharp difference between country (farming) and city (civilized) lifestyles. This difference is the theme of many stories and films. One of the most evocative is Murnau's silent film *Sunrise* (1927). *Sunrise* depicts the conflict between pastoral life and civilization amped up on modern technology. But there is also a conflict between pastoral existence and civilization before modern technology. The Roman bumpkin or the Mayan bumpkin probably met similar contempt from their more advanced civilized cousins.

But such conflict flies in the face of the obvious dependence of civilization on agriculture for food. Civilization could not exist without farming, city mouse could not exist without country mouse.[11] Even so, within civilization, the necessary support of agriculture is disparaged and devalued in favor of the multitudinous creations of a way of life with no direct connection to the ecosystem. Modern technology, by transforming pastoral farming into factory farming, replaces farming with its technologized equivalent. The predictable result is horrendous pollution of the rest of the environment, which, as we have seen, is not a concern for modern technology. The disconnect with the ecosystem becomes much greater.

An earlier unavoidable conflict between hunter-gatherers and farmers persists. By converting wild land to agriculture, farming reduces the area of land available for hunting and gathering. This type of conflict still persists in areas with significant hunter-gatherers such as the Amazon. But, unlike the later dependence between farming and civilization, there is no relation of dependence between the two ways of life: Hunter-gatherers don't need farming, and farmers don't need hunter-gatherers.

So why do city dwellers feel so superior to country folk? It is their superior lifestyle, more complex, more evolved. They are involved in purely human pursuits, distant from the barnyard or from the wild with its barbarians. What we are seeing here is a pure form of human superiority. Civilization has its advantages and its disadvantages. But the disadvantages are rarely discussed.

So far as the relation of civilization to the environment, civilization's supposed efficiencies have to be weighed against what economists call 'externalities.' That is, civilization's negative effects on the environment. Even farming has externalities which have to be considered. When the preexisting environment is replaced with cropland, the results can be disaster. In the Great Plains of North America, the result of plowing grasslands was the destructive Dust Bowl of the 30s. Deforestation in tropical rainforests yields maybe a year or two of decent crops followed by the land's having been rendered worthless, unusable for farming or forest.

But when civilization makes use of land, that land is rendered useless for any other purpose. And then the activities of civilization, even of early civilization, produce pollution. Concentrated wastes have to be disposed of. Of course, with modern technology the impact of civilization on the environment is much greater. But whether we are considering early civilization or modern technology, harm to the environment as well as the benefits of civilization or technology need to be considered.

NATURE GODS VS. MONOTHEISM

According to Elman Service (1975), all civilizations which developed on their own from preexisting societies, developed from chiefdoms headed by individuals (or classes) claiming supernatural powers. These supernatural powers derived from various types of nature gods. As civilization developed, nature gods (called pagan) were replaced by a monotheistic God. Monotheism, the currently most popular form of religious faith, appears in various forms: In Babylonia about 3500 years ago in the form of Marduk; in Egypt about 3300 years ago in the form of Aten; and in Judaism through Moses perhaps 3200 years ago.[12] It is hard to see what advantages monotheism has over nature gods. Monotheism may reflect the reality of living in civilization: The power of nature is experienced as an abstract force to which each individual is directly related. However, since this one god must contain all properties, severe problems of consistency arise which would not come up with multiple gods. They mainly concern the conflict between the properties of omnipotence and benevolence, together with the existence of evil.

There is, for example, no satisfactory way of accounting for evil spirits. An all-powerful god has to be responsible for the devil and all the devil's actions. But this is incompatible with benevolence. If there are multiple gods, some good and some bad, this is not a problem. But then these gods are not omnipotent. The Buddha drew the correct conclusion about 2500 years ago. Because there is suffering, there cannot be an all-powerful creator god that is also benevolent (Buddho Dendo Kyokai 2009). From the property of omnipotence, the philosopher Spinoza drew another, equally consistent, conclusion: God and Nature are the same thing. Spinoza's reasoning: Nature is all there is. So Nature contains God. But if there is any part of Nature different from God, God is not omnipotent. So God contains Nature. This can happen only if God and Nature are the same[13] (Spinoza 1677). But clearly this god has no claim to benevolence.

Nature gods, on the other hand, can be benevolent but not always effective. In ecosystems, there are many forces which work for the benefit of organisms in those systems. But there are also forces which work against the benefit of organisms. These forces may not be malevolent. I can recall watering a flowerbed and disturbing a nest of ants by accident. They grabbed their babies and were frantically active to escape this disaster. Similarly with so-called acts of God such as hurricanes, tornados, and volcanoes. These are acts of Spinoza's God, omnipotent but not benevolent. There are just different levels of organization in the universe that may be quite unaware of other levels.

Monotheism is a powerful force in separating civilized humans from nature and the ecosystem. The powers of nature are then subordinate to God and work through him. One's concern is then figuring out what God's will is. And since this God ultimately has contradictory properties, by logic anything at all can be ascribed to his will.[14] In the end what gets ascribed to God are just the desired actions of civilized human beings totally alienated from nature and the ecosystem.[15]

THE ACHIEVEMENTS OF CIVILIZATION

Those who regard civilization as a great advance for humanity point to achievements in the arts, sciences, technology, in public works, and in the human standard of living. It is not often noticed that many of these achievements are of value only to humans. If humans were to disappear from the earth, no artistic achievements or civic construction would mean anything to the species remaining on earth. Scientific knowledge and technology to the extent it supports science would have value insofar as they lead to actions which support the preservation of the biosphere. The value of these actions would be for the remaining species in the biosphere.

Humans sometimes justify their departures from and exploitation of the ecosystem by pointing to their accomplishments, in art, science, and technology. But there is a certain circularity in such justifications. The human species may have the average 2-million year life span of most species. (Our immediate ancestor species, *Homo erectus,* was around for almost 2 million years.). So when we go extinct, what will happen to our accumulated knowledge, our artworks, our buildings, our technology, our systems of communication, our social systems, our ethics, our spiritual connections, and so on? [16] The only thing that prevents human knowledge from dying with its possessor is other human beings who are prepared to carry on the human enterprise. What meaning can it have for non-human species? Philosophers such as Kant and Wittgenstein have noted that our knowledge is in a deep sense human knowledge, tied to the conditions of existence of human beings[17] (Kant 1787; Wittgenstein 1953). So it is problematic to think of passing the torch to other species. It is not even that other species are not intelligent enough. They did not evolve together with us as part of the complex social Web within which we can communicate with each other and share goals and values. Other intelligent species do not manipulate the environment to anywhere near as great an extent as human beings.[18] And many if not most of our accomplishments are exactly of this nature. There is not much likelihood of getting dolphins to appreciate Greek architecture or Caravaggio paintings.

A more likely alternative would be a human successor species which to some extent shares our propensity to manipulate the environment. But, given our current globalization, it is unlikely that a successor human species can evolve. The creation of a new species requires that part of the old species live separately while the genetic changes—either by natural selection or genetic drift—take place that produce a separate species. Of course over time, natural selection has been producing and will continue to produce changes to the human species. One current change in civilized human beings is reduced brain size. Civilized human beings simply don't need as much brain power to survive as did their hunter-gatherer ancestors (McAuliffe 2011).[19]

So why and for whom are we preserving whatever is worthwhile in what we have done as a species. It would be hard for me to accept that in a few million years all that would be left of humanity would be a lot of ruins and all the waste products of technology. The situation would be radically different from other extinct civilizations such as the Maya. There would be no other people, no anthropologists, to pick up the pieces and appreciate what had been going on. So the question is, what is worthwhile in what we do as a species?[20] And is it still worthwhile if there are no successors to appreciate it? We have to conclude that our accomplishments are of value only to ourselves. This is the circularity involved in using our accomplishments to justify what we are doing to the ecosystem: Civilization is a good thing because it does the things civilization is good at.

The inferior 2008 version of the science fiction film *The Day the Earth Stood Still* inadvertently reveals the important issue. In the film, an alien (played convincingly by Keanu Reeves) is about to exterminate humanity to save the rest of the planet from ecological catastrophe. [21] The alien is convinced to stop the extermination when he is shown: Bach's *Goldberg Variations*; Einstein's Theory of Relativity; and mother love. Much as I love the *Goldberg Variations,* if I were the alien, their existence would not make a difference to my decision. One must suppose it was otherwise a well-thought-out decision to exterminate humanity to save the rest of the planet. Exhibiting the Theory of Relativity was supposed to show how brainy humans are. But this should be irrelevant to the alien if that theory was part of what led humans to destroy the ecosystem. However, intellectual achievements which contribute to preserving the biosphere could be relevant. And finally mother love—the ability of humans to care intensely for each other—is the most irrelevant of all. Mother love simply leads to the preservation of the species which it is assumed is in the process of destroying the planet. The film does succeed in raising the critical issue before it obscures it with a shallow solution: We are indeed the species well on the way to severely damaging or destroying the biosphere. Doing this does not even benefit *us* in any but the shortest of time spans. So why are we doing this?

THE NET RESULT

Civilization is a completely humanized way of life. The ecosystem is experienced only indirectly, as something to be protected against or possibly enjoyed as a spectator. Even worse, civilization must do its best to control the rest of nature. The relation of civilization to the rest of nature is the same as the relation of a city to the surrounding countryside. Both civilization and city depend on what is outside for support, mainly for food and energy. But both civilization and city have some power over what happens on the outside. Thus the distinctive activities of civilization—arts, sciences, and political, social and economic organizations—have at best a tenuous relation to the ecosystem and at worst a destructive impact on it. To see this situation as unquestioningly an *advance*—as most characterizations of civilization would have it—presupposes a belief in human superiority.

The feature of the origin of civilization to be concerned about is how human beings came to accept a way of life so divorced from the ecosystem. As we have seen, the social and political sequence some anthropologists propose could easily have occurred in hunter-gatherer societies and early farm societies. The distinctive separation from the ecosystem comes about with economic specialization. When individuals or groups (classes) gain control of goods they do not produce—especially agricultural goods—that civilization begins its divorce from the ecosystem.

Society stratifies, initially into distributors and producers. Then more and more complex social forms and technology proliferate, liberated from the constraints of making a living from the ecosystem. So we now have some idea of how civilized humans became so separate from the ecosystem.

The remaining question is therefore whether civilization can embody a way of life compatible with appropriate consideration for the ecosystem. This question will be dealt with in most of the remaining chapters of this book.

REFERENCES

Adas, M. (1992). *Civilizations of the Americas: The peoples to the north.* Retrieved 1/23/12 from history-world.org/Civilizations%20Of%20The%20Americas.htm

Ament, P. (2007). Wheel. *The Great Idea Finder.* Retrieved 4/12/13 from www.ideafinder.com/history/inventions/wheel.htm

Beliaev, D., Bondarenko, D., & Korotayev, A. (2001). Origins and evolution of chiefdoms. *Revista de Antropologia*, *30*, 373–395.

Blanton, R., Feinman, G., Kowalewski, S., & Nicholas, L. (1999). *Ancient oaxaca.* Cambridge, UK: Cambridge University Press. doi:10.1017/CBO9780511607844.

Brown, C. (2011, May). El Mirador, the lost city of the Maya. *Smithsonian*. Retrieved 2/24/12 from www.smithsonianmag.com/history-archaeology/El-Mirador-the-Lost-City-of-the-Maya.html#ixzz1nMv460Y0

Buddho Dendo Kyokai. (2009). *A teaching a day: Excerpts from the teaching of Buddha*. Berkeley, CA: BDK America.

Choi, C. (2012). Huge ancient civilization's collapse explained. *Live Science.* Retrieved 5/29/12 from news.yahoo.com/huge-ancient-civilization-collapse-explained-123449804.html

Claessen, H. (1978). The early state: A structural approach. In Claessen, H., & Skalnik, P. (Eds.), *The Early State*. The Hague, The Netherlands: Mouton. doi:10.1515/9783110813326.531.

De Chant, T. (2011, August 17). Hunter gatherer populations show humans are hardwired for density. *Scientific American's Guest Blog*. Retrieved 11/29/11 from persquaremile.com/2011/08/17/hunter-gatherer-populations-show-humans-are-hardwired-for-density/

Derrickson, S. (2008). *The day the earth stood still.* [Motion Picture]. Century City, CA: 20th Century Fox Films.

Diamond, J. (2005). *Collapse: How societies choose to fail or succeed.* New York: Viking Books.

Diamond, J. (2012). *The world before yesterday*. New York: Viking Books.

Freud, S. (1930). *Civilization and its discontents*. New York: Norton.

Gill, N. S. (2000). An introduction to Sumer. *About.com Ancient/Classical History.* Retrieved 5/23/12 from ancienthistory.about.com/od/southasia/a/Sumer.htm

Hartman, P. (1996). The stone age, bronze age, and iron age. *The Hartman Website.* Retrieved 1/14/12 from www.naciente.com/essay88.htm

Infoplease. (2012). *World's ten most corrupt leaders*. Retrieved 2/3/12 from www.infoplease.com/ipa/A0921295.html

Jaynes, J. (1976). *The origin of consciousness in the breakdown of the bicameral brain*. New York: Houghton Mifflin.

Kaiser, D. (1990). *Dolphin consciousness*. Retrieved 1/30/06 from home.onemain.com/~dk1008206/html/dolph1.htm

Kant, I. (1787). *Critique of pure reason*. Riga, Latvia: Hartnoch.

McAuliffe, K. (2010, September). If modern humans are so smart, why are our brains shrinking?. *Discover*. Retrieved 1/27/12 from discovermagazine.com/2010/sep/25-modern-humans-smart-why-brain-shrinking/article_view?b_start:int=2&-C

Murnau, F. (1927). *Sunrise: A song of two humans.* [Motion Picture]. Century City, CA: Twentieth Century Fox Film Corporation.

Partch, J. (2004). *A brief history of type*. Retrieved 4/12/13 from graphicdesign.spokanefalls.edu/tutorials/process/type_basics/history.htm

Sahlins, M. (1968). Notes on the original affluent society. In Lee, R. B., & DeVore, I. (Eds.), *Man the Hunter* (pp. 85–89). New York: Aldine Publishing Company.

Service, E. (1975). *Origins of the state and civilization.* New York: W. W. Norton.

Spinoza, B. (1677). *Ethics.* Retrieved 12/6/11 from users.erols.com/nbeach/spinoza.html

Stevens, W. (1959). *Collected poems.* New York: Vintage Books.

Tainter, J. (1988). *The collapse of complex societies.* Cambridge, UK: Cambridge University Press.

Thoreau, H. D. (1854). *Walden.* Retrieved 1/1/12 from www.online-literature.com/thoreau/walden/

Welker, C. (1985). The social structure of primates. *Anthropologischer Anzeiger, 43*(2), 97-164. Retrieved 5/26/12 from www.ncbi.nlm.nih.gov/pubmed/3927823

Wittgenstein, L. (1953). *Philosophical investigations.* New York: Macmillan.

ENDNOTES

1 Civilization has its share of detractors, who will be discussed later. Notable are Thoreau (1854) and Freud (1930).

2 I use the world "culture" with some trepidation. Although widely used by social scientists, it is a murky concept. For one thing, it is not a genuine institution. No one regards themselves as a member of a culture. There are no membership rules. How would one join a culture? Or leave it? When the doctrine of cultural relativism treats culture as a genuine institution, culture then becomes quasi-totalitarian. One is stuck with obeying its rules with no recourse. And what about those who belong to more than one culture?

3 Further developments of the use of metal are: Smelting metals such as copper and bronze 6,000 years ago; and the use of iron 3,000 years ago.

4 There were numerous other domestications at this time related to farming. The horse and camel were means of transport, which was important for civilization.

5 See Diamond (1997).

6 See Chapter 5, "Humans as Hunter-Gatherers," "Prehistoric Hunter-Gatherers" and "Domesticated Dog." Pastoral nomads herding mainly sheep are not sedentary, and so are an important exception. But even in that case, it is difficult to believe that the domestication of herd animals could have taken place in a nomadic context. It is more likely that previously domesticated animals were either adopted by nomads or that sedentary farmers with domesticated animals later reverted to a nomadic lifestyle.

7 My view is similar to but more general than the Marxist view that class divisions are created by the difference between ownership of the means of production and non-ownership. I am saying that class divisions are initially created by the ability to control products one does not produce oneself. Ownership of the means of production would be an instance of that ability.

8 Service (1975, p. 307) argues that the ability to use force is not necessary for a state. But his reason, that a good government would not need to use force, clearly allows that a government must have that capability.

9 The phenomenon of choice of disembedded capital has also occurred in recent times: Washington, DC, in the United States was chosen for its neutral location between northern and southern states. It was also swamp, not prime farming land. Similarly Brasilia in Brazil, Ottawa in Canada, Canberra in Australia, New Delhi, India, and the Hague in the Netherlands (Blanton et al. 1999, pp. 65-66).

10 The unpredictability of modern technology will be discussed further in Chapter 8, "More about Modern Technology and Corporations."

11 Claude Chabrol's *Les Cousins* (1958), a more recent film than *Sunrise,* deals effectively with the same themes.

12 Jaynes (1976) argues that the shift from polytheism to monotheism signals a major change in the organization of human consciousness.

13 These views got Spinoza excommunicated by both Jews and Christians.

14 From a contradiction, any and all conclusions follow.

15 The quote from the Wallace Stevens poem "Less and Less Human, O Savage Spirit" in the Preface poignantly expresses monotheism's separation from the environment (Stevens 1959).

16 These issues will be discussed further in Chapter 14, "Extinction."

17 This is the point of Wittgenstein's enigmatic remark that if a lion could talk, we could not understand him (Wittgenstein 1953, p. 223e).

18 Dolphins and other cetaceans have much larger brains than humans but are not using them to manipulate their environment. Their brain size must be useful or else it would be selected away fairly quickly. Complex brains require a lot of energy to maintain. Human brains consume at least 33% of the individual's energy. We do not know what cetaceans are using their brains for (Kaiser 1990).

19 Similar changes have taken place in dogs. See also McAuliffe (2001).

20 This question will be addressed at greater length in Chapter 13, "Regaining Our Place as a Species among Species."

21 In the original 1951 version, the alien's warning is that humans must refrain from war, especially nuclear war, or be destroyed as a threat to the universe.

Chapter 8
More about Modern Technology

ABSTRACT

Modern technology, starting about 250 years ago, exacerbates conflicts with the ecosystem. The products and by-products of modern technology are typically either toxic or cannot be incorporated into the ecosystem. The largely petroleum-based products from about 1900 replace products more compatible with the ecosystem. Conflicts with the ecosystem therefore accelerate.

INTRODUCTION

This chapter is concerned with the development of technology, but not so much in order to chronicle human inventiveness. Rather, its aim is to uncover the growing separation of humanity from the rest of nature.

Timeline:

- **1.5 million years ago:** Complex cached tools
- **13,000 years ago:** Pottery (China, Japan) (Hirst 2011)[1]
- **11,000 years ago:** Farming (domestication of plants and animals)
- **8,000 years ago:** Pottery (Mesopotamia)
- **8,000 years ago:** Weaving
- **7,000 years ago:** Metal, mining
- **5,000 years ago:** Draft animals

DOI: 10.4018/978-1-4666-4586-8.ch008

- **5,000 years ago:** Use of concrete
- **5,000 years ago:** The wheel
- **5,000 years ago:** Writing
- **2,000 years ago:** Arches, aqueducts, bridges (Rome)
- **1,000 years ago:** Windmills
- **550 years ago:** Printing press
- **400 years ago:** Modern science
- **250 years ago:** Steam engine
- **200 years ago:** Modern technology
- **200 years ago:** Mechanization of spinning and weaving
- **200 years ago:** Asphalt road paving
- **200 years ago:** Modular steel building construction
- **70 years ago:** Postmodern technology (synthetics, bioengineering)

There were continuous important improvements in hunting technology before farming and civilization.[2] But most radical and life-changing improvements occurred with the advent of farming and civilization. In that era of what can be called "classic" technology, important technological innovations are borrowed from nature. Raw materials are derived from plants and animals. Cotton, wool, and silk come from plants and animals. Drugs are concentrated and distilled versions of substances derived from plants and animals. Almost all pharmaceuticals are derived from plants: Aspirin from willow and many antibiotics from various molds.

TECHNOLOGICAL PROGRESS

We saw in the preceding chapter, that technological progress in early civilization was rather unpredictable. Progress in modern technology is, if anything, even more unpredictable and contingent. The examples in that chapter were the inventions of the wheel and of writing. But some commentators on modern technology believe that technological progress is predetermined. The New York Times columnist Thomas Friedman calls himself a "technological determinist" (Friedman 2005, p. 374).

There are really two distinct versions of technological determinism. One version is that the development of technology, the appearance of technological advances, is determined. This version can be called *advance determinism.* The second version is that, once a technological advance appears, its widest useful application is inevitable. This version can be called *application determinism.* Friedman is an application determinist. There is no evidence one way or the other that he is an advance determinist. But application determinism is a naïve view as well as a dangerous one.

The danger in application determinism comes from the accompanying claim that technology (including the Internet) is always an improvement. This claim ignores the fact that technology is a new order imposed on an older order and can easily have deleterious side effects which cannot be prevented by due diligence in development. We saw many examples in Chapter 2, "Conflicts with the Ecosystem." The development of Chlorofluorocarbons (CFCs) is a striking example of such side effects. CFCs were inert at surface levels but highly destructive of the ozone layer necessary for life on the planet. So application determinism is false as the version that new technology is necessarily an improvement.

Advance determinism is not a particularly dangerous view, but it is also clearly false. Most technological advances were not predictable and appeared only as a result of chance factors. In this respect, technological advance is similar to the evolution of organic life forms. Important technological developments have arisen in unlikely ways, and certainly not as the result of predefined rigidly structured research programs. This is especially true of modern technology, in particular the Internet and World Wide Web. The Internet itself was originally a US Defense Department project, and such vital characteristics as no central computer were required to make the system impervious to nuclear attack. The World Wide Web was developed by Swiss physicist Tim Berners-Lee as a method of exchanging scientific information including both text and graphics. The Web browser—essential to widespread use of the Internet—was developed by graduate students at the University of Illinois (Kristula 2001). A useful peer-to-peer filesharing application (Napster) was developed by an undergraduate at Boston University (Moya 2013). So the Internet technology necessary for globalization was no more inevitable than was the rise of mammals after the age of dinosaurs.

The Napster example also shows the falsity of application determinism. Music companies used the courts to shut down the free music-sharing service in 2001. Several years later, some commentators believe the poor showing of the music industry stemmed from their failure to find a way to make money from peer-to-peer technology rather than shutting it down. However, the Napster example still supports a version of application determinism. For Friedman, the force which drives application determinism is competition within a free-market economy. He notes "if you can do it [apply the technology], you must …, otherwise your competitors will" (2005, p. 374). In the case of the music industry, they did not, and merely worked through the courts to extend their (monopoly) property rights. Thus, people can fail to exploit technologies, but ultimately the market will destroy them. This brand of application determinism is probably correct. But obviously it depends on the market functioning as a free market, without monopolistic or oligarchic impediments. Free-market application determinism is thus more-or-less the claim that a properly functioning market economy will deliver good economic results. This is not a dangerous claim and is correct.

So the conclusion is that there really is not any predetermined path that technology must take. Civilization provides the conditions for technological innovation. These conditions vary with such factors as the scientific sophistication of the civilization and the political and economic environment. These conditions by themselves make technological progress contingent. But over and above these conditions, there is extra pure contingency. No one in some group or other may happen to think about putting things together in a way that creates previously untapped value. And so the technology does not appear.

The contingency of technological progress is both a good thing and a bad thing. The unexpected proliferation of the Internet and World Wide Web has dramatically improved the lives of consumers by making a virtually unlimited inventory of items and spare parts completely accessible. And with the addition of social networking, it seems to have stimulated a social democratic revolution across the Arab world in the spring of 2011.

The bad aspect of contingency is that political and economic considerations can stifle worthwhile innovations or promote deleterious innovations. I still remember logging on to the original Napster and realizing that all music was now available to me through peer to peer sharing, with no corporate editing. Until 1998 and the Digital Millennium Copyright Act, it was not illegal to copy media for personal noncommercial use. Politics driven by shortsighted economic motives of music companies changed that. The result has been seriously impoverished popular music.[3,4]

Much worse is the efforts of fossil fuel companies to promote more use of fossil fuels. Innovation is sidetracked to projects that could well make us extinct.

THE ORIGINS OF SCIENCE AND MODERN TECHNOLOGY

"Classical" technology develops as part of civilization. Technological improvements to farming occur, to be sure. Draft animals, improved plows, and aids to irrigation are technological improvements to farming. But their goal is increased food production for civilization. Improving life for farmers is a secondary goal, worth pursuing because improved life for farmers can mean improved food production. Much "classical" technology improves both farm life and civilized life. Pottery, weaving, metal objects, and the wheel are useful both to farmers and city dwellers. But technology such as writing and more elaborate mechanical devices are primarily of use within civilization.

More about Modern Technology

We have just seen that the development of technology is highly contingent. That is, as civilization develops, the progress of technology is by no means predetermined. Its development depends on chance bursts of ingenuity on the part of individuals. Modern technology especially depends on chance bursts of technological innovation. The steam engine, the mechanization of manufacturing and related inventions—the 'industrial revolution'—finally begin to apply the methods of science to improving technology. Science abstracts from values and so does the application of science in modern technology. But just as science aims to give us a complete account of the world, so too does modern technology come to believe it applies to everything in the world.

Julian Young correctly notes that *all* technology treats the world as resource. What distinguishes modern technology is that it treats the world as *nothing but* resource (Young 2006, pp. 49-50). Thus science in the service of technology becomes modern technology.

An alternative view is that modern technology[5] arises primarily because people are living more densely. Three economists, Raouf Boucekkine, Dominique Peeters, and David de la Croix, argue that modern technology arises because of a great increase in literacy. And this increase in literacy was produced by greater population density (De Chant 2011).

The rise of literacy was real and may have helped the Industrial Revolution take place. Literacy in England skyrocketed between the 1720s and 1740s. In just two decades, literacy rose from 58 percent to 70 percent among men and from 26 percent to 32 percent among women. This rise was due to unprecedented school construction. In the 40 years beginning in 1700, 988 schools were founded in Britain, nearly as many as had been established in all previous centuries (De Chant 2011).

The economists argue that the reason for the greater number of schools was increased population density. The result was that schools were closer and hence better attended. Hence, literacy increased. And, *voila,* the Industrial Revolution.

If one is thinking of the Industrial Revolution as just an accumulation of inventions, this account has some plausibility. But, as we have seen, a major change in world view is taking place. Increased literacy can enable more inventions but it cannot account for the change in worldview characteristic of modern technology, which is the importation of science and the consequent view of everything as resource.

But if modern technology stems from science, we then need to ask how science arises when it does. As we saw in Chapter 3, "Science, Modern Technology, and Corporations," modern science arose about 1600, and modern technology arose about 1800. From the history of the origins of science, it is clear that its invention was the work of a few individuals. Thus, science originated in the same haphazard way most technology originates. Of course, like many useful inventions, once it originates, it spreads widely. Its later adaptation to technology is also a contingent event. Evidence for this is that there were not multiple inventions of science, nor were there multiple inventions of modern technology.

MODERN AND POSTMODERN TECHNOLOGY

Like 'now' and 'contemporary,' 'modern' applies to items close in time to our present.[6] But the term 'modern' violates its own meaning when it continues to be applied well beyond the present. This has happened in the arts. 'Modern art' now means roughly 'art produced in the first three-fourths of the 20th century. New York's Museum of Modern Art is not getting rid of art produced in the first half of the 20th century because it no longer fits its name taken literally. So now we see 'postmodern' art, which is nonsense in the literal meaning of the term modern: How can something be later than now? I will follow contemporary usage and use modern to indicate a particular style rather than what is current at the time of writing. So I will feel free to describe a recent development of modern technology as *post*modern technology.

Ecologists such as Barry Commoner have noted that many products developed after 1945, the end of World War II, have a much greater impact on the environment than the more natural products they replaced. Detergents replace soap. Synthetic fibers replace wool and cotton. Pesticides replace earlier methods of pest control. Plastics replace metals and wood. These are *postmodern technology.* Commoner emphasizes the contrast between the older, pre-World-War II technologies and the postmodern technologies adopted since then: Pasturing animals vs. factory farming; maintaining soil fertility through crop rotation vs. chemical fertilizers. Production technologies with intense impact on the environment have displaced those that did not. These new technologies are more profitable for the producers—the farmers, cleaning product manufacturers, textile producers. Commoner's solution is simply to stop using these technologies. But the solution may be (as happened in 1965) to produce instead bio-degradable detergent molecules with benzene components making them even more toxic to the environment (Commoner 1971, pp. 153-154). Commoner is certainly correct about the destructiveness of these technologies. But the question is, what in this technology leads in this direction?

Part of the answer has to be the profitability of the new products or techniques, together with the failure of economic institutions to account for offsetting environmental costs. Environmental costs are of the kind called 'externalities' by economists, meaning they are costs external to the regular workings of a market economy. Accounting for externalities within a market economy is a challenging problem, which will be addressed in the next two chapters.

But the question underlying the issue of externalities, is why did these dramatically more environmentally destructive products and techniques arise when they did? Here is a timeline of some of the major postmodern technologies:

- **1900:** Gasoline-powered automobile
- **1909:** Synthetic fertilizers
- **1909:** Plastics (Bakelite)
- **1926:** PVC (polyvinyl chloride)
- **1927:** PCBs (polychlorobiphenyl, banned 1979)
- **1930:** Chlorofluorocarbons (discontinued 1994)
- **1933:** Synthetic detergents
- **1939:** Nylon
- **1944:** Insecticides (organochlorine compounds; DDT banned 1973)
- **1949:** Jet engines for aircraft
- **1949:** Plastics (polystyrene)
- **1953:** Plastics (polyester)

The automobile and jet engines use petroleum-based products. All but one of the other products are petroleum derivatives—PVC, PCBs, chlorofluorocarbons, detergents, nylon, insecticides, polystyrene, polyester. The remaining product—synthetic fertilizer—is produced from natural gas. About 70% of petroleum is used for fuel, and about 10% for plastics. So postmodern technology seems to be fossil fuel technology—primarily petroleum. Because petroleum and its derivatives are highly toxic, it is not surprising that the products of postmodern technology have serious effects on the environment. And when these products can be disseminated with no regard for their environmental costs, they can cause major environmental disruption. This combination of factors—toxic products and market externalities—are important in causing conflicts with the ecosystem.

THE POINT OF VIEW OF MODERN TECHNOLOGY

In Chapter 3, "Science, Modern Technology, and Corporations," the views of the philosopher Martin Heidegger on modern technology were presented. Heidegger's important insight was that *modern technology has its own point of view* which is completely separate from any other structure of human aims and purposes. The critical feature of modern technology is its willingness to treat anything as a resource to be reordered in the furtherance of other aims, especially its own. Heidegger concluded that modern technology is an independent force in human existence. It builds a new and incompatible order on top of what was there, primarily in order to extract and store energy for later uses (Heidegger 1955, pp. 14-17).

The point of view of modern technology[7] is to regard *everything* as potential resource, as "standing reserve" to be reused later in other related processes. Forests become timber resources. Land becomes a resource for the building industry. Human beings themselves become human resources or consumers. Electric power is always standing reserve, on hand for potential use. Genetic engineering treats DNA, the basis of the continuation of all living species, as just another resource to be manipulated. For modern technology, the biosphere is just another collection of resources.

It is also a critical part of the point of view of modern technology that it presents itself to us as a mere means, an enabler for our other ends. But it is impossible to deal with modern technology correctly without realizing that *modern technology has its own ends,* which are to reorder everything as standing reserve for further uses.[8]

Julian Young makes the important point about the relation of the point of view of modern technology to the environment. The point of view of modern technology requires us not to see things as they are in themselves. Rather, objects become just means to whatever ends the institutions of modern technology require. Young remarks that since nothing shows up as anything other than their utility as a resource, nothing presents itself as a limit on the ability of modern technology to manipulate and exploit everything as resource. Therefore, modern technology is oblivious to the need to do anything but treat the ecosystems as merely another collection of resources. For example, the point of view of modern technology does not have the ability to treat a forest as a self-sustaining ecosystem. For modern technology, the only consideration of the forest as ecosystem might lie in its being more valuable for timber if preserved for a longer period (Young 2002, p. 53).

In Chapter 3, "Science, Modern Technology, and Corporations," it emerged that the reason why technological advances so often have adverse side effects lays in the nature of modern technology. Modern technology is an entirely different order imposed on the environment which contains ecosystems. And so it regularly conflicts with ecosystems. But if one has adopted the point of view of modern technology, one might say, so what? The reply might be, that a substantial portion of the ecosystem is required for human survival.[9] And the modern technologist could reply, OK, we will set limits on ecosystem destruction to insure human survival.[10] There are many problems with this approach. Most will be discussed in the next chapter. One important problem is that the ecosystem is extremely complex and loaded with hard-to-predict feedback effects. So there is a good chance of serious mistake.

Another important problem has to do with the place of human beings in this world. Treating everything in the world as something to be exploited ignores completely where we came from. This point of view is human superiority at its most grandiose.

We are saying that we can determine the course of everything on the planet—plants, animals, land, sea, atmosphere. And that we have absolutely no respect for the being of anything else—actually including ourselves. We ourselves are a product of Being, of the coming to be and passing away of all things. Our presence here as humans was not determined by us, and our leaving this earth is not determined by us. Therefore, we should respect our place in the natural order over and above the aim of merely exploiting and for the most part destroying all that exists around us.

MODERN TECHNOLOGY AND THE ACCUMULATION OF CIVILIZED OBJECTS

Technology in the service of civilization does not merely improve existing objects. It creates new objects to such an extent that almost the entire environment is human-created.[11] Of course, humans have been creating new objects for millions of years.

First, our ancestors *Homo habilis* and *Homo erectus* created ad hoc tools. Then *Homo erectus* created more complex tools, which were likely stored. *Homo erectus* discovered and used fire and likely passed it on to us *Homo sapiens.* But at first fire was used for heat, cooking, and protection from predators. It was not until after 7,000 years ago that *Homo sapiens* began using it to smelt metal and make metal objects. *Homo erectus* did, however, build shelters. But as hunter-gatherers, these shelters were not permanent. *Homo sapiens* later developed and wore clothing.

But before farming and civilization, the number of objects a typical human would accumulate was not large. A nomadic hunter-gatherer cannot accumulate that many possessions because he or she has to remain mobile. Hunter-gatherer women had to space children every four years because children had to be carried until they were old enough to walk with the rest of the band. However, hunting tools improved enough for humans to cause the extinction of the megafauna 60,000 years ago.

When a sedentary agricultural lifestyle became possible, it also became possible to accumulate possessions. Important technological changes before civilization were in the domestication of animals. Farming implements also improved. But it is within civilization and city life that human beings live almost entirely with man-made objects. Modern technology eventually creates a consumer economy, where great effort is spent in getting people to throw things away and buy new things. 'Consumer' is very much a modern technology term. A consumer is defined as a person who uses things up.

But such a person is also one who produces waste. So consumerism clearly has the potential for conflict with the ecosystem. Consumerism encourages people to replace things with newer versions and therefore to discard things that are still usable. There is also a tendency on the part of corporations to produce items with merely cosmetic changes, solely in order to get consumers to buy something new. Or worse, the strategy of 'planned obsolescence' actually results in items designed to wear out more quickly than they need to. American automobile manufacturers followed this strategy until the 1980s, when more durable Japanese cars captured a large portion of American automobile market share. It was rare to see an American car that lasted for much more than 100,000 miles; whereas Japanese cars lasted 200,000 miles or more.[12]

These consumerist strategies conflict with the ecosystem not only in generating extra waste, but also in wasting resources. The considerable effort and resources spent on producing multi-blade razors, for example, could have been redirected toward research on cures for cancer or preserving the environment. And planned obsolescence deliberately increases the amount of waste and the cost and effort to deal with it.

Consumerism, especially in America, is defended as a necessary part of a free market economy. But the extra waste and wasted effort are clearly economic externalities. Exactly how free markets deal with externalities is an important part of dealing with causes of conflict with the ecosystem. We will discuss this topic in the next chapter.

Systems of modern technology have a tendency to generate waste which is not handled by the environment. This tendency is acknowledged by the existence of recycling programs. The efficacy of these programs in dealing with technological waste will be discussed in a later chapter. The immediate question is, what portion of the waste can be dealt with through recycling? The follow up questions are: Are there satisfactory means for dealing with the waste not handled by recycling? Also, can a product be recycled in a way that reduces the need to produce more of the item. As we saw earlier, plastic bags cannot be recycled into more plastic bags. So 'recycling' them does not solve any waste management problems.[13]

CORPORATIONS

Corporations are only one manifestation of capitalism, and in some ways an unbelievably destructive one. An institution with the single overriding purpose of maximizing profits, cannot help but be in conflict with nature. For profits, measured in money, are a purely human institution. If humans vanished, profits and money would be meaningless. But capitalism need not be solely about profits. The business theorist Peter Drucker once said that profits are a cost of doing business (Malik 2012). He meant that profits enabled one to keep an enterprise going, but that one was in business for reasons other than profits. Nobel Peace Prize winner Mohammed Yunus of Bangladesh has developed the idea of a *social business* (Yunus 2003). Such a business uses the mechanisms of capitalism, including market competition, to make profits. But the profits are used to sustain the business, thus avoiding the nonprofits' chronic continual problem of fundraising. And investors do not get the profits. At best, their investment is returned. All of this is in the service of worthwhile social goals.

But such ideas do not have as much currency as the doctrines of pure selfishness and single-minded pursuit of profit attributed to Milton Friedman and his followers. These doctrines themselves are based on views about the economic efficiency of free markets. There is good evidence that free markets are efficient only in highly idealized circumstances. So pure free market capitalism as the singleminded pursuit of profit is destructive even on its own terms. It will be examined in the next chapter whether unrestricted free markets are a contributing cause to our separation from the ecosystem.

In an earlier chapter, *knowledge pollution* was discussed. Knowledge pollution is often practiced by corporations and in non-democratic regimes such as China. The corporation has the power to enforce false beliefs because of its enormous wealth. An autocratic or totalitarian state also has the power to enforce false beliefs. When knowledge pollution is practiced, it is not so much that a state or corporation actually knows that what they are promulgating is false. Rather, they seem to have a *pragmatic* view of truth: Truth is only what works, that is, truth is what produces their desired results. A less distorted version of truth is that truth about things reveals things as themselves. Things are not hidden, they are unconcealed.[14] The truth is that which is unconcealed. Truth is necessarily transparent. This version of truth works better than the pragmatic view because it connects us to the reality of the environment around us.

Of course deliberately promulgating falsehood is unethical. But the wrongness of knowledge pollution goes beyond the wrongness of lying. The ethical basis for the wrongness of lying is that lying cannot be a cooperative principle. In order for the lie to succeed, the person being lied to has to believe in the truth of what is being said. And if lying were made a universal principle, this could not happen. No one would believe what was said.[15] Even though a corporation is not an ethical individual, this universal principle consideration still applies. However, since a corporation exists to maximize profits, it often regards its profitability as taking precedence over the cooperative principle of telling the truth.

But because of the impact of corporations on the environment and on environmental policy, there is an additional consideration, discussed in Chapter 3, "Science, Modern Technology, and Corporations." Corporate behavior aligns itself only with whatever will maximize profits. Being ruled by profit maximization, a corporation is not using language to reveal the truth. Thus, it can have no relationship to the reality of the world through its use of language. Whatever happens to the world as the result of its actions cannot be something it is concerned with. The environment is a major expression of this reality. Thus corporate knowledge pollution shows that it has no relation to the reality of the environment. The real danger is that corporations distort reality to serve their own end of profitability.[16]

REFERENCES

Adas, M. (1992). *Civilizations of the Americas: The peoples to the north.* Retrieved 1/23/12 from history-world.org/civilizations%20of%20the%20americas.htm

Chirping Bird. (2012). *Plastic ban bag report.* Retrieved 5/27/12 from plasticbagbanreport.com

De Chant, T. (2011). *Population density fostered literacy, the industrial revolution.* Retrieved 11/29/11 from persquaremile.com/2011/11/11/population-density-fostered-literacy-the-industrial-revolution

Friedman, T. (2005). *The world is flat.* New York: Farrar, Straus and Giroux.

Hansen, J. (2009). *Storms of my grandchildren.* New York: Bloomsbury.

Heidegger, M. (1927). *Sein und Zeit.* New York: Harper & Row.

Heidegger, M. (1955). The question concerning technology. In *The Question Concerning Technology and other Essays.* New York: Harper & Row.

Hirst, K. K. (2011). *The invention of pottery.* Retrieved 1/18/12 from archaeology.about.com/od/inventions/a/pottery_invent.htm

Kristula, D. (2001). *The history of the internet.* Retrieved on 4/3/08 from www.davesite.com/Webstation/net-history.shtml

Malik, H. (2012). *Managerial accounting - Profit planning.* Retrieved 1/27/12 from hafeezrm.hubpages.com/hub/Managerial-Accounting-Profit-Planning

Mason, M. (2013). *Early history of Napster.* Retrieved on 4/13/13 from www.moyak.com/papers/napster-history.html

Olesen, A. (2006, July 22). Stephen Hawking warns about global warming. *USA Today.*

Schultz, R. (2006). *Contemporary issues in ethics and information technology.* Hershey, PA: IGI Global.

Schultz, R. (2009). *Information technology and the ethics of globalization.* Hershey, PA: IGI Global. doi:10.4018/978-1-60566-922-9.

Seabrook, J. (2012, March 26). The song machine. *New Yorker (New York, N.Y.).*

Young, J. (2002). *Heidegger's later philosophy*. Cambridge, UK: Cambridge University Press.

Yunus, M. (2003). *Banker to the poor*. New York: Public Affairs.

ENDNOTES

1 Even though farming did not begin in China until about 10,000 years ago, these early potters must have had a sedentary life style likely with the benefit of farming. It is hard to see how nomadic hunter-gatherers could have carted around lots of fragile items.

2 The ability of humans to carry out the megafauna extinction of 60,000 years ago is a testament to the quality of their hunting tools.

3 Current "top 40" songs are better called "music product" than music. Pieces determined, probably by focus group, to please inexperienced audiences are combined in ways that will cause people to buy the songs with no regard to meaning. There is a bizarrely sympathetic account of the process in Seabrook (2012).

4 Precisely because of the element of contingency in technological progress, intellectual property laws should not grant excessively long copyright or patent periods. Innovations must be available to build upon. Corporations apparently ignore this, even though it is to their long-term benefit. The long term benefit they have obtained is 120-year copyright periods in the Bono Copyright Extension Act of 1998. Short-term profits for as long as possible seems to be their objective. The US Supreme Court has shown itself oblivious to these considerations in the case *Eldred vs. Ashcroft* (2003).

5 These economists argue in terms of the Industrial Revolution. The Industrial Revolution is the same as modern technology.

6 These words and words such as "here" and "there" and "this" are called *token reflexive*. This means that they refer to the time and/or place when they are spoken or written.

7 Heidegger has a one-word German neologism, *Gestell*, for "the point of view of modern technology." William Lovitt, a translator, uses his own neologism "Enframing." Commentator Julian Young leaves the word in German. I prefer the longer phrase "the point of view of modern technology" because one does not have to keep reminding oneself of the meaning.

8 The most chilling image of the film *The Matrix* (Wachowski Brothers, 1999) is of banks of human beings as a pure standing reserve.

[9] The question whether humanity can survive without the ecosystem will be discussed in Chapter 9, "Causes of Conflict with the Ecosystem."

[10] This is exactly the approach of the author Mark Lynas in his *The God Species* (2011). Lynas' position will be discussed in Chapter 9, "Causes of Conflict with the Ecosystem," and Chapter 10, "Environmentalism and Sustainability."

[11] My thanks to John Dittmeir for his observations resulting in this topic.

[12] From my own experience, I had a 1974 Ford, which did not quite make 100,000 miles. I replaced it with a 1983 Toyota which had reached 280,000 miles at last report from its third owner.

[13] It is good news that a significant number of cities and counties in 10 states ban disposable plastic bags. Such bags are banned in the entire state of Hawaii. A number of cities elsewhere have also banned disposable plastic bags, including Mexico City and Dharamsala, India (Chirping Bird 2012).

[14] This version of truth was used and promoted by Heidegger (1927, p. H33). It stems from the Greek word for truth, aletheia (ἀλήθεια) which means not-hidden. The popular idea of truth as correspondence is often incoherent. If the correspondence is supposed to be between what is in your head and what is outside of your head, this is impossible.

[15] This is Kant's universal principle argument for the wrongness of lying (Kant 1785).

[16] Who but a conservative blinded by money and power could ever think that a corporation was the same as a person?

Chapter 9
Causes of Conflict with the Ecosystem

ABSTRACT

Potential causes of humanity's conflict with the ecosystem are: human institutions misusing technology, technology itself, population growth, economic growth, and human attitudes. The main causes are modern technology together with attitudes of human superiority and human exceptionalism. Population growth and economic growth are secondary causes, but they do make environmental problems worse.

INTRODUCTION

Conflicts between human activity and the ecosystem have been recognized for quite some time. Since the beginning of the industrial revolution, groups have been working to end industrial pollution. And a number of individuals have advocated respect for and conservation of nature—Henry David Thoreau, Gifford Pinchot, John Muir, and others. The proliferation of postmodern technology since the 1940s has accelerated the conflict in much more severe ways. So there has been more attention to the conflict and its causes. Heidegger's analysis of modern technology from 1955 is one of the first such analyses—and also perhaps the deepest. Many different analyses have appeared since the 1970s and continue to appear. Many are still worth some attention; almost all offer solutions based on their conception of the cause of the problem of the conflict of human activity with the ecosystem.

DOI: 10.4018/978-1-4666-4586-8.ch009

I will consider five major views about the causes of human conflict with the ecosystem:

- Technology is not the cause. Rather, various human practices and institutions utilizing technology are the cause.
- Population growth is the main cause.
- Economic growth is the main cause.
- Human attitudes are the main cause.
- Humanity is a species doomed to destroy itself.

I will consider each of these in turn. These views are not mutually exclusive, and some commentators believe more than one. But usually one cause is singled out as the primary cause.

TECHNOLOGY IS NOT THE CAUSE

Those holding this view believe it is social and economic institutions which are misusing technology. The authors to be examined are Barry Commoner (1971), E. O. Wilson (2002), and Mark Lynas (2011). Each has a different view about how technology is being misused.

As we saw in the preceding chapter, Commoner calls attention to the contrast between the older, pre-World-War II technologies and the newer technologies adopted since then such as: Maintaining soil fertility through crop rotation vs. chemical fertilizers; biodegradable soap vs. detergents; natural vs. synthetic textiles. These are *postmodern* technologies. Production technologies with intense impact on the environment have displaced those that did not. These new technologies are more profitable for the producers but tend to cause great damage to the ecosystem.

Commoner's solution is to change public policy to jettison postmodern technologies. Commoner's solution would eliminate much of the damage of modern technology, but it may turn out to be difficult to sort out harmful postmodern technology from beneficial postmodern technology. Computer applications such as the Internet may be beneficial, but they depend on computer hardware whose manufacture and ultimate disposal do harm to the environment.

The prior question is, what in technology leads it to be so destructive? The answer has to be modern technology's tendency to have its own ends and its own goal of remaking the entire world by utilizing it as resources for its own applications. So it will be tricky to separate harmful from beneficial modern technology or postmodern technology.

The eminent biologist E. O. Wilson in his *The Future of Life* (2002) is largely concerned with defending biodiversity. The major threat to biodiversity is habitat destruction, and the major cause of habitat destruction is human activity, often economic activity. Wilson lists a number of projects to preserve, protect and restore existing habitat. NGOs are now undertaking many of these projects, but Wilson notes eventually governments will have to take over a large part of this. He also calls for projects to make biodiversity contribute more to the world economy and make conservation profitable. He wants *all three* of business, government, and technology to cooperate in preserving biodiversity and engaging in conservation.

First of all, why would or should business engage in unprofitable activities? There is a real danger in justifying concern for biodiversity and conservation by profitability. This justification concedes the priority of economic growth. Wilson, to be fair, does offer non-economic justifications for biodiversity.[1] His ethical justification is this: ". . .life forms around us are too old, too complex, and potentially too useful to be carelessly discarded" (2002, p. 131) Although Wilson then spends some time arguing that the biosphere deserves respect for itself, that last "potentially too useful" in his remark completely vitiates respect for the biosphere for itself. We might very well *carefully* discard any number of life forms that prove not to be useful to us after careful examination.

A better argument (which Wilson almost makes) is that not only cannot we tell what parts of the biosphere will be useful to us, we cannot even tell which parts will turn out to be essential for our survival. As radical environmental activist Dave Foreman puts it, the biosphere is essential for our survival. Near the end of this chapter claims that humans can do without it will be considered.

Wilson fantasizes that corporations will simply come to cooperate with environmentalists in protecting the biosphere. It is possible when individual corporate CEOs have epiphanies and guide their enterprises to a balance between profitability and environmental responsibility that compromises neither. There is the well-documented case of Ray Anderson, the CEO of Interface, a leading carpet manufacturer, who had an epiphany about the environment.[2] He transformed his company from an environmental hazard into one that was not only environmentally sustainable, but also, as he puts it, environmentally "restorative." Interestingly enough, it was not necessary for Interface to alter its goal of maximizing profits. Rather, a carefully-thought-out strategy yielding both sustainability and profitability was possible. From a corporate point of view, it would have made no difference whether the CEO of Interface had chosen an environmentally terrific plan or a plan producing large amounts of pollution, provided both were just as profitable. But it made a huge difference to Anderson, and he was able to get his workers and supporters to take his view—not as corporate citizens but as (eco-) ethically responsible human beings. (Arena 2004)

By contrast, Pacific Lumber, a California company, until the 1980s owned and preserved most major old-growth redwood forests. Pacific Lumber was bought out by leveraged buy-out king Charles Hurwitz with the aid of Michael Milken and Ivan Boesky, who later served jail time for related felonies. The previous managers practiced sustainable lumber harvesting. Hurwitz liquidated the old growth to pay off the debt he incurred in taking over the company, often resorting to illegal cutting when US Forestry Rangers were not on the job (Worldwatch Institute 2008, p. 146). So even if a corporation is run by a management *personally* committed to sustainability, the corporation's commitments can change when management changes through hostile buyouts.

Ultimately, the aim of preservation and conservation has to take priority over corporate profitability. And the practical question is how this priority might come to be. The radical environmental slogan 'Earth First!' may ultimately be the correct view.

TECHNOLOGY FIRST

At least Wilson wants both environmental protection and corporate profitability. An English writer, Mark Lynas, could be said to subscribe to the slogan 'Technology first.' Lynas describes himself as a former environmentalist. But in his 2011 book, *The God Species,* after carefully laying out the serious conflicts between technology and the environment, he proceeds to endorse many environmentally problematic technologies as the solution to the conflict.[3]

Lynas applies the 2009 work of Johan Rockström and a panel of environmental scientists who attempted to determine 'planetary boundaries,' ecological limits to human development. If these limits are crossed, there is a risk of irreversible and abrupt environmental change. The boundaries are in the following areas: climate change, biodiversity loss, biogeochemical (nitrogen, phosphorus), ocean acidification, land use, freshwater, ozone depletion, atmospheric aerosols, chemical pollution. The aim of these scientists was to "determine a safe space for sustainable human development" (Rockström 2009).

Lynas uses these boundaries in a way cautioned against by some of the project's critics. Environmental policy analyst Bill Clark feels:

Tipping points in the earth system are dense, unpredictable... and unlikely to be avoidable through early warning indicators. It follows that... 'safe operating spaces' and 'planetary boundaries' are thus highly suspect and potentially the new 'opiates' (Clark 2011).[4]

Lynas takes the boundaries as a license to develop technology so long as one stays within the limits. Of course, the stated aim of the boundaries project permits this interpretation. The sustainability aims of this project will be discussed in the following chapter. This chapter will be concerned with Lynas assumptions, stated or implied, about the causes of conflict with the ecosystem. Of special concern are those of his assumptions which lead him to believe that almost unlimited development of technology is what we should be doing.[5]

One of his assumptions is human superiority together with cheerleader superiority: Humans are the "conquering army" of earth. Also, we are "completely natural" because we arose by evolution. But, as pointed out in Chapter 2, "Conflicts with the Ecosystem," we would not be in conflict with nature if we turned the entire planet into a lifeless desert. In fact, this outcome is exactly what planetary boundaries are supposed to avoid.

Lynas has two main reasons why more technology is the solution to technology's conflict with the ecosystem. First, he has an almost religious belief in technology. His blind trust in technology is a form of human superiority. Second, he believes any other possible courses of action will be either ineffective or politically impossible. In particular, he believes population reduction or economic growth reduction are not viable solutions. His unreasoning faith in technology will first be discussed and then we will return to why he thinks other courses of action besides technological development are not feasible.

Discussing environmental problems with the atmosphere, he mentions radically changing the atmosphere by pumping sulfur compounds at higher levels. Lynas remarks that "the unintended consequences of geoengineering would be better than six degrees of global warming" (2011, p. 197). But how can he possibly know this? In a later chapter on ozone depletion Lynas makes it clear that absolute disaster was narrowly averted by blind luck. If the inventor of CFCs—the ozone destroyers—had used bromine as their base rather than chlorine, the ozone would have been gone before we were able to determine what had happened (p. 218). An international agreement to phase out CFCs—the Montreal Protocol—was reached with great difficulty (p. 221). The result of that protocol is that the ozone layer is expected to recover by about 2050. We should notice that there was no way of telling in advance that implementing CFCs would have severe unintended consequences. Also, the political action necessary to fix the problem was by no means guaranteed.

As adverse effects of global warming accumulate, technologists will push hard for technological fixes. More sensible technologists such as Ken Caldeira point out that if one is going to try to cool the planet by injecting sulphur or some other extraneous compound into the atmosphere, it should be done a little at a time (Specter

2012). Better than all at once, to be sure, but once again we really don't know what the longer term unintended consequences would be. But in a strange way it could be worse. Peter Eisenberger, President of the company Global Thermostat, is developing a relatively inexpensive method of removing CO_2 from the atmosphere. The method sounds promising and less disruptive to the ecosystem. But here is Eisenberger's description of the ecosystem: "[It is] violent, amoral, and nihilistic. If you look at the history of this planet, you will see cycles of creation and destruction that would offend our morality as human beings" (Specter 2012). This flabbergasting distortion of the history of life on this water planet coming from someone who may be able to save it is mind-wrenching. Of course there have been and will continue to be disruptive events on the planet (volcanoes, asteroids). But evolved life mostly proceeds under a stable and delicately balanced set of conditions called the ecosystem.[6] To use the catastrophes as an excuse for trashing the ecosystem is insane.

Lynas' reasons for discounting other causes for conflict with the ecosystem are based on a belief that the ideal world would be one with unlimited population all enjoying the lifestyles and gadgets of the current first world. He discounts any possibility that current first world lifestyles may be not as fulfilling as less consumerist lifestyles. A New Economics Foundation (NEF) study showing that people were happiest in Costa Rica, Dominican Republic, and Jamaica is dismissed out of hand,[7] as is the NEF claim that it would take three earths to provide the resources to bring the entire world up to first world standards (pp. 68-69). Yet the first world uses 80% of the world's resources with 20% of the people (Hall & Ng 2004). So to bring the other 80% up to first world standards, an easy calculation indicates that this would take four earths. Also missing from Lynas' discussion is any consideration of the impact on the environment of wastes and other pollutants produced by a first-world economy. Pollutants such as plastic bags are mentioned but not quantified as part of the nine boundaries.

But his argument is really that attempts to control population or rein in the economy will not help in dealing with the ecological boundaries. He claims population reduction cannot be done fast enough to help with global warming. Ironically, Joseph Bookstein (who will be discussed shortly) argues in favor of population reduction for exactly the opposite reason: Technological fixes cannot be done quickly enough. Additionally Lynas regards economic reduction as politically impossible.

These considerations are enough to show that Lynas is wrong about technology not being the cause of the conflict with the ecosystem. The next chapter will return to a discussion of his specific proposals in favor of genetic engineering, pesticides, and nuclear power as part of a sustainable world, as well as his opposition to organic farming.

POPULATION GROWTH IS THE MAIN CAUSE

Paul Ehrlich in his 1968 book *The Population Bomb* presented the view that population increase was the root of the problem. In his later 2000 book *Human Natures*, he modified his view. His belief in 2000 was that, although overpopulation was important, defective human institutions and practices were the main cause. Overpopulation is to be handled by better institutions.

Others such as Joseph Bookstein have argued that population growth is a more important cause of global warming than burning fossil fuels. (2008) Bookstein points out that any carbon dioxide reduction from compact fluorescents or improved automobile gas mileage will be nullified by population increase in about two years. 75-year lifetime savings are as shown in Table 1.

As previously mentioned, Bookstein feels that technological solutions to climate change can't be implemented quickly enough, are expensive, and would not reduce carbon dioxide enough. So, whether or not population growth is the cause of the conflict with the ecosystem, population reduction for Bookstein is the only viable remedy.

Bookstein thinks that the population reduction necessary for preventing catastrophic climate change can be achieved voluntarily. A number of developed countries already have fertility rates low enough to cause their populations to decline. And some developing countries such as Iran and Mexico have managed dramatic voluntary reductions to their fertility rates. Iran's fertility rate went from 5.5 in 1989 to 1.7 in 2006. Mexico's rate went from 6.8 in 1970 to 2.4 in 2006 (Bookstein 2008). China also had a dramatic reduction in fertility but that was, of course, not voluntary. A fertility rate of 2.05 would stabilize world population at about 7 billion. A fertility rate of 1.85 would reduce world population to about 4 billion by 2300. Since the world average fertility rate is about 2.47, the necessary reduction in fertility is well within a possible range, given what countries like Iran and Mexico were able to achieve. A population of 7 billion would result in a temperature rise of about 3 degrees.

Table 1. CO_2 savings[8]

Policy	Lifetime CO_2 Saved
Low-wattage lights	11 tons per person
Doubling auto gas mileage	75 tons per person
Reduce home thermostat	93 tons per house
Three less flights per year (2 US, 1 overseas)	480 tons per person
80 kw home renewables	800 tons per house
One less child	2130 tons per child

As we saw, Lynas' discussion assumes technology will be able to provide a first-world standard of living for not only the current global population but also an indefinitely growing population. There is a serious question whether this is possible within the limits of the ecosystem. But leaving that objection aside, there is an important ethical issue here concerning population: Is it really better to have extra billions of people? It is pretty obvious that if the life prospects of the extra people were less, the answer is no, it would not be better. The ethical theory of utilitarianism has something to say on this issue.[9] Utilitarianism says to produce the greatest good for the greatest number. But consider situation #1 with 100,000 people who have 10 units of good, versus situation #2 with 10,000 people who have 20 units of good—twice as much good. A utilitarian would say that situation #2 is better than situation #1, because what counts for each individual is the *average* amount of good. So even though the total good in situation #1 is 1,000,000 units and the total good in situation #2 is 200,000 units, situation #2 is better[10] (See Table 2).

What if the average amount of good in two situations is the same, but in situation #3 there are 7 billion people and in situation #4 there are 1 billion? Is there any value simply to having more people if their life prospects are the same? These are the approximate populations for the year 2012 (situation #3) and the year 1800 (situation #4). That is, there are seven times as many people in situation #3 (2012) as in situation #4 (1800).

In 1800 there was Mozart. Do we now have seven Mozarts? In 1800, there were David Hume and Immanuel Kant. Do we now have fourteen Humes and Kants? Even in the area of technology, do we have seven technologists producing inventions as innovative as the steam engine?[11] My point is, there is no particular value in proliferating the number of human beings just to proliferate them. The main institutions in favor of human proliferation for its own sake are the Catholic Church and corporations: In the case of the Church, to increase the number of souls; in the case of corporations, to increase the number of customers. There is probably a minimum population below which creativity might suffer, but the 1800 number of one billion is a plausible workable minimum.

Table 2. Utilitarianism and population

	Population	Average Individual Good	Date	Total Good
Situation #1	100,000	10		1,000,000
Situation #2	10,000	20		200,000
Situation #3	7,000,000,000	20	2012	140,000,000,000
Situation #4	1,000,000,000	20	1800	20,000,000,000

Of course, my considerations ignore the fact that increased numbers of human beings will most likely not improve the average life prospects. Lynas says nothing about how a first-world standard of living would be achieved for more than six billion extra people. But we know that achieving that standard of living for the existing first world takes 80% of the world's resources. So somehow we have to come up with an extra 360% of the world's resources. Can this be done without smashing through all the environmental boundaries? Clearly not. The predictable result is "irreversible and abrupt environmental change" instead of everyone's moving up to first world standards.

So population growth is a major contributing cause to the conflict with the ecosystem. It is clear that it didn't cause the conflict, any more than population growth caused farming or civilization to arise.[12] Also, the major threats to the ecosystem arose within countries with low to moderate population growth. However, there is no question that population growth makes solutions to conflicts with the ecosystem much more difficult.

ECONOMIC GROWTH IS THE MAIN CAUSE

Some commentators believe that economic growth is the primary culprit in the conflict of technology with the ecosystem. The authors to be examined are E. F. Schumacher (1973), Al Gore (1992), and Robert Nadeau (2006). Nadeau's main argument has some weak points, but can be made stronger with the aid of some research by Joseph Stiglitz (2010).

E. F. Schumacher was an early proponent of this view in 1973, with his *Small is Beautiful.* Schumacher sees technology and social structures moving in tandem to produce the conflict with the ecosystem. Both require things to be done on a larger scale than necessary, and produce detrimental results. For Schumacher, the solution is simply to scale down expectations, both social and economic as well as technological. Small applications are more likely to be sustainable, and are less likely to contain inefficiencies destructive of the ecosystem. Indeed, his "small" applications look very much like instances of technology which are not termed modern or postmodern. For Schumacher, smallness is a spiritual correlate to a technology appropriate to human needs and to the environment. Ultimately Schumacher is calling for a change of heart to a more human scale as the basis for the relevant downsizing to sustainability. But his solution does not deal directly with ecological problems. He does indicate that smaller scale, more responsive systems are less likely to cause environmental damage, but these consequences seem to be only a desirable by-product of smallness. And so he really does not have a diagnosis of the cause of the conflict with the ecosystem.

Gore (1992) believes we know what is causing the conflict with the ecosystem but have just failed to take appropriate action. He believes that the cause is unsustainable consumption. It is destroying the planet. But we just have to be firm about not consuming so much and achieving sustainability; and we should alter our social and economic institutions and practices accordingly. Although this will clearly help and some individuals will be motivated to reduce consumption, these actions will not help when the damage is more indirect, as with global warming. A society can be efficient in the sense of not consuming wastefully and still be destructive to the environment. Gore's message that we just need to get the economy back on track in working to avoid wasteful consumption does not address important aspects of the problem.[13]

Nadeau (2006) thinks that classical economics is to blame. He finds the idea of equilibrium achieved by market economies faulty (the "invisible hand" of Adam Smith). According to Nadeau two other faults contribute: The inability of classical economics to recognize part/whole relationships[14] and the inability of classical economics to include environmental considerations properly. Classical economics does recognize environmental considerations as 'externalities.' Nadeau's argument is that there is no good way to internalize these considerations, to incorporate them into classical economics in a reasonable way.

Classical economics makes much use of the so-called 'invisible hand.' This is the theory of Adam Smith that self-interested buyers and sellers competing in a free market will arrive at a mutually beneficial state of affairs without any additional outside actions. The buyer wants to buy, but preferably at the lowest price. Sellers compete by lowering their price, but in order to keep selling they must make some profit, so there is a limit to how low the seller can go. Buyers in a competitive market will therefore receive the lowest price consistent with the seller continuing in business. This situation is said to be *Pareto optimal,* which means that no one can be made better off without someone else being made worst off.[15] The major question is what are the conditions under which this optimal result happens.

Nadeau seems to be saying that any application of mathematics to human affairs is wrong. His reasoning is that the human motivations driving economics are just "too complex" to be captured by mathematics. Thus Adam Smith's 'invisible hand' or Pareto optimality fail just because human reality has complexity mathematics cannot possibly capture. Now it is known that free markets do not optimize when there are externalities or monopolies. And Nadeau has some interesting things to say about why the externality of the environment cannot be easily incorporated into classical market economics. His points will be returned to shortly.

But the claim that mathematics cannot be applied to economics is close to bizarre. For the physical world itself is also complex and physical laws are also idealized. Falling bodies do not completely obey Newton's laws—air resistance complicates the results. Nadeau considers it a criticism to say that economics borrowed from physics. This is normally called technology transfer. Within mathematics, it is common that new results are proved in one area by using results from a seemingly unconnected area. The economist John Nash[16] proved that there were solutions to non-zero-sum games using the Brouwer Fixed Point Theorem from topology. Nash's result has implications for economics, since many economic transactions can be modeled as non-zero-sum games. The Brouwer Fixed Point Theorem has many other non-abstract applications. For example, it is a matter of mathematics that every human head of hair must have a cowlick. Like it or not, mathematics does apply to reality.[17]

It is one thing to criticize the doctrine that free markets will solve every social problem. This doctrine, *market fundamentalism,* is more of an ideology or religion than science and is demonstrably wrong.[18] But it is equally wrong to make Nadeau's criticism. Free market economics is not wrong simply because it applies mathematics to human behavior. The mathematical discipline of economic forecasting of expected supply and demand applies mathematics to human behavior. And such forecasting is essential for any business other than the smallest.

The criticisms of Nobel Prize winning economist Joseph Stiglitz (2010) of the classical economic equilibrium are less general and have more plausibility. Stiglitz's main criticism is that the classical "invisible hand" equilibrium holds only under ideal conditions. Markets generally have imperfections, and some common imperfections cause great departures from invisible-hand equilibriums. When conditions for a perfect efficient market are not satisfied, there are always government interventions which make everyone better off (2010, p. 243). For example, imperfect information changes the equilibrium. With perfect markets (with perfect information), there is never unemployment. With imperfect information, there can be unemployment. Stiglitz also points out that markets are not efficient in decisions about how much to spend on research for innovation.

Waiving Nadeau's implausible blanket criticism, we can now turn to his plausible discussion of some important difficulties with classical economics. These are how to include the environment within classical economics; and the use of Gross Domestic Product as a measure of economic well-being (These are also concerns of Stiglitz.). In classical economics, environmental problems are external to the market system. So the question is how to include them within the market system. Government regulation is one way of handling them, but it is not within the market system. Three ways of handling them within the market system are: By a tax on the activity producing the negative environmental effect ('emissions tax'); by issuing tradeable permits to establish a certain amount of environmental pollution ('cap and trade'); or by including environmental effects within cost/benefit analyses (2006, p. 126). Nadeau's criticisms of these approaches are mainly of how they are implemented, rather than of the approaches themselves. For example, the whole point of emissions taxes as opposed to government regulation is to preserve market functioning. So taxes will never be set high enough to remove a company which should not be in business at all because of the severity of the environmental damage it is causing. Nadeau criticizes cap and trade as being expensive and difficult to administer as well as not working very well (2006, pp. 128-129). Some say that cap and trade worked well against acid rain. However, climatologist James Hansen believes cap and trade is useless against carbon dioxide emissions. First, it sets a floor for emissions; emission cannot go lower than the cap. Second, if one buys a fuel efficient car, it simply helps someone else buy an SUV[19] (2009, pp. 213-219).

So, it seems that classical economics cannot deal with externalities very well. Additionally, its measure of when an economy is doing well is deeply flawed. Nadeau points out that Gross Domestic Product (GDP) and Gross National Product (GNP) take no account of environmental losses incurred in production (2006, p. 140). Stiglitz points out that since GDP is a bad measure of economic well being, talk about tradeoffs between economic growth and the environment are seriously misguided: If we correctly measured output—including worsening the environment—there would be no tradeoff (Stiglitz 2010, p. 284).

So what can we conclude about economic growth as the main cause of conflict with the ecosystem? It is not obvious that economic growth is by itself the cause of conflict. But it is not clear that sustainable economic growth is possible. Sustainable economic growth is by definition economic growth compatible with maintaining the ecosystem. The fact that current economic growth largely stems from modern technology immediately raises issues of compatibility with the ecosystem. And even if we ignore this problem, classical technology still uses up resources. So in that respect it too is not sustainable. The economy will be discussed further in Section 3, "Possible Outcomes."

HUMAN ATTITUDES ARE THE MAIN CAUSE

The authors Dave Foreman (1991) and Bill McKibben (1989, 2010) explicitly regard human attitudes as central to the conflict with the ecosystem. They would both require a radical change of heart in how we view our relation to the earth. McKibben says ". . . we could limit ourselves voluntarily, *choose* to remain God's creatures instead of making ourselves gods" (1989, p. 214). While the attitude McKibben favors is probably the correct one, it does not seem to be a *choice*. As we saw in Chapter 1, "Humans as Masters of the World," Bill McKibben tacitly accepts human exceptionalism. In his best-selling *The End of Nature* (1989), he tells us that human beings are no longer part of nature, they have *ended* nature. But this is not correct because we *are* still part of nature. Injecting large amounts of greenhouse gases into the atmosphere has consequences because of the effects *in nature* of increased concentrations of gas in the atmosphere. This is a part of nature over which we have limited control once the gases are in the atmosphere.

McKibben clearly thinks of nature as *non-human* nature. Human beings essentially stand outside the world. But if human beings have, as he thinks, become completely separate from nature, the question is untouched of the proper place of human beings in the rest of nature and how we have come to be in conflict with it.[20] Perhaps the better view is that of the biologist E. O. Wilson: "It is in the nature of all biological species to multiply and expand heedlessly until the environment bites back. The bite consists of feedback loops—disease, famine, war, and competition for scarce resources" (2002, p. 76). Such feedback loops spell the end of any claim for superiority over nature a species might have had. And that is what we humans are now experiencing.

Dave Foreman, one of the founders of Earth First!, has a clear conception of the proper place of human beings in nature. Earth First! is a movement rather than an organization. Its slogan is 'No compromise in the defense of Mother Earth!' Foreman came to his position after many years of working with establishment environmental organizations such as the Wilderness Society. They compromised nature in favor of human beings. Foreman says: "Human beings are not paramount. Earth is not for Homo sapiens alone; human life is but one life form on the planet and has no right to take exclusive possession" (Foreman 1991).

Their views both challenge human superiority. But both challenge human superiority on the basis of *feelings.*[21] McKibben is rhapsodic about untrammeled nature. Foreman's praise of unspoiled nature is more robust:

The oceans of the Earth course through my veins, the winds of the sky fill my lungs, the very bedrock of the planet makes my bones. I am alive! I am not a machine,... a cog in the industrial world.... When a chain saw slices into the heartwood of a two-thousand-year-old Coast Redwood it's slicing through my guts. When a bulldozer rips through the Amazon rain forest, it's ripping into my side. When a Japanese whaler fires an exploding harpoon into a great whale, my heart is blown to smithereens. I am the land, the land is me (1991, pp. 4-5).

And it is true that if their feelings against human superiority and human exceptionalism and the priority of nature were accepted, this could be a solution to the problem of conflict with the ecosystem. But there is a difficulty in using *feelings* as a basis for action.[22] Basing any obligation to the ecosystem on how you feel about it, makes it seem more like a special interest or hobby than a genuine requirement.[23] Also, in considering principles of justice governing society, it is important that justice require only that the basis of social cooperation be shared. Social requirements cannot be based on the adoption of deep-seated comprehensive views about the nature of reality because they cannot realistically be shared as the basis of social cooperation.[24] Foreman's and McKibben's views about putting human beings in a less exalted place are correct about the cause of the conflict. Indeed, they are the only views about the cause of the conflict which reject human superiority. However, how to use their insights as the basis for a solution is problematic. Ultimately, though, a spiritual conversion may be exactly what is required to end the conflict between human beings and the ecosystem.

HUMANITY IS A SPECIES DOOMED TO DESTROY ITSELF

The paleontologist Peter Ward formulated what he calls the Medea hypothesis. This is the hypothesis that a ruthlessly selfish species that competes too successfully will eliminate itself (Flannery 2010, p. 23). Since humans are ultimately subject to natural selection, it remains to be seen whether the destructive forces at work in human society will prevail over the forces working to resolve our conflicts with the ecosystem. The ecosystem is, after all, the system that supports all of us and all life. For most of the remainder of this book, it will be assumed that it is possible for humans to resolve their conflicts with the ecosystem. This is, however, just an assumption.

PRIORITIES BETWEEN THE CAUSES

Although economic growth and population growth exacerbate problems of conflicts with the ecosystem, they are not by themselves causes of the problem. Economic growth without proper attention to impacts on the ecosystem will likely be fatal in a number of ways. So regardless of any other solution, economic growth must be suitably constrained by the needs of the ecosystem. However, the market system itself is not at fault. But "market fundamentalism" causes grief wherever it is applied.

Although population growth tends to worsen any conflict of technology with the ecosystem, it is not by itself a cause of the conflict—any more than population growth was by itself a cause of farming. Or that population growth was by itself a cause of modern technology. Again, some environmental problems may not be soluble without reduction in population.

The causes of human conflict with the ecosystem are modern technology coupled with attitudes of human superiority and human exceptionalism. Without attention to these causes there is little chance of resolving the conflict with the environment. Section 3, "Possible Outcomes," will discuss suggestions about how to do this.

The issues that must be confronted are: Whether it will be possible to continue to employ modern technology at all; and whether rejection of human superiority and exceptionalism can be given a firmer basis than strong personal feelings about the environment.

Of course, the possibility that humans will self-destruct is entirely different. Unfortunately, it cannot be ruled out.

CAN WE EXIST APART FROM THE ECOSYSTEM?

There is another possibility to consider, suggested by Hans Moravec (1988) and other technologists. Could the human species survive without the ecosystem? The idea would be to eliminate the ecosystem in favor of some substitute supplied by technology. This possibility depends on the efficacy of human technology as compared to nature. Human technology regularly has unanticipated conflicts with the ecosystem which if not corrected produce great damage. So technology is simply not in a position to replace the ecosystem. Besides, as E.O. Wilson notes, it is impossible to put together a functioning ecosystem from scratch. Most of the necessary microorganisms are still unknown. And species have evolved to fit together in ways biologists are only beginning to understand. Wilson concludes, "To synthesize ecosystems on bare ground or in empty water is no more practicable than the reanimation of deep-frozen human corpses" (Wilson 2002, p. 130).

But what gives us the right even to consider this possibility? Without the still unjustified belief in human superiority and human exceptionalism, we have no right to destroy or even significantly damage the ecosystem which has supported us and all our fellow living beings.

ORIGINS OF HUMAN SUPERIORITY

In the Preface, it was noted that the prevalent belief that we are in charge of nature, has a history. In this section of the book, the attempt was to trace the development of technology and human superiority. Modest separation from the ecosystem begins with our ancestor *Homo erectus* storing tools, thus creating a dedicated human space distinct from the environment. *Homo erectus* later controls fire and constructs shelters, again instancing the control of a human environment distinct from the rest of the environment. Our species, *Homo sapiens,* continues the development of a distinct human environment by the invention of clothing. Body ornaments and pigments are precursors of symbol systems. Ceremonial burials may or may not indicate belief in another world.[25] But the representations of cave paintings clearly demonstrate the human ability to create surrogates for the actual world, culminating in writing and digital worlds.

But by far the greatest contributions to the separation of human beings from the ecosystem are farming and civilization. Farming is a limited separation. Although domestication of plants and animals entails a belief in the superiority of human intervention in the ecosystem, humans are still dependent on the ecosystem for the growth of plants and animals. Civilization is, however, almost completely removed from the ecosystem. The origin of these two dramatic and relatively sudden changes in human life is not completely understood. Current understanding of the origin of farming, outlined in Diamond (2012), is that the sudden appearance of abundant plant food in nine different areas caused people to become sedentary. Once humans settled down, the low hunter-gatherer birthrate became much higher, causing the need for yet more food available only through farming. It is possible that the ending of the last glacial period was what produced the new plant abundance which set the whole practice in motion.

Like farming, civilization was clearly not an accident because civilization arose independently in three completely different locations. It is clear that farming is a necessary condition for civilization, because civilized humans don't grow their own food. The distinctive separation from the ecosystem comes about with economic specialization. When an abundance of food allows individuals or groups (classes) to gain control of agricultural goods they do not produce, civilization begins its divorce from the ecosystem. Society stratifies, and more and more complex social forms and technology proliferate, liberated from the constraints of making a living from the ecosystem.

We can see how farming and civilization contributed to human superiority and exceptionalism. With the domestication of plants and animals, humans take over some of the work of natural selection. To that extent, they have control of nature. With civilization, humans create a totally human world. One of its features is improving on the way the natural world does things. This is called technology, and its premise is that humans can do better than nature. Its final illusion is that nature can be completely ignored or destroyed.

REFERENCES

Achbar, M., Abbott, J., & Bakan, J. (2004). *The corporation.* [Motion Picture]. New York: Zeitgeist Films Ltd.

Arena, C. (2004). *Cause for success.* Novato, CA: New World Library.

Bookchin, M., & Foreman, D. (1991). *Defending the earth*. Boston, MA: South End Press.

Bookstein, J. (2008). *Global warming: The population connection*. [Broadcast]. La Jolla, CA: UCTV. Retrieved 2/4/12 from www.uctv.tv/search-details.aspx?showID=14886

Clark, B. (2011). *Grand challenges of sustainability science*. Paper presented at the 2011 Conference on Resilience, Innovation, and Sustainability: Navigating the Complexities of Global Change. Phoenix, AZ. Retrieved 2/11/12 from www.hks.harvard.edu/var/ezp_site/storage/fckeditor/file/pdfs/centers-programs/centers/cid/ssp/docs/about-us/in-the-news/2011/Clark_Grand_Chal_Sust_Science_110314.pdf

Diamond, J. (2005). *Collapse*. New York: Viking.

Diamond, J. (2012). What makes countries rich or poor? *The New York Review of Books*, *59*(10).

Ehrlich, P. (1968). *The population bomb*. New York: Ballantine Books.

Ehrlich, P. (2000). *Human natures*. New York: Penguin Group.

Foreman, D. (1991). *Confessions of an eco-warrior*. New York: Harmony Books.

Foreman, D. (2011). *Manswarm: The killing of wildlife*. Durango, CO: Raven's Eye Press.

Gore, A. (1992). *Earth in the balance*. New York: Houghton Mifflin.

Hall, M., & Ng, S. (2004). *Overpopulation vs. overconsumption*. Retrieved 2/14/12 from www.odec.ca/projects/2004/ngsi4s0/public_html/indexx.htm

Hansen, J. (2009). *Storms of my grandchildren*. New York: Bloomsbury.

Heidegger, M. (1955). The question concerning technology. In *The Question Concerning Technology and Other Essays*. New York: Harper & Row.

Jackson, T. (2008). *The challenge of sustainable lifestyles. In 2008: State of the World*. New York: Norton.

Klein, N. (2007). *The shock doctrine*. New York: Henry Holt.

Lynas, M. (2011). *The god species*. Washington, DC: National Geographic Society.

McKibben, B. (1989). *The end of nature*. New York: Doubleday.

McKibben, B. (2011). *EaArth*. New York: St. Martins Griffin.

Mill, J. S. (1848). Principles of political economy. *Library of Economics and Liberty*. Retrieved 2/28/12 from www.econlib.org/library/Mill/mlP.html

Moravec, H. (1988). *Mind children*. Cambridge, MA: Harvard University Press.

Nadeau, R. L. (2006). *The environmental endgame*. New Brunswick, NJ: Rutgers University Press.

Reports, N. (2009).. . *Climatic Change*. doi: doi:10.1038/climate.2009.95.

Rockström. Johan, et al. (2009). Planetary boundaries: Exploring the safe operating space for humanity. *Ecology and Society, 14*(2). Retrieved 2/11/12 from www.ecologyandsociety.org/vol14/iss2/art32/

Schumacher, E. F. (1973). *Small is beautiful*. New York: Harper & Row.

Specter, M. (2012, May 14). The climate fixers. *New Yorker (New York, N.Y.)*.

Stiglitz, J. (2010). *Freefall*. New York: Norton.

Wilson, E. O. (2002). *The future of life*. New York: Vintage Books.

Worldwatch Institute. (2008). *2008 state of the world*. New York: Norton.

ENDNOTES

1 These were discussed in Chapter 2, "Conflicts with the Ecosystem," "Species Extinction."

2 Ray Anderson, CEO of Interface, Inc., details his epiphany in the film *The Corporation* (Achbar, Abbott, & Bakan, 2004).

3 For those knowledgeable about environmentalism, this is a difficult book to read. Its mixture of environmental sensitivity, pro-corporate propaganda and preposterous predictions is bewildering.

4 The Wikipedia entry on planetary boundaries has relevant excerpts from criticisms of the boundaries by environmental scientists. Nature Reports (2009) devotes a full issue to those criticisms.

5 It should be noted that the developers of planetary boundaries are neutral about how respect for the boundaries are to be implemented. In particular, they subscribe neither to human superiority nor a policy of technology ueber alles.

6 The 20% oxygen in the atmosphere necessary for animal life is produced by plants. The processes which produce rain necessary for food crops depend on

complex causes and conditions. Even irrigation depends on processes in the ecosystem to produce water. Irrigation does not produce water out of nothing.

7 See Jackson (2008) for a more comprehensive discussion.

8 This table is modified from Bookstein (2008). 80kw rather than 10kw is used as the basis for home renewables because 80kw is the amount necessary for significant home power generation.

9 After writing this section, I discovered that the founder of utilitarianism, John Stuart Mill, also believed that having more people was not better. His eloquent reasons are different from those given here. See Mill (1848, book 4, chapter 6, section 2). This passage is reprinted as the foreword to Foreman (2011).

10 Leftists tend to recoil in horror at the idea of reducing human population. See, for example, Steve Chase's Introduction to Bookchin and Foreman (1991). Chase observes, "the goal of many deep ecologists [including Earth First!ers] is not transforming society but rather drastically depopulating the earth as if human numbers were all that mattered and the various kinds of societies that people can create are of little or no relevance to the ecological question" (p. 20). Many leftists like Chase regard human societies as having primary value. But, as we saw in Chapter 7, "Civilization," such claims about the superior value of human institutions are basically circular. The undoubted value of human societies is no justification for destroying the ecosystem, any more than was the undoubted value of human arts, sciences and technology.

11 My thanks to the late Alan Ross Anderson for this observation.

12 Although, as we saw in Chapter 6, "Humans as Farmers," population growth may have been the reason farming continued after it originated.

13 More recently, Gore has made important proposals to end short-sighted short-term corporate planning. His proposals will be discussed in Chapter 10, "Environmentalism and Sustainability."

14 This criticism will be ignored because I cannot see how it applies.

15 Pareto optimality is named for the Italian economist Vilfredo Pareto, who formulated this criterion for everyone's benefiting.

16 John Nash's life was well reported in Sylvia Nash's *A Beautiful Mind.* The film of that name presents a dumbed-down inaccurate version.

17 The reality of mathematical objects has always been somewhat mysterious. Provably they are not just human constructs. But since they are also not physical objects, the only alternative is that they are abstract objects existing in their own realm. This view is called "Platonism."

18 For a good discussion of a number of horrendous cases of the application of market fundamentalism, see Naomi Klein, *The Shock Doctrine* (2007).

19 Hansen's positive proposal of fee-and-dividend will be discussed in Chapter 10, "Environmentalism and Sustainability."

[20] McKibben's more recent book, *EaArth* (2010), continues his theme of separation. It also seems to espouse a doctrine, not of human superiority, but of *Vermont* human superiority. My criticisms are of McKibben's theoretical positions. His practical work in ameliorating climate change through his organization 350.org is beyond reproach.

[21] E. O. Wilson's *biophilia,* love of nature, is clearly a related feeling but with no power (2002, p. 134).

[22] See Schultz (2009), Chapter 2, "Background in Ethical Theory," "Theories of Right: Intuitionist vs. End-Based vs. Duty-Based."

[23] Personal or organizational recycling, or Earth Day activities like cleaning beaches, are praiseworthy activities. The only danger is to think that if we recycle and pick up after ourselves, the problem of conflict with the environment will thereby be completely solved. More on this issue in Chapter 10, "Environmentalism and Sustainability."

[24] See Schultz (2009), Chapter 5, "Domestic Theories of Justice," (p. 68).

[25] Domestication of dogs and the megafauna extinction do not contribute to increasing human separation from the ecosystem.

Section 3
Possible Outcomes

Chapter 10
Environmentalism and Sustainability

ABSTRACT

Sustainability is maintaining equilibrium with the ecosystem. Four accounts of sustainability are examined. Unlimited economic and population growth will create more environmental problems and thus not produce sustainability. Reducing population and dealing with climate change will help but do not deal with modern technology's threat. If humans come to realize that the ecosystem is primary, that realization will produce sustainability, but it needs a firmer basis than emotional commitment.

INTRODUCTION

Conflicts between technology and ecology have complex causes, and any prediction of outcomes must therefore be tentative. My original title for Section 3 was "Solutions." But on reflection that title would have been not only inaccurate but also arrogant. In this chapter, proposals to make changes to our way of life that are not drastic will be examined. These changes may not be sufficient to resolve the conflict. In the three following chapters, Chapters 11, 12, and 13, further proposals to help achieve equilibrium with the ecosystem will be examined. This involves determining what parts of our way of life we can keep, consistent with resolving the conflict. Finally, in Chapter 14, "Extinction," the worst-case outcome if we are unable to resolve the conflict will be discussed.

DOI: 10.4018/978-1-4666-4586-8.ch010

In this Section, there are no detailed prescriptions of what humanity needs to do to regain equilibrium with the ecosystem. In our current situation, detailed policy recommendations would be pointless. Some of the most powerful corporations on the planet think it is in their interest to discredit sound science concerning climate change. It is impossible to understand how they could have come to think that their desire for profits could change what is happening in the environment. But as long as they do, specific policy recommendations may be a waste of time. At some point, environmental reality will overcome corporate knowledge pollution and then sensible policy to produce equilibrium with the environment will be possible. One can only hope that environmental reality harsh enough to overcome corporate knowledge pollution will not be horrendous. It is also important not to take too limited a view of what changes might be required for equilibrium, even if those requirements may now seem totally unattainable or even not quite sane.

The question to be answered in this chapter is to what extent environmentalism and sustainability can solve problems about conflicts with the ecosystem. The goals of environmentalism and sustainability—to achieve equilibrium with the ecosystem—are correct. But it is another matter to achieve those goals in the context of powerful economic institutions and in the context of dominant attitudes of human superiority. The goals of environmentalism and sustainability can themselves become unsustainable. These goals can present themselves as merely one more political interest, to be balanced and compromised against other political and economic interests.

There are a number of important concerns about sustainability that are rarely discussed. First, exactly what are we trying to sustain? Lester Brown's Plan B aims to save civilization. But can we save civilization and the ecosystem at the same time? Even if we can, do we want to? Second, recommendations for saving the environment are mostly presented piecemeal. Some recommended actions are going to be more efficacious than others. How do we compare the efficacy of actions, especially actions with different aims. For example, actions making contraceptives available and thus reducing population, versus actions to promote renewable energy? Given that we have limited time to achieve sustainability, would it be better to ignore actions with low efficacy such as recycling plastic bags?[1] I will call this problem, the "Piecemeal Problem."

FOUR PROPOSALS CONCERNING SUSTAINABILITY

I will first examine four different proposals about sustainability. Each has its own account of what features of the environment are worth saving or are necessary to save—whether human or nonhuman. Consequently, each has a different account of what is harmfully unsustainable. And then each also has a correspondingly different account of sustainability, of how we humans must modify our actions in order save valuable or necessary features of the environment.

The four accounts are:

- Mark Lynas'(2011) technophile goal of raising the entire population of earth to first-world levels.
- E. O. Wilson's (2002) vision of cooperation between environmentalists and corporations.
- Lester Brown's *Plan B 4.0* (2009), a detailed plan for attaining sustainability.
- Dave Foreman's (1991, 2011) agenda for protecting the earth.

SUSTAINABILITY WITH UNLIMITED ECONOMIC AND POPULATION GROWTH

As we saw in the preceding chapter, Mark Lynas' goal is actually *not* sustainability. He believes humans can and should expand current levels of first-world economic activity to everyone on the earth. He believes public opinion will demand this level of economic activity and there is nothing to be done about that. He believes this will be OK for the environment provided the nine planetary boundaries developed by Rockström et al. (2009) are not crossed.

As we also saw in the preceding chapter, critics of the planetary boundary approach warn against using the boundaries as a license for technological development right up to the edge of the boundaries. Treating them as early warning indicators is also unlikely to work (Nature Reports 2009). Leaving aside for the moment Lynas' possibly illegitimate use of the boundaries, the usefulness of these boundaries in achieving sustainability will first be addressed.

The planetary boundaries are intended to set ecological limits to human development: If these limits are crossed, there is a risk of irreversible and abrupt environmental change. The boundaries are in the following areas: climate change, biodiversity loss, biogeochemical (nitrogen, phosphorus), ocean acidification, land use, freshwater, ozone depletion, atmospheric aerosols, chemical pollution. The aim of these boundaries scientists was to "determine a safe space for sustainable human development" (Rockström 2009). Setting boundaries does seem to be an improvement over just randomly minimizing human impacts on the planet. The boundaries are an attempt to address the Piecemeal Problem.

There is an obvious omission from this list, namely population.[2] Several of the boundaries reflect population size directly and some less directly. The boundaries directly connected to population are: Biodiversity, nitrogen and phosphorus, land use, and freshwater. The biodiversity boundary is stated in terms of extinction rate. And extinction primarily occurs because of habitat loss. And habitat loss occurs because of human encroachment. Population increase clearly impacts the land use boundary, even if humans chose to live in ever more condensed cities as Lynas proposes. Nitrogen and phosphorus affect water quality by stimulating algae growth which can be toxic. Nitrogen pollution occurs because of the use of artificial fertilizer. More population will mean more use of artificial fertilizer, since even now the current population is fed with intensive use of artificial fertilizer. The freshwater boundary is also sensitive to population increase. There is a limited amount of freshwater, and although technology can help in purifying water, that requires the expenditure of large amounts of energy. The boundaries indirectly connected to population are climate change, ocean acidification, ozone depletion, atmospheric aerosols, and chemical pollution. All will increase with population increase.

It is also clear that the work on the boundaries is incomplete. Both atmospheric aerosols and chemical pollution have no quantifiable boundaries set, apparently because of their complexity. So progress on the piecemeal problem is limited. The creators of the boundaries also acknowledged that interaction between the boundaries could lower the boundary values. This is another defect in the ability of the boundaries to help with the piecemeal problem. But there may be no way around these defects.[3]

If one looks at Table 1, Planetary Boundaries, the preacher in the woodpile[4] is that all the unsustainable increases which we are trying to control are due to the industrial revolution. As we have seen, the industrial revolution *is* modern technology. And for modern technology, everything is available for transformation by its processes. There is no natural world whose limits have to be respected. Unfortunately there are those who believe the uncontrolled destructive processes of modern technology are the answer to all our problems, even though modern technology—with all its benefits—has also created problems severe enough to destroy many of us humans and even all life.

So what force could be powerful enough to oppose modern technology? Although modern technology may not be an addiction, it mimics many features of an addiction. As with alcohol in alcoholism, modern technology seems to be the solution to our problems. And sometimes it is. But both alcohol and modern technology can drag us to undreamed-of depths of despair. Continuing the analogy, we would really like to use technology in moderation and in harmony with the health of our support system, the ecosystem. We can do this if and only if we are not addicted to it. The addict has lost the possibility of control and thus the possibility of moderate use.

Table 1. Planetary boundaries

Earth-System Process	Control Variable	Boundary Value	Current Value	Boundary Crossed	Pre Industrial Value
1. Climate change	(a) Atmospheric carbon dioxide by concentration (ppm volume)	350	387	yes	280
	(b) Or: Increase in radiative forcing(W/m2) since the start of the industrial revolution (1750)	1.0	1.5	yes	0
2. Biodiversity loss	Extinction rate (number of species per million per year)	10	> 100	yes	0.1–1
3. Biogeochemical	(a) anthropogenic nitrogen removed from the atmosphere (millions of tons per year)	35	121	yes	0
	(b) anthropogenic phosphorus going into the oceans (millions of tons per year)	11	8.5–9.5	no	-1
4. Ocean acidification	Global mean saturation state of aragonite in surface seawater (omega units)	2.75	2.90	no	3.44
5. Land use	Land surface converted to cropland (percent)	15	11.7	no	low
6. Freshwater	Global human consumption of water (km3/yr)	4000	2600	no	415
7. Ozone depletion	Stratospheric ozone concentration (Dobson units)	276	283	no	290
8. Atmospheric aerosols	Overall particulate concentration in the atmosphere, on a regional basis	not yet quantified			
9. Chemical pollution	Concentration of toxic substances, plastics, endocrine disruptors, heavy metals and radioactive contamination to the environment	not yet quantified			

From Wikipedia (2012)

There are signs of addiction in Lynas' proposals, in his case for unlimited economic growth through (of course) unlimited use of technology. The supposed fact which leads him to adopt this position is that most everyone in the world wants a first-world lifestyle. But how can he know this? The one study he cites but dismisses fails to support this so-called fact. The study by the New Economics Foundation (2009) showed that people were happiest, not in first world countries but in Costa

Rica, the Dominican Republic, and Jamaica. Once Lynas' supposed "fact" is accepted, it follows that we have no choice but to use putatively harmful technologies. For Lynas, organic farming is useless if we have to feed a burgeoning population. So we must ramp up our use of synthetic fertilizers, pesticides, and genetically engineered crops.

There are two points here: First, organic farming is not inefficient. Lynas cites Professor Tim Benton of Leeds University who concluded from his own research that it takes twice as much land to get the same yields organically than with chemicals (Lynas 2011, p. 100). Professor Benton is an officer in a biotechnology organization and might be expected to favor technologically intensive agriculture. Brian Halweil (2012) of the World Watch Institute reports in a meta-study that organic techniques used in wealthier countries tend to yield 80% to 95% as much as artificial techniques. Whereas, in poorer countries, organic yields are often 20% higher and as much as 93% higher than the yields of artificial techniques. Halweil gives reasons for the differences: In wealthy countries, higher yields are achieved with copious amounts of synthetic fertilizer and pesticides, which produce nitrogen pollution and toxic pesticide pollution (I will return to this issue shortly.). Whereas farmers in poorer countries often cannot afford synthetic fertilizers and pesticides. Overall, they have less resources, including water, and organic techniques of cover crops, compost, and manure help them to conserve water.

Addiction also involves denial. Getting greater yields from artificial agricultural technologies is going to increase nitrogen pollution and pollution from pesticides. Lynas believes these problems are going to be solved by (future) genetic engineering.

As remarked in Chapter 2, "Conflicts with the Ecosystem," genetic engineering of species in the environment cannot be safe because it has unforeseen consequences which do not play out within individual life spans.[5] In nature, genetic changes are honed over time against the existing environment by natural selection. Genetically engineered changes are not. We already have cases in which genetically altered individuals cannot be safely released into the environment.[6] Genetic engineering is a prime example of modern technology. There is no natural world, there are only items available to manipulate as resources to serve human aims.

But, even before genetic engineering eliminates the threats of nitrogen pollution and pesticide pollution, Lynas reassures us that neither is that bad. Nitrogen runoff fertilizes forests. And, "if you don't mind the dead zones," nitrogen fertilizes plankton (Lynas 2011, pp. 93-94). To be fair, he does say that with his pro-nitrogen pollution statements, he is "get[ting] carried away." Nitrogen pollution does reduce biodiversity and cause eutrophication and a lot of dead spots in the ocean, some quite large, where nothing can live. However, according to Lynas, it does not matter whether the nitrogen comes from artificial fertilizers or manure. But just how often have you seen a field over-manured? Except as part of factory farm operations.

Here is Lynas' take on pesticide pollution. Monsanto has genetically engineered a number of crops—soy, rapeseed, flaxseed and sugar beet—to be resistant to its pesticide Roundup, their trademarked version of glyphosate.[7] Numerous questions continue to be raised about the safety of glyphosate (Barrow 2006). Of course Monsanto vehemently defends its safety. Given Monsanto's track record, this defense is exactly worthless.[8] Lynas' defense is: Glyphosate is less toxic than other commonly used pesticides (p. 104). So a parallel defense of carbon dioxide pollution is that carbon dioxide is a much less powerful greenhouse gas than methane. Other problems with the use of this pesticide include substantial ecosystem damage and biodiversity loss. Also, other plants and weeds become resistant to the pesticide, so more and stronger pesticides have to be used to get the same results. The upshot is more and more damage to the surrounding ecosystem. As of spring 2012, Monsanto was seeking FDA approval to combine glyphosate with dioxin for use with its pesticide-resistant genetically engineered crops. Glyphosate alone was no longer working. Dioxin is one of the most toxic substances ever produced. It has been banned for any use since 2004 (Energy Justice Network 2012).

Concerning the issues just discussed, Lynas is misled by his technophilia. But concerning nuclear power, he is at best dishonest and at worst deliberately deceptive. Whatever credulity Lynas might have had vaporizes when we read on pages 71-72 that the severity of the Fukushima accident is due to "hyperbolic media coverage" and resulted in "panicky measures" in many countries (2011).

Lynas argues that there is absolutely no basis for being concerned about the safety of nuclear power. The exclusion zones following accidents around Chernobyl and now Fukushima are unimportant. Radiation is not that dangerous, he claims: Although several thousand children got thyroid cancer from Chernobyl, thyroid cancer is treatable and only 15 children died. This figure is from the International Atomic Energy Agency, so it could easily be knowledge pollution by the atomic power industry. If, as Lynas claims, there were "only" 65 fatalities at Chernobyl, why did the normally not-that-caring Soviet Union evacuate 350,000 people? The WHO, a somewhat more neutral source, estimates 4,000 extra cancer deaths. For Lynas, the connection between increased radiation and increased cancer levels must be media hype.

We also discover from Lynas that nuclear waste is pretty harmless: He says most nuclear waste is more-or-less harmless after 40 to 50 years (Lynas, p. 179). Spin is spin, but this borders on criminality. Spent fuel rods typically need to be stored for 1000 years, and some fractional radioactive parts of fuel rods have half-lives of several million years. Besides, some forms of plutonium, a by-product, are also extremely poisonous and need to be sequestered permanently from the biosphere (Lai 2013).

If Lynas had not chosen to peddle nuclear industry untruths about the complete harmlessness of nuclear power, there would still be an argument that with all its flaws, nuclear power may still be needed to reduce carbon emissions. Respected climatologist James Hansen almost makes that argument (2009, pp. 192-204). Hansen also dismisses the serious problems with nuclear power because he believes there is a technological fix—what he calls "the new fast-reactor." Nuclear waste is accumulating and there are still no good solutions for needed long-term sequestration for hundreds of thousands of years. But according to Hansen, the fast reactor will burn it all up. Hansen must also deal with the inherent safety problems in the process. Nuclear chain reactions must run at about 2100 degrees to be sustained. This is done to boil water at 800 degrees to produce steam for turbines. 2100 degrees melts most metals and structurally damages all metals. Consequently, an outside mechanism must cool the area around the fuel so that it does not escape its confines. This is never a fail-safe operation. When the cooling fails, we get a *meltdown.* This is what happened at Chernobyl and Fukushima. Other means of heating water just stop heating when they run out of fuel. Hansen believes technology will allow reactors to shut down harmlessly. This is impossible. When the safety technology malfunctions, we have a meltdown. Third, when the inevitable but hopefully infrequent accident occurs, radiation creates damage over a widespread area, including contamination of large chunks of the biosphere. Hansen's belief that the nuclear industry has an "enviable safety record" indicates that he has not seen the records of individual reactors such as San Onofre in Southern California. In some cases at San Onofre essential safety equipment has been inoperative for periods of up to five years during which the plant was operating. It is just luck that there have not been more major accidents[9] (Hirsch 2012).

Unfortunately Hansen's argument that nuclear power is essential for reducing carbon emissions in a timely manner is weakened by his technophile arguments about nuclear safety. He notes that coal, the worst of the fossil fuels, tends to replace nuclear. He also notes that there is not much renewable energy generation and it is growing very slowly. Finally, he says, "The difficult part is that we must make the transition [from fossil fuels] with extraordinary speed if we are to avert climate disaster" (2009, p. 209). Whether nuclear power can be used to reduce carbon emissions will be discussed further in the next chapter.

The excuses of technophiles like Lynas (and Hansen) are somewhat like the excuses of the overfond parent of a badly behaved child: "He didn't do it, it was those environmentalists," or "If he did do it, he won't do it again," or "What he did wasn't so bad anyway." But technophilia even goes beyond the indulgence of a doting parent. It becomes a faith impervious to rational criticism: "If technology causes problems, technology can fix them. We can depend on technology even if we don't know the solution yet." The religious quality of this faith is underscored by the fact that the products of modern technology are almost certain to have unpredictable consequences for the biosphere, with no guarantee that a technological solution may not make things worse.

PROTECTING BIODIVERSITY THROUGH COOPERATION WITH CORPORATIONS

The eminent biologist E. O. Wilson in his *The Future of Life* (2002), characterizes humanity's current situation as expending and destroying the natural capital necessary for economic development—land, water, forests, fisheries, petroleum (pp. 149-150).

The problem is that this natural capital is finite and the human population continues to grow. Wilson is concerned with one consequence of this situation, namely the threat to biodiversity. The major threat to biodiversity is through habitat destruction, and the major cause of habitat destruction is human activity, often economic activity. Wilson lists a number of projects to preserve, protect, and restore existing habitat:

- Salvage hotspots (habitats at greatest risk with numerous vulnerable species)
- Keep intact the five remaining wilderness forests (Amazon, Congo, New Guinea, Canada, and Russia + Scandinavia)
- Stop all logging of old-growth forests
- Preserve lake and river ecosystems
- Complete inventory of world's biodiversity
- Develop a global conservation strategy
- Make conservation profitable
- Use biodiversity to benefit world economy (new drugs and plant species)
- Restore up to 50% of land to nature
- Have zoos breed more endangered species
- Support human population planning (pp. 160-164).

According to Wilson, governments will eventually have to take over a large part of these projects. Two of the projects concern making biodiversity projects profitable, and thus are more likely to be accomplished. However, coming up with those profitability ideas is not automatic. So we cannot depend on this happening, although it would be nice if it does.

At least five of the projects involve removing potentially economically productive land or water area from the economy. Wilson actually has little to say about slowing down our rapacious resource-intensive economy. But unless this happens, it is hard to see how any of five of his projects will happen: saving hotspots, wilderness forests, old-growth forests and water ecosystems; and returning land to nature. Of course, the project of reducing population will help reduce the economy. Of the remaining three projects, the two planning projects are fine—that is, to inventory the world's biodiversity and to develop a global strategy. But who will enforce the strategy? The last remaining project, to have zoos breed more endangered species, seems quite feasible.

Wilson sees *all three* of business, government, and technology cooperating in preserving biodiversity and engaging in conservation. There is a real danger in justifying concern for biodiversity and conservation by profitability. This justification concedes the priority of economic growth. Wilson's ethical justification is this: ". . . life forms around us are too old, too complex, and potentially too useful to be carelessly discarded" (2002, p. 131). But we still might very well *carefully* discard any number of life forms that prove potentially not profitable for us.

A better justification (which Wilson almost gives) is that all life is interdependent in complex ways. One indication—and this is important for the preservation of as many species as possible—is that ecosystems cannot be recreated artificially by humans.[10] So we do not know what is important in the ecosystem for our survival and we do not know what we can remove without potentially disastrous consequences for us. Not only cannot we tell what parts of the biosphere will be useful to us, we cannot even tell which parts will turn out to be essential for our survival.

Corporations may not cooperate with environmentalists in protecting the biosphere. Sometimes individual corporate CEOs balance profitability and environmental responsibility in a way that compromises neither.[11] From a corporate point of view, it makes no difference whether a CEO chooses an environmentally terrific plan or a plan producing large amounts of pollution, provided both are just as profitable. Sometimes the results are environmentally horrible. Pacific Lumber owned and preserved most major old-growth California redwood forests. A leveraged buy-out was financed by clear cutting the old growth redwoods.[12] So even if a corporation is run by a management *personally* committed to sustainability, the corporation's commitments can easily change drastically.

The real difficulty with approaching sustainability through biodiversity is that most of the threats to biodiversity stem from human economic activity. The impact of the economy on biodiversity must be addressed. And, as we have seen in the discussion of Lynas' views, expanding technology in the service of exploding population will not sustain the biosphere.

PLAN B 4.0: A DETAILED PLAN TO ACHIEVE SUSTAINABILITY

Lester Brown's *Plan B 4.0* is a coherent account of sustainability problems and a comprehensive program to solve those problems. As Brown sees it, the two main threats to sustainability are: Increasing population, causing resource degradation and competition; and climate change, with implications for energy and the food supply.

Plan B has four components:

1. Cut CO_2 emissions 80% by 2020
2. Stabilize population at 8 billion or less
3. Eradicate poverty
4. Restore earth's natural ecosystems (pp. 23-24)

These goals are completely interdependent. We cannot stabilize population without eradicating poverty; we can't restore ecosystems without cutting emissions and stabilizing population; we can't cut emissions without stabilizing population; and we can't eradicate poverty without restoring ecosystems (2009, p. 25). Logically, this implies that to accomplish any of the components, you have to do all of them. Brown makes the important comment that "the ambitiousness of this plan is not driven by perceived political feasibility but by scientific reality" (p. 24). Unfortunately it is far too common to think of environmental problems in the reverse order. Some contributors to the planetary boundaries report think that setting a boundary on carbon dioxide concentration is "too ambitious" for political reasons (Morris 2010). This policy of letting the agenda for human survival be based on short-term political calculations does not seem rational.

The goals of Plan B 4.0 are to be achieved largely by the judicious application of technology. Stabilizing the climate (by cutting CO_2 emissions 80% by 2020) is to be achieved by increasing energy efficiency, shifting to renewable energy, redesigning cities (to reduce energy use), and localizing food production (also to reduce energy use). Population stability is to be achieved through availability of contraception. Eradicating poverty cannot be done without stabilizing population. Poverty reduction also requires education, health, and food. In order to produce enough food to alleviate poverty, land and water productivity need to be raised, and protein production needs to become more efficient. In general, it helps to eat lower on the food chain. The main parts of earth's ecosystems to be restored are: Forests, soils, fisheries. Also biodiversity is to be protected. Forests aid in carbon sequestration and in protecting soils. Soil conservation and repair is essential. Climate and population stabilization are also essential for preserving biodiversity.

Plan B 4.0 is a theoretically feasible plan for ameliorating much conflict with the ecosystem. Brown had the right idea to determine what must be done rather than what is politically feasible. Even so, it is not irrelevant to consider how Plan B 4.0 could actually be realized. In the final part of the book, Section 3, "The Great Mobilization," Brown asks, "Can we mobilize fast enough?" The action items and budget are in two tables (see Tables 2 and 3).

To provide evidence for the practicality of this plan, Brown compares these annual costs to the annual US military budget. The annual US military budget is $607 billion; whereas the cost for all of Plan B other than carbon emission reductions is $187 billion.

Brown may not have provided costs for carbon emission reduction because he thinks it has to be done regardless of cost. But then since Brown believes all goals of Plan B must be accomplished, why provide costs for three of the goals but not carbon emission reduction?

The tipoff to the unreality of Brown's view of the implementation of Plan B is the comparison with the US military budget. It is not that the size of the military budget has escaped everyone's notice. A few progressives always try to reduce it and are met with howls from conservatives. It is necessary to ask why conservatives think this much money is necessary for the US military. They may be frightened that someone somewhere may come and take away their *loot*. Conservatives correctly conclude that since they got their loot by aggressive self-interest, someone

Table 2. Annual cost for Plan B carbon dioxide emissions reductions and sequestration in 2020

Action	
Energy Restructuring	**Amount (Millions of Tons Carbon)**
Replacing fossil fuels with renewables for electricity and heat	3,210
Restructuring the transport system	1,400
Reducing coal and oil use in industry	100
Biological Carbon Sequestration	
Ending net deforestation	1,500
Planting trees to sequester carbon	860
Managing soils to sequester carbon	600
Total Carbon Dioxide Reductions in 2020	7,670
Carbon Dioxide Emissions in 2006	9,350
Percent Reduction from 2006 baseline	82%

From Table 10-1 in Brown (2009, p. 254)

Table 3. Added annual costs for social goals and earth restoration

Goal	
Basic Social Goals	**Funding (in Billions of Dollars)**
Eliminating poverty	57
Stabilizing population	20
Subtotal	77
Earth Restoration Goals	
Planting trees for carbon sequestration	17
Cropland soil protection	30
Restoring range land and fisheries	22
Stabilizing water tables	10
Protecting biodiversity	31
Subtotal	110
Grand Total	187

Condensed from Table 10-2 in Brown (2009, p. 263)

else also aggressively self-interested could take it away. Unfortunately, these views have led to a series of completely unnecessary wars that were extremely costly in innocent lives and in money.

Funds spent on converting Iraq into an unstable state with warring Muslim factions, could have funded Brown's social and environmental goals for several decades.

This kind of budget transfer will never happen.[13] In making this kind of proposal, Brown has strayed from his announced basis of scientific reality into the realm of political feasibility. Conservatives are single-minded in defending freedom as they understand it, which is the unencumbered freedom to make money. Note that conservatives have been fighting to have regulations removed banning known poisonous pollutants such as mercury. How can we expect them to get enthusiastic about reducing carbon dioxide? A recent study showed that conservatives are not as intelligent as liberals, and conservative statements about carbon dioxide confirm this (Waugh 2009). Michelle Bachman and Rick Santorum, 2012 Republican presidential candidates, argue that we do not have to worry about carbon dioxide because it is not a pollutant. The task of getting action on carbon emissions is almost insuperable when a fair number of politicians lack the brain power to understand the situation.

Brown outlines three models of social change: "Pearl Harbor," where a catastrophe prompts action; "Berlin Wall," where long-simmering grassroots unhappiness finally erupts; and "sandwich" where grassroots and national leadership are congruent. Brown, writing shortly after Obama's election, thinks that grassroots and the

Obama administration are congruent. Four years later, absolutely no legislation on carbon emissions has passed, and Obama contributed to the failure of the climate talks in Copenhagen in 2009. No sandwich.

"Pearl Harbor" may be the only realistic option. It will take massive climate-related fatalities in the United States to prompt action. The 90,000 heat-related deaths in Europe did not cause a stir in the US. One can imagine a senator from Oklahoma saying, "Too bad that those European guys didn't have the sense to put in air conditioners." A real senator from Oklahoma, Jim Inhofe, declared a few winters ago that a heavy snowstorm in Washington DC refuted global warming[14] (Wolf 2010). Once again, what can one do with those lacking brain power?

Corporations who largely fund conservative politicians are worse. They have the brainpower to perpetrate massive knowledge pollution. And the situation regarding fossil fuel corporations is worse still. Mark Lynas' (2011) account of how the Montreal Protocol saved the ozone layer is a valuable lesson about corporate behavior. Dupont, who manufactured the chlorofluorocarbons destroying the ozone layer, fought tooth and nail to keep making them. Eventually a safe profitable replacement was discovered. Lynas implausibly believes it was the threat of strong international regulation that caused the manufacturers to change their course and not the appearance of profitable alternatives (p. 222). The Montreal Protocol itself, which saved the ozone layer, was touch-and-go. Numerous countries fought against a ban to protect their own chemical industries. So it is hard to see how a strong threat of international regulation would have much effect.[15]

It is natural to think of the repair of holes in the ozone layer as analogous to the repair of holes in a roof. You fix the hole in the roof, and that is it. No more rain comes in. However, there turn out to be other causes of holes in the ozone layer. More powerful storms activated by global warming have begun to punch holes in the protective barrier called the *tropopause* (The characteristic anvil shape of thunderstorms occurs when storms reach the tropopause and spread out just beneath it.). But stronger storms go right through this barrier into the ozone layer, and the additional water vapor greatly increases the activity of the remaining chlorofluorocarbons in the ozone layer. Scientists say this process was a surprise and are mainly concerned for two reasons: Since the injection of water vapor into the ozone is caused by global warming, it is irreversible. Second, holes in the ozone layer produced by injection will occur over the densely populated areas that experience thunderstorms, rather than just over the poles (Fountain 2012). So fixing the technology causing an ecological problem does not guarantee that some other technology will later cause the same problem.

These are the lessons to apply to the effort to reduce carbon emissions. The fossil fuel companies are the largest and most powerful corporations on the planet. There is no reason to believe they would be cowed by any international threat. By contrast, it was the US that led the way in saving the ozone layer. There is no hope whatever that the US, the second largest emitter of carbon in the world, will take any action on carbon emissions in the foreseeable future. This reinforces my case that it will take a sizable catastrophe to cause any change. Of course, by then the consequences are going to continue to be bad. But perhaps the goals of Plan B will be achieved in a brutal way.

Four related questions about Plan B will be discussed:

1. Why is there no discussion of limits on economic growth?
2. The plan makes extensive use of technology. Is it possible to use technology only in a way that does not conflict with the ecosystem? How much modern technology is needed?
3. The only item in the plan which is not primarily for human benefit is biodiversity. Is the rest of the plan consistent with protecting biodiversity?
4. Where does the plan stand with respect to human superiority?

1. Plan B's Limits on Economic Growth

My best guess is that Brown believes that if the economy is restructured so as not to produce gross damage to the ecosystem, and that if population is stabilized, economic growth can continue without other limits. This position is somewhat similar to Mark Lynas' position on the economy, except Lynas does not believe that we need to do anything to limit population, nor that we need to restructure the economy. But regardless of the large difference between the two positions, both ignore the assault on the ecosystem by technology in the service of the economy. The economic reforms in Plan B are primarily directed at reducing carbon emissions by switching to renewable energy. Plan B does reject nuclear power, but for economic reasons, not environmental ones (pp. 111-112). The words 'pesticide' and 'pollution' occur nowhere in *Plan B 4.0,* but Brown does want markets to be reformed to reflect "ecological truth," that is, ecological externalities.

I conclude that Plan B would at best resolve some current conflicts with the ecosystem. Without a more radical restructuring of the economy, resource depletion will remain an ongoing problem. And without some modification of our deployment of technology, unexpected side effects are likely to continue causing environmental crises. There will be more said about technology in the answer to the next question.

2. Plan B's Use OF Technology AND Modern Technology

The critical feature of modern technology is its inherent propensity to produce unintended side effects which cause serious damage to the ecosystem. This is not a correctible flaw in modern technology. For modern technology, everything is resource, including the entire ecosystem and all its processes. It has no respect for the ecosystem. The entire conception of genetic engineering is ample evidence that modern technology has no place for natural processes, even those as important as natural selection.

On the other hand, modern technology is indispensable for the scientific instruments necessary for understanding our place in the ecosystem and for appropriate action concerning the ecosystem. The sophisticated instruments necessary for measuring our impacts on the ecosystem are products of modern technology. Also, at current levels of population, our food supply is dependent on modern technology.

Sorting out the role of technology in a truly sustainable world is a complex task which will be attempted in the next chapter. In this chapter, there is a brief sketch of the technological requirements for Plan B. Stabilizing the climate by increasing energy efficiency and renewables, requires the most advanced technology possible. All the remaining components of Plan B require either low-tech equipment or policy changes or both. Policy changes are what is needed to implement redesigning cities, localizing food production, stabilize population, or eradicating poverty. Some technological assistance would be needed to implement the education and health components of eradicating poverty. For increasing food production, raising land and water productivity requires low-tech techniques such as double cropping. Policy changes would produce more efficient protein production—less beef and pork, more poultry, and aquaculture of herbivorous fish. Also using soy rather than grain for animal feed would be a more efficient policy for producing protein. Restoring earth's natural ecosystems would require conservation policies and low tech techniques, such as metering water pumped from nonrenewable aquifers,

3. Does Plan B Sufficiently Protect Biodiversity?

Brown states that the essential steps necessary to protect diversity are stabilizing population and stabilizing the climate. These are clearly necessary conditions, but they are not sufficient. Both E. O. Wilson and Dave Foreman believe it is necessary to set aside much more land for non-human use. If population is stabilized at 8 billion and those 8 billion are raised from poverty, there will be increased pressure for humans to make use of more land. Mark Lynas also believes more reserves are necessary to preserve biodiversity, but he believes market forces can be tweaked to accomplish this. There will be more on this topic in the following discussion of Foreman's views.

4. Where Does Plan B Stand ON Human Superiority?

Plan B clearly assumes human superiority. Plan B's treatment of biodiversity includes the view that the planet is for human use. If we reduce our numbers and avoid wrecking the climate, that is all we have to do for the living things we share the planet with. This version of sustainability is not sustainable. As was pointed out previously, we do not fully understand our dependence on the ecosystem and are likely never to understand it enough to be able to manipulate it with impunity.

Considering purely hypothetical scenarios is usually a waste of time. *One* purely hypothetical scenario will be considered: That humans are *not* dependent on the ecosystem and therefore need have no interest in preserving it. In this case, to save the ecosystem we would have to fall back on the feeling that earth comes first and that humans do not have priority over everything else on the planet. As noted in the preceding chapter, there are difficulties basing policy on such feelings. But fortunately this scenario is purely hypothetical and we can base policy on ecological science rather than feelings. Feelings can, however, reinforce the correct attitude against human superiority.

One of the advantages of a scientific worldview, including evolution, is that one avoids the delusional view that humans are not part of nature, that humans are not basically animals like other animals. So-called intelligent design theories seem to have human exceptionalism as an ulterior motive. As Richard Dawkins (2006) has pointed out, intelligent design explains absolutely nothing, because there are no coherent details about how God designed anything.[16] Of course, if God is a big guy in the sky, like us but more powerful and not in an earthbound location, it is possible to make some statements about his actions and intentions. But this God is obviously a human projection. This God did not make man in his own image; rather, man made this God in his image.

OVERALL ASSESSMENT OF PLAN B 4.0

On the basis of the answers to the questions, Lester Brown's Plan B 4.0 is most concerned with *immediate* threats to human civilization. He apparently believes that civilization can survive just fine if we meet immediate challenges and institute appropriate reforms. Altering markets to reflect environmental externalities is an important reform. But this reform does not touch the problems involving humans overwhelming the rest of nature. It is also clear that Brown does not see the danger in technology at all. So if we manage to stabilize the climate and the ecosystem, we are still completely vulnerable to create still further environmental crises.

Without taking steps to give up beliefs in human superiority, we are still asking for ecosystem collapse. Plan B is perhaps weakest in its provisions for biodiversity, probably because of its tacit acceptance of human superiority. Views which give priority to the rest of the ecosystem will now be considered.

EARTH FIRST! AND SUSTAINABILITY

In the preceding chapter, the strong feelings of Dave Foreman, a founder of Earth First!, were noted. The slogan of the movement is: 'No compromise in the defense of Mother Earth!' Foreman feels that establishment environmental organizations such as The Wilderness Society compromise nature in favor of human beings. Foreman asserts: "Human beings are not paramount. Earth is not for Homo sapiens alone; human life is but one life form on the planet and has no right to take exclusive possession" (Foreman 1991). Some proponents of animal rights make somewhat similar statements. They maintain that humans should leave other animals completely alone. This claim and other animal rights issues will be discussed at length in Chapter 13, "Human-Animal Relations."

It is tricky to propose that human livelihood and welfare be compromised in favor of the rest of the ecosystem. Loggers, for example, tend to be hostile to environmental concerns. Recently, in Mexico, fishermen were bought out so they wouldn't fish and exterminate the world's smallest porpoise in the Sea of Cortez. One fisherman asked, how could the government prioritize a porpoise over a human being? (Elton 2011).

The problem is that human beings actually *are* an animal species and must make demands on the ecosystem in order to survive. Striking an appropriate balance is not that easy. But if the guiding principle is that human development is unlimited and takes precedence, there is no question that the rest of the ecosystem will lose out. The current massive extinction by humans is proof. An appropriate balance must include two components: Segregating functioning ecosystems from human encroachment; and putting limits on human development.

But segregation and development limits will have no force unless human population is stabilized. Virtually all commentators on the environment see that population stabilization is necessary if any other environmental goals are to be achieved. Lester Brown in Plan B suggests 8 billion as a good target. Joseph Bookstein believes 4 billion would be the right number for stabilizing the climate.[17] In the preceding chapter it was argued that there was no value in merely increasing human numbers just for the sake of numbers. The 1800 human population of 1 billion was certainly at least as creative as our 7 billion. It is hard to see that any value would be added to a global population of more than one or two billion. Two billion maximum is the number favored by Dave Foreman.

A reduction to 4 billion could probably be achieved by voluntary means together with education, incentives, and propaganda.[18] For this reason, it is a possible target—doable, and yet sufficient to enable environmental goals.[19] In a later book, *Man Swarm* (2011), Dave Foreman discusses the issue of population limitation. He spells out the near-impossibility of achieving biodiversity goals without population stabilization. He also outlines how the goal of population stabilization has been derailed by leftists, feminists, and technophiles. He quotes one feminist as saying that a mother has the right to have as many children as she wants (Foreman 2011, pp. 128-129). This feminist would find most vermin such as roaches or mice in complete agreement (At least baby mice are cute.). Of course such a mother of a big brood will have appropriate instincts to care for her brood, but not the intelligence to realize they are likely not to have good lives.

But then near the end of his discussion, Foreman goes a bit off the rails. The issue is immigration. According to Foreman, the US was well on the road to a stable population when the Immigration and Nationality Act of 1965 dramatically increased population (McFayden 2012). If one reads commentary on the 1965 act, there was no stated intention to increase the numbers of immigrants. Rather, the act tried to remedy the previous bias in favor of Northern Europeans. However, Foreman notes that after 1995, the number of immigrants to the US increased from 200,000 per year to 1,000,000 per year. This increase prevented population stabilization from happening. So this significant population increase was unplanned, unintended, and unnoticed.

Later illegal immigration was definitely noticed. Foreman correctly acknowledges an important cause of illegal immigration from Mexico, namely NAFTA's consequence that small Mexican farmers could no longer compete with *subsidized* American corn. This is not greater American efficiency, it is greater American farm industry political clout. In fact, American farm subsidies are illegal according to the World Trade Organization.

So we have illegal immigrants who come here because of illegal subsidies. At very least, if we take action, we should do something about both. Foreman proposes a reasonable course of action for legal immigrants: Reduce the number. However, for illegal immigrants, his recommendations range from draconian to out-and-out fascist. First, he calls for absolutely no amnesty because that will encourage more illegals. Have all employers make sure their employees are legal. Jail employers who hire them. Further, no drivers licenses, post office boxes, business licenses for illegals. Deport them immediately and bar them permanently from the US. Foreman believes draconian penalties will be effective, "make lawbreaking hurt" (p. 232). Well? The US has draconian drug penalties and a resulting per capita jail population five times that of any other country. Making it hurt is just a redneck sadist fantasy.

It is hard not to notice the difference in suggested treatment for legal and illegal immigrants. Both contribute to population growth. Most studies show that both have a positive effect on the economy. Yet illegals are to be treated worse than child molesters? What have they done but try to get out of intolerable home conditions? In any case, Foreman is in no position to call for punishing lawbreaking till it hurts. His favored method for dealing with bad environmental actions was called 'monkeywrenching,' deliberately breaking the law. Monkeywrenching is explicitly not civil disobedience, which is an act to address the sense of justice of the public. Instead, monkeywrenching is a deliberate individual illegal act to damage something that produces bad environmental actions[20] (1991, chs 11-13).

For the purposes of this book, a solution to the illegal immigrant problem is not necessary. But Foreman's negative solutions clearly indicate he is thinking in terms of the US as an isolated fortress. No environmental problems can be addressed successfully that way. To be fair, he does recommend; 'mak[ing] homelands better than going to the US' (p. 193). In the case of Mexico and Central America, that would involve doing something about our supplying weapons to the drug cartels. Mexico internally has strict gun controls—each family is allowed one gun for hunting or protection. Along the Rio Grande border with Texas, There are 3 gunshops on the Mexican side. On the Texas side, there are over 3000 (Gay 2010). With its gun situation, the US is not a civil society. As the social theorist Hobbes (1651) pointed out, if everyone can use lethal force, you have a war of all against all, where human life is nasty, brutish and short. The US more and more fits this description.

WHAT IS SUSTAINABILITY?

At the beginning of this chapter, sustainability was characterized as humans maintaining equilibrium with the ecosystem. Other prevailing conceptions of sustainability characterize it in human terms. A popular definition of sustainability was given at a 1987 UN conference. Sustainable developments were defined as those that "meet present needs without compromising the ability of future generations to meet their needs" (WCED, 1987). Robert Gillman elaborates on this definition as follows: "sustainability refers to a very old . . . concept (The Golden Rule)...do unto future generations as you would have them do unto you."

Architects Bartuska, Kazimee, and Owen (1999) expand the UN definition as follows:

Sustainable developments are those which fulfill present and future needs (WCED, 1987) while [only] using and not harming renewable resources and . . .human-environmental systems . . . [involving air], water, land, energy, and human ecology . . . or . . . other sustainable systems (1999).

If we bring human activity into equilibrium with the ecosystem, then "resources' will be available for future generations. But if resources are available for future generations, will human activity then be in equilibrium with the ecosystem? It is likely but not certain. If resources are not squandered and continue to be available for future generations, protecting and preserving the ecosystem would be the easiest way to keep resources available over a long period of time. But perhaps we decide to go all nuclear power, maybe setting aside a continent to bear the brunt of radiation problems. Whatever continent we pick, its ecosystem is damaged or destroyed.

The mention of resources in the definition should be a red flag. In this context, one question is how many generations do we need to preserve resources for? If indefinitely, then protecting and preserving the ecosystem has got to be the way to go. On earth, after all, there are limited amounts of lots of resources. Besides air, water, and soil, fossil fuels are finite. So are many of the metals and minerals we employ. There are technophiles who counter we can begin exploiting other worlds in space![21] But the background problem is the view of everything as nothing but resource. We meet here again modern technology with its single-minded view. From the point of view of modern technology, the ecosystem is just another collection of resources. And then we proceed to destroy the biosphere because our dependence on it is not visible to modern technology. So long as modern technology holds sway, sustainability as equilibrium with the ecosystem is impossible.

Although modern technology is clearly not sustainable, nevertheless we cannot blithely give it up. This issue will be discussed in the next chapter. It is also clear that economic growth is not sustainable. Exactly what to do instead is a difficult question addressed in the following chapters. But when economic growth depends upon nonrenewable resource extraction, we eventually run out. And in later stages of resource extraction, more complex and risky procedures make environmental catastrophes more-or-less inevitable. It is easy to see how this works with oil. BP's Gulf of Mexico disaster involved drilling one mile below the surface of the sea. What happens if something goes wrong? Almost impossible-to-control oil contamination of the ocean. Similarly shale oil and gas extraction ('fracking') easily leads to drinking water contamination.

Could it also be that civilization is not sustainable? Discussion of this question will be postponed to Chapter 13, "Regaining Our Place as a Species among Species." Some of the material necessary to answer that question will not be available until then, especially the material in the next chapter on the role of technology in sustainability.

However, defining sustainability in terms of what we owe to future generations is a valuable step away from modern technology. Kids are not resources, they are our extensions over time because we are mortal.[22] Modern technology has no way of taking notice of mortality and the meaning of descendants. Corporations, its agents, do not have children. This is a fundamental way in which they are not people. So corporations cannot allow for future generations. Because corporations as agents of modern technology cannot have any reason to plan over long periods of time, and because their aim is profitability to shareholders, they tend to plan for short periods of time such as the next quarter or fiscal year. Any longer planning must be imposed on them.

Al Gore has recently proposed a way to accomplish this. Together with David Blood, former CEO of Goldman Sachs Asset Management, he has formulated a blueprint for 'sustainable capitalism.' This blueprint includes:

- Ending the default practice of issuing quarterly earnings guidance, which results in managing and investing for the short-term.
- Identifying and incorporate risk from stranded assets, whose value would dramatically change when large externalities are taken into account—for example, subprime mortgages. Appropriate adjusted pricing for carbon or water could produce stranded assets..
- Aligning compensation structures with long-term sustainable performance. Hold asset managers and corporate executives accountable for the ramifications of their decisions over the long-term and pay out financial rewards over the long term.
- Incentivizing long-term investing with loyalty-driven securities that reward investors for holding securities for at least three years.
- Mandating integrated reporting. Companies are encouraged to integrate both their financial and Environmental, Social, and Governance (ESG) performance into one salient report. If companies refuse to issue such reports, they may be thought to have problems with stranded assets (Gore & Blood 2012).

The two principles directly relevant to the environment are the ones about reporting externalities and environmental reporting. These are clearly helpful. All these principles encourage more rational and hence more effective behavior on the part of corporations. Compliance with these principles will help the environment. But their adoption is more-or-less independent of any other sustainability actions. There is nothing wrong with this. Independent actions are one solution to the Piecemeal Problem. We mostly do not have to worry about how such actions fit into any broader plan of action for sustainability or saving the environment. The next set of actions to be discussed, recycling, is sometimes this kind of independent action.

RECYCLING AND INDIVIDUAL ACTION

Processes of modern technology produce substances such as plastic which do not easily return to the environment. Many of these products are said to be non-biodegradable. If left to persist in the environment, they cause large amounts of damage of various kinds. In Chapter 2, "Conflicts with the Ecosystem," a number of examples of toxic substances and environmental damage associated with these products will be discussed. One way to ameliorate these problems is through recycling. The idea is to reclaim the original resource materials from discarded items, so as to reduce the burden on the planet in producing these items. The success of recycling is usually measured in terms of the energy reduction in producing new products. This is a modern technology point of view: Energy is a resource available for utilization for other processes of modern technology.

However, reduction of the need for primary materials also typically means reduced ecological damage. Extraction and refinement of metals and minerals nearly always requires severe damage to habitat for almost all species, as well as the release of highly toxic chemicals to the environment. Most metals are in principle completely recyclable (Biello 2012b). So a lower percentage of a metal that is recycled is a measure of unnecessary damage to the ecosystem. Fourteen metals have a recycling rate above (but not much above) 50%: Aluminum, cobalt, chromium, copper, gold, iron, lead, manganese, niobium, nickel, palladium, platinum, rhenium, rhodium, silver, tin, titanium, and zinc. There are 46 other metals used in making products which have low to nonexistent recycling rates (UNEP 2011). So even though these fourteen have much better recycling rates than the remainder, there is in principle a lot of room for improvement.

An example is neodymium, a rare earth metal used in batteries, is almost never recycled. This, despite the fact that neodymium is involved in the so-called 'rare earth crisis' with China. As David Biello notes,

Mining for neodymium is not benign (which is why the world lets China monopolize its production). And it's not just neodymium. Mining waste or tailings, leach ponds, slurries and the like are among the world's largest chronic waste problems. North

America alone produces 10 times as much mining waste as it does the municipal solid waste . . . from all the neighborhoods in the U.S. (Biello 2012b).

There is not that much an individual consumer can do about this kind of recycling, although it does no harm for them to reduce their contribution to municipal waste. Table 4 shows energy and related savings per recycled consumer item.

All but plastic containers are biodegradable products. The metals would eventually return to the earth, but it would be better to recycle them. As we noted earlier, recycled plastic bags do not become plastic bags. Neither do plastic containers. Both become lower-grade plastic which eventually ends up in landfills. Plastic container manufacturers have succeeded in defeating laws requiring them to use some recycled material in their products (Ecology Center 2012).

So although recycling helps reduce primary resource consumption and energy consumption, 94% of materials in production processes become waste before we even see the product (Citizens' Clearinghouse on Waste Management 2004). A major benefit of consumer recycling is reducing the waste stream in landfills. This is a benefit to the environment even if recycling costs more. New York City in 2002 terminated recycling glass and plastic because on their analysis it cost more than dumping. When dumping costs rose, they began recycling again (Earth Talk 2012). New York City may have considered more than just costs when they stopped recycling glass and plastic. The market for recyclables had diminished to the point where the recycled items were ending up in landfills anyway. This is an excellent reason for stopping recycling. Cost has to be a secondary issue.

It is clear that most waste is not recycled. Waste that is not recycled goes into landfills or is incinerated. Incineration can add to air pollution and carbon emissions. Also, it helps when products are recycled so that the need is reduced to use more resources to produce more of the item. As we saw, plastic bags or containers cannot

Table 4. Saving through recycling[23]

Item recycled	Energy Saved by Recycling	Pollution Reduction	Waste Reduction
Aluminum cans	96%		
Glass bottles	80%		
Cereal boxes	50%		
Paper		75% air 35%water	60% less water
Scrap iron to steel		86% air 76% water	97%
Plastic containers	Don't recycle	0%	0%

be recycled into more plastic bags or containers. So 'recycling' does not solve many waste management problems, although it is usually clearly better to recycle than not.

Companies also believe consumers are aware of such differences. Baja Fresh Mexican Grill lines its serving trays with a statement that they are "saving the planet one burrito at a time." They mention their biodegradable plates, bags made from 50% recycled materials, and unbleached compostable burrito wraps (2011).

Other individual choices can matter. Researchers at the Potsdam Institute for Climate Impact found that by 2055 the emissions of methane and nitrous oxide from agriculture could be cut by more than eighty percent. Methane and nitrous oxide comprise about 15% of greenhouse gases. This reduction could be achieved by reducing the consumption of meat and dairy products and improving agricultural practices (ScienceDaily 2010). The effect of reducing consumption of meat and dairy was much more significant than improving agricultural practices. A reduction of 15% per decade or 1.5% per year would be enough to achieve the 80% result.

Individual action by itself, whether recycling or eating lower on the food chain, may not have that much effect. So, what should the individual do when his or her individual actions to help the environment may have an effect, but a small one? This brings us back to the Piecemeal Problem.

THE PIECEMEAL PROBLEM

There seems to be no choice besides approaching the environment and sustainability piecemeal. The planetary boundaries project at least addresses the problem but the project has internal problems. Plan B 4.0 attempts to set out a coherent plan of action. But there too problems of prioritizing actions arise. Perhaps the best we can do is to separate out actions which have the greatest claim to be done.

Three kinds of actions worth considering are these:

1. Actions necessary to prevent biosphere collapse;
2. Actions toward goals to prevent crossing negative thresholds; and
3. Actions that would help but are independent of other actions.

Examples of actions necessary to prevent biosphere collapse are reducing carbon emissions, taking steps to prevent an asteroid collision, and preserving biodiversity. Reducing carbon emissions to ameliorate climate change is also a goal to prevent crossing a negative threshold. Beyond a certain point, nothing can be done to prevent radical climate change. Examples of actions that would help no matter what else is done are recycling and Al Gore's revision of corporate governance away from only short term considerations.

For the first category, actions necessary to prevent biosphere collapse, the appropriate decision rule is the maximin rule. The maximin decision rule rather than an expected value decision rule is appropriate when there is a possibility of a seriously bad alternative, such as, for an individual, their death. The maximin rule directs you to choose the alternative which yields the best payoff when the worst happens. Whereas, expected utility directs you to choose the greatest expected payoff. Suppose you decide to rob a bank. Your payoff matrix is shown in Table 5.

Don't rob the bank gives you the best payoff if the worst happens. The greatest expected utility payoff depends on the probability of being caught. But using expected utility, there will be some probability and some dollar amount large enough to give robbing the bank the greater sum of utility even if the other alternative is death. The maximin is rational only when the choices make a major difference. For more usual choices, expected utility is usually the most rational choice(see Appendix A).[24] Therefore, when a choice is necessary to prevent biosphere collapse, it is irrelevant to consider the payoff. And therefore, actions in this category should have highest priority.

Actions in the second category (to prevent crossing negative thresholds) may or may not be as serious as collapse of the biosphere. For most biological species in the wild, there is a threshold below which the species will go extinct. Captive breeding can be tried, but again if there are not many individuals left, it is not clear that even captive breeding can sustain the species. Climate change is like this too. There is a threshold which, once passed, there is nothing to be done to prevent radical climate change and its effects. Where there is a threshold and it has been passed, then the piecemeal prescription is to cease working on those goals and move on to something else. However, in some cases the negative effects may prevent most actions.

The final category of actions, those that are helpful pretty much regardless of whatever else happens, are those which individuals are free to do if they have enough spare time. Recycling is an action in this category. So would be buying a lower-mileage or electric car. The hitch is that most independent actions won't impact large goals unless a large number of people do them. And often we have no way of telling this. So, from an eco-ethical point of view, it is good to keep doing these actions unless they interfere with higher-priority actions.

Table 5. Payoff matrix, maximin

	Get Caught	Don't Get Caught
Rob Bank	Shot to death	Get $1,000,000
Don't Rob Bank	Security guard waves to you	Life as usual

CONCLUSION

True sustainability is hard to come by. We can do things to make sustainability more likely. But equilibrium of human activity with the ecosystem will require more than just sustainability of current human activities. In the following chapters, we will see what more than sustainability is required, and how much of that is compatible with human activity as we now practice it.

REFERENCES

Barnofsky, A. et al. (2012, June 7). Approaching a state shift in Earth's biosphere. *Nature*, 486.

Barrow, J. (2006). Glyphosate and cancer. *Australian Plants Online*. Retrieved 4/17/12 from anpsa.org.au/APOL20/dec00-3.html

Bartuska, T., Kazime, B., & Owen, M. (1999). *Defining sustainability*. Retrieved 3/14/12 from www.arch.wsu.edu/09%20publications/sustain/defnsust.htm

Biello, D. (2012a, June 13). Walking the line: How to identify safe limits for human impacts on the planet. *Scientific American*. Retrieved 6/13/12 from news.yahoo.com/walking-line-identify-safe-limits-human-impacts-planet-120100550.html

Biello, D. (2012b, August). Recycling reality: Humans set to trash most elements on the periodic table. *Scientific American*. Retrieved 8/10/12 from news.yahoo.com/recycling-reality-humans-set-trash-most-elements-periodic-220200185.html

Bookstein, J. (2008). *Global warming: The population connection*. [Broadcast]. La Jolla, CA: UCTV. Retrieved 2/4/12 from www.uctv.tv/search-details.aspx?showID=14886

Brown, L. (2009). *Plan B 4.0*. New York: Norton.

Citizens' Clearinghouse on Waste Management. (2004). *Environmental problems from municipal solid waste*. Retrieved 3/17/12 from www.citizenswasteinfo.org/A559CA/ccwm.nsf/5155d8f53ce25d2785256cc300567828/c050375504a2a7508 5256e7d005a325f?OpenDocument

Dawkins, R. (2006). *The god delusion*. New York: Houghton Mifflin.

Diamond, J. (2002). *Collapse*. New York: Viking.

Earth Talk. (2012). *Do the benefits of recycling outweigh the costs?* Retrieved 3/18/12 from environment.about.com/od/recycling/a/benefit_vs_cost.htm

Ecology Center. (2012). *Seven misconceptions about plastic and plastic recycling*. Retrieved 3/17/12 from www.ecologycenter.org/ptf/misconceptions.html

Elton, C. (2011). Safety net. *Audubon Magazine, 113*(6).

Energy Justice Network. (2012). *Dioxins & furans: The most toxic chemicals known to science*. Retrieved 4/17/12 from www.ejnet.org/dioxin/

Foreman, D. (1991). *Confessions of an eco-warrior*. New York: Harmony Books.

Foreman, D. (2011). *Manswarm: The killing of wildlife*. Durango, CO: Raven's Eye Press.

Fountain, H. (2012, July 26). Storms threaten ozone layer over U.S., study says. *New York Times*. Retrieved 8/10/12 from nytimes.com/2012/07/27/science/earth/strong-storms-threaten-ozone-layer-over-us-study-says.html

Gay, M. (2010, December 13). Texas biggest gun source for Mexican drug war. *Huffington Post*. Retrieved 4/17/12 from www.aolnews.com/2010/12/13/report-texas-biggest-gun-source-for-mexican-drug-war/

Gillam, C. (2012, March 10). Scientists warn EPA on Monsanto corn rootworm. *Reuters*. Retrieved 3/10/12 from news.yahoo.com/scientists-warn-epa-monsanto-corn-rootworm-193441402.html

Gore, A., & Blood, D. (2012). *A manifesto for sustainable capitalism*. Retrieved 3/17/12 from www.generationim.com/media/pdf-wsj-manifesto-sustainable-capitalism-14-12-11.pdf

Halweil, B. (2012). Can organic farming feed us all? *Worldwatch Institute*. Retrieved 2/12/12 from www.worldwatch.org/node/4060

Hansen, J. (2009). *Storms of my grandchildren*. New York: Bloomsbury USA.

Hirsch, D. (2012). *Nuclear power: Ethical and policy implications*. Retrieved 4/17/12 from www.uctv.tv/search-details.aspx?showID=23504

Hobbes, T. (1651). *Leviathan*. Retrieved 5/20/04 from darkwing.uoregon.edu/~rbear/hobbes/leviathan.html

Lai, L., & Morrison, K. (2013). *Nuclear energy fact sheet*. Retrieved 4/13/13 from www.wagingpeace.org/menu/issues/nuclear-energy-&-waste/nuclear-energy-fact-sheet.php

Lynas, M. (2011). *The god species*. Washington, DC: The National Geographic Society.

McFadyen, J. (2012). What is the immigration and nationality act? *About. com Immigration Issues.* Retrieved 3/13/12 from immigration.about.com/od/immigrationlawandpolicy/f/INA.htm

Morris, A. (2010, April). Climate change in Foley, J., "Boundaries for a Healthy Planet. *Scientific American.*

Nadeau, R. (2006). *The environmental endgame.* New Brunswick, NJ: Rutgers University Press.

New Economics Foundation. (2009). *The happy planet index 2.0.* Retrieved 3/4/12 from www.happyplanetindex.org/public-data/files/happy-planet-index-2-0.pdf

Rockström, J., et al. (2009). Planetary boundaries: Exploring the safe operating space for humanity. *Ecology And Society, 14*(2). Retrieved 2/11/12 from www.ecologyandsociety.org/vol14/iss2/art32/

ScienceDaily. (2010). Conscious choice of food can substantially mitigate climate change, research finds. *ScienceDaily.* Retrieved 3/18/12 from http://www.sciencedaily.com /releases/2010/06/100628075744.htm

Sewell, A. (2012, June 18). Environmental group demands full San Onofre review. *LA Times.* Retrieved 7/3/12 from latimesblogs.latimes.com/lanow/2012/06/san-onofre-friends-of-the-earth-embargoed-until-7-am-monday.html

United Nations Environmental Programme (UNEP). (2011). *Recycling rates of metals: A status report.* Retrieved 8/12/12 from www.unep.org/resourcepanel/Portals/24102/PDFs/Metals_Recycling_Rates_110412-1.pdf

Walsh, J. S. P. (2012, March 2). The final environmental frontier: Space development and its consequences. *Scientific American.* Retrieved 3/21/12 from news/yahoo.com/final-environmental-frontier-space-development-consequences-160000428.html

Waugh, R. (2012, February 8). Right-wingers are less intelligent than left wingers, says study. *Daily Mail Online.* Retrieved 4/17/12 from www.dailymail.co.uk/sciencetech/article-2095549/Right-wingers-intelligent-left-wingers-says-controversial-study—conservative-politics-lead-people-racist.html

Wikipedia. (2012). *Planetary boundaries.* Retrieved 3/4/12 from en.wikipedia.org/wiki/Planetary_boundaries

Wilson, E. O. (2002). *The future of life.* New York: Vintage Books.

Wolf, Z. B. (2010, February 11). Inhofe uses blizzard to refute global warming. *ABC News*. Retrieved 4/17/12 from abcnews.go.com/blogs/politics/2010/02/inhofe-uses-blizzard-to-refute-global-warming/

World Commission on Environment and Development. (1987). *Report of the world commission on environment and development: Our common future*. Retrieved 3/16/12 from www.un-documents.net/wced-ocf.htm

ENDNOTES

1 Plastic bags are singled out because they do not recycle to more plastic bags. They do recycle to usable lower-grade plastic. But this fact means that recycling plastic bags does not reduce the amount of plastic bags in circulation. And plastic bags are serious pollutants.

2 A less obvious omission is biotechnology, Clearly those producing the boundaries did not think biotechnology with mentioning. Biotechnology is discussed further in Chapter 11, "The Role of Science and Technology."

3 Biello (2012a) contains further criticisms of the Planetary Boundaries along these lines. Barnofsky et al. (2012) take a somewhat different approach. They are looking for signals that radical environmental change is about to happen. However, the complexity of interactions leads them to conclude that the best course is to reduce human effects on the biosphere. These include reducing population growth and resource consumption, reducing fossil fuel use, reducing human use of land, and maintaining biodiversity.

4 I hope this is an acceptable substitute for the traditional racist simile. It is surprising that there does not seem to be a totally different phrase with the same intent. The early 20th-century comedian W. C. Fields substituted "An Ethiopean in the fuel supply," but this harks right back to the original racist phrase.

5 My criticism here is of agricultural applications of genetic engineering. Medical applications of genetic engineering will be discussed in Chapter 11, "The Role of Science and Technology."

6 Salmon engineered for increased size are also sterile. Wild female salmon prefer to mate with the larger engineered salmon and thus will produce no offspring. If engineered salmon are introduced into the wild, salmon extinction would be highly probable. More recently, Monsanto's corn seed engineered to make the plants toxic to the pest corn rootworm is rapidly losing its effectiveness. Scientists say that continuing to plant a failing technology increases the risk that resistance will increase. This is happening because of non-engineered corn

seed is scarce. "Using insecticides along with the biotech corn as Monsanto has advised is not a good approach, according to the scientists, because it elevates production costs for farmers and masks the extent and severity of the building insect resistance" (Gillam 2012). Also this incurs the risk from pesticide this genetic engineering was supposed to avoid. Two lessons are repeated here. First, genetic engineers apparently do not understand the role of natural selection in living things. Second, Monsanto's response was entirely predictable: Their spokesperson said there is no problem here at all.

7 Other problems with this use of pesticide are ecosystem and biodiversity damage. Monsanto's genetically engineered Roundup-resistant flaxseed crossbred by natural means with regular flaxseed in adjacent farms. Monsanto then sued the farmers for patent infringement. The fact that US courts granted patents to genetic material show that they regard the world as corporate property. Any possible harm to the ecosystem and its genetic basis are completely irrelevant. See Chapter 2, "Conflicts with the Ecosystem," "Indirect harm to humans,"

8 See Chapter 2, "Conflicts with the Ecosystem," "Indirect harm to humans."

9 San Onofre was shut down in early 2012 until the causes were determined of unexpected wear in the steam generation facilities. See Sewell (2012).

10 See Chapter 2, "Conflicts with the Ecosystem," "Species extinction."

11 See the discussion of Ray Anderson, CEO of Interface, Inc. in Chapter 9, "Causes of Conflict with the Ecosystem." Anderson details his epiphany in the film *The Corporation* (Achbar, Abbott, & Bakan, 2004).

12 Details of this case appear in Chapter 9, "Causes of Conflict with the Ecosystem," Human attitudes are the main cause.

13 This is different from letting the agenda for dealing with the environment be based on short-term political consideration. Here the issue is means to fulfilling this agenda and the political considerations are long term.

14. In case you missed it, warmer air holds more water vapor which supplies more energy for all sorts of extreme weather, including snow.

15 See also Nadeau(2006, pp. 39-41).

16 An employee in the space program at JPL was recently fired because he was proselytizing for Intelligent Design. For some reason, he thought that Intelligent Design explained complexity better than evolution. The space program requires intelligent people, and this employee was clearly not one.

17 See Chapter 9, "Causes of Conflict with the Ecosystem," "Population Growth is the Main Cause."

18 See the discussion in Bookstein (2008).

19 Obviously methods such as genocide and forced sterilization are unacceptable. As human beings, we must remain ethical. That is, we must act on cooperative principles good for all human beings. There would be nothing ethically wrong

with China's one child edict if it had been adopted by a democratic regime. But in present-day China, the state acts by issuing autocratic edicts which must be obeyed.

20 Because monkeywrenching has no social implications, it amounts to a physical struggle between an individual and a state agency or a corporation. The agency or corporation will win. Possibly as a result, monkeywrenching has pretty much disappeared as a tactic.

21 For an excellent discussion of why exploiting resources on other worlds is an irresponsible fantasy, see Walsh (2012). Walsh dissects the untenable assumptions of an exhibit promoting exploitation of space resources at the American Museum of Natural History in 2012.

22 Not only that, our species itself is mortal, it will some day no longer exist. This and related issues are discussed in Chapter 14, "Extinction."

23 This table is based on information found in Citizens' Clearinghouse on Waste Management (2004).

APPENDIX

If the choice is between a restaurant with highly variable quality of food and one that always has decent but unexciting food, your payoff matrix might be Table 6:

The maximin rule would say always to choose the decent restaurant—you come off best if the worst happens. Expected utility would allow you to pick the uneven restaurant if the spectacular meal has higher utility for you than the decent meal. With the numbers above, the expected utility of the uneven restaurant would be:

(20% x 80) + (60% x 20) + (20% x (-10)) = 16 + 12 - 2 = 26.

The expected utility of the decent restaurant would be 100% x 20 = 20. Thus the choice of the uneven restaurant has greater expected utility.

Table 6. Payoff matrix, expected utility

	Best Happens (20%)	**Ordinary (60%)**	**Worst Happens (20%)**
Uneven Restaurant	Spectacular meal (80)	Decent meal (20)	Inedible meal (-10)
Decent Restaurant	Decent meal (20)	Decent meal (20)	Decent Meal (20)

Chapter 11
The Role of Science and Technology

ABSTRACT

The role of modern technology needs to be reduced because of the ongoing threat of catastrophic environmental consequences. Regardless, some modern technology needs to be employed to monitor the ecosystem and to deal with potential celestial collisions. Other parts of modern technology that do not contribute to survival need to be reduced as quickly as possible without causing any more damage than necessary. Economic growth needs to be rethought with environmental costs properly included.

INTRODUCTION

Our examination of the role of science and technology in reaching equilibrium with the ecosystem starts with two conflicting facts: First, modern technology can never be compatible with the ecosystem because it does not acknowledge the ecosystem as something to be preserved. Second, science with its modern technological apparatus is essential for telling us where we stand with respect to the ecosystem, especially about threats to humans and to the ecosystem. For example, killer asteroids could wipe out both us and the ecosystem, completely resetting the clock of evolution, as happened when the dinosaurs died.

If modern technology has compatibility problems with the ecosystem, what about earlier technology? Classic agricultural technology is a paradigm for humans of nature itself: Pastoral scenes *are* nature for us. However, the classic technology

DOI: 10.4018/978-1-4666-4586-8.ch011

of civilization has always collapsed, normally through practices which resulted in the inability to sustain their populations (Diamond 2002). So it seems that even this technology (agriculture practiced for the benefit of civilization) has compatibility problems—although collapses of classic technology were local and did not threaten entire ecosystems.

Because science does not regard human activity as the highest reality, it may play the role of a saving power. For it is largely science and its technological apparatus that are responsible for our knowledge of our position in the environment. And it is because of science and technology that we can even begin to have any hope of overcoming the difficulties in our current position. If our intelligence has failed us in leading us in the direction of destroying our own ecosystem, it has not failed us in revealing this very situation to us.

I speak of science as "our intelligence," but powerful forces—especially in the US—disparage science for religious or economic reasons. Republican presidential candidates state their disbelief in evolution, presumably because of their religious beliefs. In any case, their reliance on religious belief in only some matters is suspect. It is not religious beliefs which keep planes in the air and automobile engines from exploding. It is the same scientific methodology that establishes evolution and climate change. So there does not seem to be any rational justification for picking and choosing between science you like and science you do not like. Either do not drive cars and do not fly, or start believing in evolution and climate change. Beliefs against evolution and climate change as a basis of policy make it impossible to deal with environmental problems in any effective way.[1] Although freedom of religion is a basic human right, the separation of church and state in the US means that religious beliefs should not impact public policy. Religious beliefs determining public policy can only result in human extinction.

The alternative to evolution is creation by a mythical humanlike being whose actions are rather arbitrary. Human beings on this account are not part of nature but are not physical beings at all, but rather souls which on death move to another much better world. Thus on this alternative, this world is a throwaway world, and we do not have to grant it much respect. The discovery of DNA in chromosomes as the basis for all life should have rendered creation by a large human being in the sky unbelievable. But, after all, the discovery that the world is not flat and therefore not a three-storied affair should have rendered unbelievable the idea of souls going to an afterlife in a world in the sky. Belief in an afterlife is, however, relatively harmless. It does not impact public policy in any direct way. However, disbelief in evolution can impact public policy by ignoring or downplaying the impact of human action on the ecosystem. Faith-based beliefs also support climate change denial. Climate change denial as of now is the most serious threat to the continued existence of human beings and even the continued existence of the ecosystem.

Attempts to disparage science for economic reasons are in some ways more disturbing. When fossil fuel corporations spend upwards of thirty million dollars to disparage climate change science, they do not have alternative correct ideas. They are motivated solely by removing a threat to their profits. In their case, the disconnect between the science, they accept and the science they reject is even greater: Their business of producing fossil fuels requires sophisticated science.

Conservatives may distrust science for ideological reasons. Recent polls show that conservatives have been losing trust in science. Only 35% of conservatives regard science as a trustworthy source of knowledge (Pappas 2012). Some commentators believe the source of this distrust is recent scientific claims calling for global collective action on climate change.[2] Conservative ideology does not believe in the reality of collective action and cooperative benefits. Margaret Thatcher famously declared that society does not exist, there are only individuals. This view has been clearly false for human beings and their ancestors for at least 4 million years. Human beings and their ancestors have always been social animals. Single human beings are not able to kill large animals. Single human beings did not develop and implement the technology necessary for civilization. The conservative illusion that actions of individuals alone are what produce value is so far from reality that it deserves to be called a mental disorder.

MODERN TECHNOLOGY, R.I.P.?

Is it possible to eliminate just the dangerous part of modern technology? In this section it will be determined what needs to be done to protect the environment. But technology for protecting the environment or for human economic requirements may not be compatible with protecting the environment. In the next sections What kinds of compromises we need to make will be considered, as well as their ultimate impact on the ecosystem.

There are two features of modern technology which make it incompatible with long term human survival. First, modern technology's tendency to produce unintended and unpredictable side effects. Second, modern technology's treating everything (including the ecosystem) as resources to be exploited. The two features are connected. Because modern technology has no respect for the integrity of the ecosystem in its processes, it is no wonder that unintended and unpredictable side effects for the ecosystem occur with some frequency.

There is also no way of sidestepping the problems caused by these features. Unintended and unpredictable consequences are just that. A Dilbert comic strip featured the clueless boss demanding to be notified in advance of unplanned outages. The case of the ozone layer is sufficient by itself to show that we need to minimize modern technology. It is worth repeating that chlorofluorocarbons, apparently useful and unique substances produced by modern technology, were destroying the ozone layer. We (and most of the biosphere living on land) need the ozone layer to survive. The substances in question were deemed completely safe. Their creator actually drank a glassful of chlorofluorocarbons at a press conference. It turned out that chlorofluorocarbons are completely inert except in the upper atmosphere, where they destroy ozone. It was touch and go to get the international agreement[3] which phased them out. Also, if the chemist who created chlorofluorocarbons had happened to have based his product on bromine rather than chlorine, the ozone layer would have been gone before we knew what hit us.[4]

Modern technology is *just not safe.* Climate change is another instance. Carbon dioxide is not a pollutant at ground level. In the upper atmosphere, it functions as a greenhouse gas, trapping more of the sun's heat. One of the reasons why people cannot accept the straightforward science of climate change is that they see that they must preserve modern technology. To do this, they have to reject the idea that human technology is unintentionally causing conditions which can do serious harm to humans and living things. Of course, as it gets harder and harder to deny the damage, it gets harder and harder to preserve modern technology as is.

But we are now also dependent on modern technology for many aspects of our survival, especially in the developed world. Therefore, it is likely that modern technology must be reduced rather than eliminated. But given the unpredictability of bad consequences, it still needs to be reduced as quickly as possible without causing other serious damage. Some parts of modern technology probably should be terminated immediately, whereas other parts may have to remain if they can be employed without serious risk. And some, for example those needed to monitor the state of the climate, must be retained even if there is some risk. Using computers, we can not only monitor but also model complex systems involving living beings, weather, and climate. Thus we have a much better idea of how our actions affect the other beings we share the ecosystem with. Clearly, these technologies should be retained.

It would be tidier if we did not need to use modern technology to help solve the problem of our relation with the ecosystem. But this seems unavoidable. We can, however, monitor the modern technology we must use for damage to the ecosystem. This is not completely safe because of modern technology's unpredictable consequences, but it seems to be the best we can do.

One part of modern technology which has its own problems is genetic engineering. As explained in earlier chapters, genetic engineering treats genetic material as just one more resource to be exploited. It is also the ultimate short-term procedure. Genetic material is treated as a means of creating living things with definite characteristics serving immediate human needs. The longer-term effects of natural selection of living things in the environment are ignored. Therefore, all long-term effects of genetic engineering are unpredictable and unintended. Especially because genes act in combination in ways which are again unpredictable, the chances of good long-term effects are low. Two examples previously mentioned were large sterile salmon preferred as mates by female salmon unaware of genetic engineering. And corn engineered to be pest-resistant has lost useful resistance because of a more rapidly evolving pest. Genetic engineering is actually of limited usefulness. Lester Brown points out that genetically modified crops have not increased yields. The potential of most crops has been exhausted by conventional breeding (2009, p. 8). So the uses of genetic engineering are either for minor improvements in crop yield or improvements in Monsanto's profits by breeding crops resistant to its proprietary pesticides such as Roundup. Not surprisingly, planting pesticide-resistant crops and then dousing the field with pesticide has horrendous effects on the rest of the ecosystem. In the spring of 2012, a coalition of over 2000 farmers sued to get the US Department of Agriculture to conduct an independent study on the effects of releasing the banned chemical dioxin into the environment. Dioxin is a contaminant of 2,4-D, the new pesticide Dow and Monsanto are engineering crops to resist. These companies counter that the stronger pesticide is needed *because earlier less toxic pesticides have begun to fail* (my italics, Gillam 2012). Modern technology again ignores the ecosystem.

However, genetic engineering has other uses than modifying food crops and livestock. Organisms have been modified to produce new drugs (biopharmaceuticals). Typically, organisms are modified so that they will produce new drugs, or older drugs more cheaply and easily. One of the first bioengineered drugs was insulin, produced from modified *E. coli.* Insulin was previously collected from pancreases of dead animals. Less positively, the possibility of creating of treatment-resistant bacteria is acknowledged by genetic engineers. Organisms have also been modified to help clean up the environment (bioremediation). In fact, the engineered organism which led the Supreme Court to allow granting patents to living things was engineered to break up crude oil and thus help with oil spill cleanup. The Supreme Court in this case took the point of view of modern technology. Genetic material is merely another resource. These oil-eating bacteria were used in helping to clean up the 2010 BP oil spill, but some sources report the bacteria caused mutations in other bacteria with harm to humans and the environment (Benson 2011).

Both these applications of biotechnology present exactly the same dangers as the agricultural applications. All completely ignore the role of natural selection in calibrating the living thing with the ecosystem. Bioengineering is *not* the same as traditional breeding, only faster, as commentators Robert Krulwich and Jad Abumrad (2008) would have it. Traditional breeders use natural selection to get their results. They deal with whole animals, which have to interact with the environment.

Misfires like the big sterile male salmon could not happen. There is the possibility of unintended consequences. Krulwich and Abumrad ask, "What happens if these things, once built, don't do exactly what they are told?" Even worse, the impact of human genetic manipulation on gene pools will be much like that of strong destructive bursts of radiation. From the point of view of the ecosystem, human aims and purposes are completely arbitrary.

There is some evidence that the antibiotic-resistant *E. coli* superbug killing people in Europe since 2011 is a genetically engineered strain. But, even if this is paranoid nonsense, in 2012 we have Dutch researchers producing a bird flu virus transmissible between humans! Of course everyone is always super careful. One can imagine the next bird flu epidemic starting when the cleaning lady dumps a Petri dish into the trash by mistake. The arrogance here is staggering. Whatever is done in a research lab is completely under "our"control.

One important difference between biopharmaceutical applications and agricultural applications is that biopharmaceutical engineering has produced beneficial drugs which could not have been produced otherwise.[5] But even if we grant this, it is very hard to quantify and compare the risks and rewards, because of the unknown and unanticipatable risks of serious consequences. The maximin reasoning developed at the end of the preceding chapter is appropriate: We pick the alternative, which leaves us best off if the worst happens. That means to stop using biopharmaceuticals even if they have clear benefits.

However, it is doubtful whether these considerations would cause biotechnology or biopharmaceuticals to cease being practiced. Acceptance of the maximin argument depends on how serious the worst-off alternative is believed to be. Technologists will generally admit that mishaps occur. But they generally believe that more technology can fix the problem. As pointed out earlier, this belief is an article of faith rather than a rational conclusion.[6] But it is unlikely that a technologist would accept the possibility of an unforeseen disaster, however great, as a reason to stop producing lifesaving pharmaceuticals.

Agricultural and remedial uses of biotechnology are not so defensible. Certainly, the efforts of corporations like Monsanto to introduce products with toxic side effects and then vigorously defend them deserve no privileged status. If they were made to pay anything like the actual amount of damages they cause, they would go out of business. And the biotechnology applications of companies like Monsanto only seem to benefit their own profitability. Because remedial uses of biotechnology involve introducing modified organisms into the environment on a widespread basis, if there is any mistake, the consequences would be horrendous. So the maximin argument applies here.

Even modern technology applications for agriculture with apparent wonderful consequences are not completely beneficial. Frank Borlaug, the father of the "green revolution," using artificial fertilizer and voluminous pesticides, claims that without these technologies, one billion people would have died (McKibben 2007, p. 66). Or maybe they would not have been born. Technologists seem to assume that our goal is to keep producing more and more human beings. This is the strategy of a cancer.[7] In any case, as we have seen, careful studies have shown that organic, that is, non-modern-technology farming, is probably more productive than technologically enhanced agriculture. And it avoids side effects such as nitrogen pollution (a serious threat to the oceans) and biodiversity damage.[8]

The next uses of modern technology to consider are consumer products. Some consumer products with no real uses other than those produced by marketing are also uses of technology which do not contribute to the ecosystem. It would probably be best to regard them as valueless by-products of the functioning of a capitalist market economy. The market economy is a purely human institution. Like all such institutions, it would disappear if there were no humans. But the products we are now considering can use up a huge amount of resources from the ecosystem. If we want to stop doing human activity which makes superfluous demands on the ecosystem, we must consider eliminating these products.

For example, the amount of energy and resources involved in producing and distributing different varieties of flavored sugar water is staggering. These products (Coke, Pepsi, and others) have global reach and negative health benefits. Their persistence should be considered a flaw in the current practice of capitalism. Also in this category are essentially useless elaborations of products solely to produce market differentiation. Sometimes accompanied by what is called 'planned obsolescence,' they again absorb resources for no other reason than the workings of a flawed market economy. If environmental costs are factored in, these items would most likely have negative value.

The background issue here is that modern economies are designed to consume more and more resources with no discernible payoff except what is called 'economic growth.' And economic growth amounts to consuming more and more resources with no discernible payoff. We will return to this issue after determining what we need to preserve from technology in order to protect ourselves and the planet.

So it will be necessary to cut back on many applications of modern technology. Biopharmaceuticals have the same types of risk as other biotechnology applications, but they also have clear benefits. So biopharmaceuticals should be continued with strong monitoring for any modified organisms escaping or any other unintended side effects.

SCIENCE AND TECHNOLOGY AIDING ECOLOGY

It is modern science supported by modern technology that has revealed our position in the ecosystem to us. Climatology requires elaborate equipment as well as resources for making lots of observations, sometimes at difficult-to-reach corners of the world or from space. This, in turn, requires the technology of modern transportation and satellites. So how can we determine what is 'good' technology to be encouraged and what is 'bad' technology disturbing the ecosystem to be discouraged?

It is important to remember that technology is not needed only to remedy other problems caused by technology. Much of the technology needed to monitor the climate is also necessary to detect potential collisions with asteroids or comets. And more technology not yet developed would be needed to deflect such an asteroid or comet. It is worth anticipating the objection that since modern technology is so harmful, we should forgo the technology for avoiding the asteroid collision and let nature take its course.

This objection misses an important point, important both now and in the following chapters. The universe is composed of different levels which are largely unaware of each other and whose interests conflict. I became aware of this situation when I was attempting to water some plants in my front yard. I severely disturbed a nest of ants, who began furiously running around and rescuing their larvae from this unforeseen threat. I had absolutely no intention of causing this crisis. Asteroids and comets are simply another level of activity in the universe. Their behavior in striking the earth and destroying life has nothing to do with anything they or we intend. So our deflecting the asteroid has nothing to do with not following nature. We are following nature in deflecting the asteroid just as much as the ants were in protecting their larvae. Another instance is the relation of predators to prey. The lion has no concern whatever for the gazelle's concerns. That is how things have been organized by natural selection.

It is also true that modern technology and the ecosystem are in effect different levels in our portion of the universe. So, like lion and gazelle, their interests will conflict. And, in situations like this, each of their interests has nothing to do with the interest of the other. The analogy may be useful in considering the question of whether nuclear power can be used to reduce carbon emissions. Nuclear power per se has nothing to do with preserving the environment. In fact, its drawbacks include a propensity to widespread and persistent environmental damage. So its use should be confined only to situations where there is no other way of reducing carbon emissions. In the actual situation, discussion is distorted by knowledge pollution. But the current situation makes no sense. We subsidize nuclear power by amounts orders of magnitude greater than subsidies for renewable energy.[9] Perhaps the lions in this case are the nuclear power companies and the gazelles are the rest of us.

I now return to the question of what technology is necessary to monitor the state of the ecosystem. That includes climate and animal movements.[10] Table 1 presents the equipment necessary to monitor climate change and the technology necessary for that equipment.

As you can see from Table 1, electronic and computer equipment are needed to record continuous data. The few low-tech non-electronic instruments require manual monitoring, which is not only labor-intensive but somewhat impractical even at ground or sea level. Also, virtually all planet-wide measurement of climate variables is now done by satellite. Even local weather forecasting now uses satellite data. It would be impossible to duplicate this data without the use of satellites. We need this level of data accuracy and immediacy to be able to handle unanticipated effects on climate brought about by human activity. As we saw just above, if the chemist who produced chlorofluorocarbons had chosen bromine as his base instead of chlorine, we would have needed almost instantaneous knowledge of the effect on the ozone layer in order to deal with it. So this technology is essential.

Electronic instruments, satellites and launch rockets, require pretty much the whole apparatus of modern technology. Electronics and satellites require precision manufacturing using rare metals and minerals. Rockets require sophisticated metal and composite technology for the rocket and its engines, for maximum strength and minimum weight. Current common rocket fuels are liquid oxygen plus liquid nitrogen, or some variant of hydrazine.[11] But fuel is not the major cost with rockets. Midrange satellite launch rocket hardware can cost from $50 million to $150 million (Benson 2011). These technologies and more will also be required to deal with the ongoing threat of asteroid collisions.

Basically, we need sophisticated manufacturing technology supported by sophisticated extraction and refining to deliver the materials required. Materials needed for electronic equipment include fiberglass, copper, tin, lead, nickel, cadmium, beryllium, organobromides, and gold. However, since according to the Environmental Protection Agency, only 15-20% of electronic waste is recycled (2011) unless this percentage increases drastically, we will need to continue to mine for all these metals and minerals and then process them. All mining involves some destruction of land. Processing of ore almost always involves as by-products toxic air, water, and/or land pollution. Characteristically mining companies employ various strategies to avoid the costs of restoring land. The simplest is to deny responsibility for the pollution. The Taganito Mining Corporation, a large Filipino nickel-mining firm, claimed that the obvious pollution in a bay adjacent to its operation was caused by rain (Caliguid 2012). More devious is the strategy of gold miners in the west. Jared Diamond describes how mining corporations avoid cleanup costs by declaring bankruptcy and transferring assets to other corporations owned by the same individuals. Pegasus Gold, Inc. opened the largest gold mine in Montana. Its processing continu-

Table 1. Weather monitoring technology

Component Measured	Low Tech Version	High Tech Improvement	Nature of Improvement	Background Tech Requirements
Temperature	Thermometer	Electronic sensor	transmits info max&min temps	Electronics
	Radiosonde	Computer communication, GPS use	records continuously, stores data	Radio technology Electronics Satellites for GPS
		Satellites—Geostationary (10) Polar orbit (10-15)	measures sea temperature & salinity	Electronics Satellite technology
Precipitation	Rain gauge	Self-adjusting rain gauge		Electronics
		Weather radar		Electronics
		Satellites	all environmental information	Electronics Satellite technology
Biomass	Insect surveys	Satellites	map extent of insect devastation	Electronics Satellite technology
	Effects on plant species range	Satellites	map plant ranges & climate degradation	Electronics Satellite technology
		Soil monitors	no low-tech equivalent	Electronics
Sea level	Tide gauges, barometers, Wind gauges thermometers	Electronic tide gauges connected to satellite altimeters	higher tech can correct for surface tide measurement	Electronics Satellite technology
Solar activity		Pyrheliometer Pyranometer	measure solar energy at earth	Electronics Internet
		Sunphotometer	measure particles in atmosphere	Electronics
Volcanic eruptions		4 instruments to measure gases	chemicals in eruptions	Electronics. First used to measure industrial pollutants
	Gas collect bottles		delivered for gas analysis	
Chemicals in atmosphere		Plane & satellite instruments	measures of 6 gases	Electronics Plane & Satellite technology
Chemicals in water	Buoys	Buoy + satellite	gases in ocean and acidification	Satellite technology
		Satellite	salinity	Satellite technology

Data in table derived from Global Greenhouse Warming.com (2010) ally leaked such contaminants as cyanide. After settling $36 million in lawsuits, the company directors voted themselves $5 million bonuses and declared the company bankrupt. Another mining company, Galactic Resources, declared bankruptcy and closed its mine a week later and abandoned the mine, halting essential environmental maintenance. An overflow a month later poisoned 18 miles of river with cyanide. Cleanup costs were $180 million, of which $28 million came from the bankruptcy settlement. The remainder was paid by US taxpayers (Diamond 2004, pp. 455-457).

The problems here have as much to do with corporations as with conflicts between technology and the ecosystem. What needs to be done to prevent this dysfunctional corporate behavior? Unregulated corporations are the perfect expression of modern technology. But unintended and unpredictable consequences are a part of modern technology. If we are going to minimize such life-threatening consequences, we need to regulate corporations.[12]

What exactly do we need to do with corporations to ensure they will not severely damage the planet in their quest for profits? At this point in the history of corporations, it would be difficult to add back in the value and ethical elements that were removed from corporations back in the 19th century. Corporations have had a lot of practice at seeming to be ethical and yet actually following the bottom line. Therefore, external constraints are necessary. If corporations obey external constraints which do not maximize profits, that should not bother shareholders nor detract from their single goal of maximizing profits.

1. Make the continued existence of corporations subject to their passing a periodic ethical and environmental review.
2. Prevent corporate knowledge pollution. This requires prohibiting any corporate influence on public policy in any form. Our experience with the ozone layer and global warming shows that corporations should not be able to prevent appropriate action to deal with unexpected consequences of technology.
3. Require corporations to use longer time periods for planning and profit reporting (Al Gore's suggestion).[13]
4. Prior testing of any new substances to be released into the ecosystem.[14]

Even though we cannot completely avoid unintended and unpredictable consequences from modern technology, we have the means for detecting obvious conflicts when they occur, assuming we can end corporate knowledge pollution. We have scientific methods for determining causes. So if bees suddenly start dying in large numbers, we can determine why, without the knee-jerk response and delays from the manufacturer of the likely pesticide. Or worse—an expensive effort to discredit the science that uncovered the problem, as fossil fuel industries have done with respect to climate change.

SCIENCE, TECHNOLOGY, AND ECONOMIC GROWTH

If we curtail modern technology, how can we deal with negative effects on economic growth? Technology has an impact on the economy. Reducing technology to the extent necessary also looks to produce a drastic reduction in GNP. But to some extent this is an illusion. Studies have shown that people with much lower levels of economic well-being are much happier than those with more income.[15] But this will not be much of a consolation to those well off.

One of GNP's illusions is that it only counts money expended without considering results. Joseph Stiglitz remarks that US GDP includes the enormous amount spent on health care, without distinguishing the roughly 33% which goes to private insurance overhead rather than health care expenses. This overhead actually detracts from health care output rather than adding to it. The result is that countries having much better health outcomes with lower expenditures, are ranked lower in GDP productivity in health than the US (Stiglitz 2010, p. 284).

But perhaps the most important illusion is that Gross Domestic Product (GDP) and Gross National Product (GNP) take no account of environmental costs. In the preceding chapter, the views on this issue of both Stiglitz and Robert Nadeau were discussed (2006, p. 140). Stiglitz points out that since GDP is a bad measure of economic well being, talk about tradeoffs between economic growth and the environment are seriously misguided: If we correctly measured output—including worsening the environment—there would be no tradeoff (Stiglitz 2010, p. 284). Therefore, if we eliminate that part of economic "growth" that worsens the environment, we have not eliminated any real economic growth.

All too often, people think growth can be unlimited. As Figure 1 illustrates, there are only two possibilities for growth.

Figure 1. Two possibilities for growth: a) sustainable growth, b) growth crash collapse

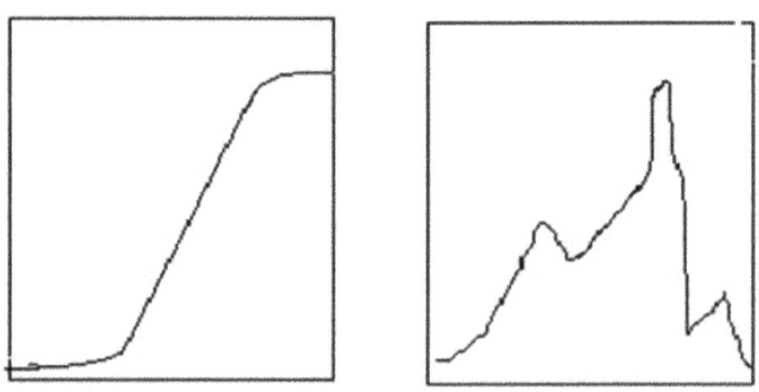

Figure 11.1a
Sustainable growth

Figure 11.1b
Growth Crash Collapse

Figure 1a shows sustainable growth. Some process or entity has an opportunity to expand in the environment and expands until the environmental resources are fully exploited. At that point, the process or entity reaches a new, higher, sustainable equilibrium. Animal and plant populations as well as products in new markets often follow the curve in Figure 1a. The curve in Figure 1b includes economic bubbles—rubber in London in 1909, gold and silver in 1979-1980, US housing in 2006-2008. Such booms feed on themselves until their irrationality is exposed. Animal populations sometimes follow the crash curve. Locust and lemming populations are examples. Introduced animals can also sometimes experience a population boom before they exhaust resources and crash.

Clearly economic growth since the advent of modern technology and civilization has been following the growth-crash-collapse curve. Even if modern technology had not produced substances capable of destroying the ecosystem, its exploitation of the environment is in itself not sustainable. A recent study done by the Forrester Institute at MIT indicates that the developed world will simply run out of resources by the year 2030. One of the scientists involved in the study, Graham Turner, says that perhaps the most startling finding from the study is that the results of their computer scenarios were nearly identical to those predicted in the computer scenarios used as the basis for 1972 study "The Limits to Growth" issued by the Club of Rome. "There is a very clear warning bell being rung here," Turner said. "We are not on a sustainable trajectory" (Pfeiffer 2012).

Our economic goal cannot be unending growth. The planet is finite. There are those who think this is not a problem. We just exploit other worlds or asteroids the same way we exploited earth! Because that's what we do! That is our destiny! These people do not notice that this is the destiny of a cancer cell. Cancer cells are no doubt delighted that their growth is going so well, that major organs have been won over to their view. Maybe those organs aren't functioning so well since they were converted to the cancer lifestyle. But no matter. Everything is fine until the host expires and all the gung-ho cancer cells are buried with him.[16]

To move away from continuing economic growth as a basis for life will not be easy. This is why I am inclined to believe that it will take a few serious crashes in environmental infrastructure and resources to convince policy-makers that continued economic growth is impossible. Current difficulty in making gradual changes to avert climate catastrophe make it very unlikely that gradual voluntary means will prevail. However, if I am wrong, Lester Brown's *Plan B 4.0* (2009) and Bill McKibben's *Deep Economy* (2007) provide sensible gradualist approaches to redoing the economy.

ECONOMIC DEVELOPMENT OF SPACE

A recent analysis of the development of space by Professor Justin St. P. Walsh makes it clear that such development is not a good idea (2012). The proposal in the exhibit *Beyond Planet Earth*[17] is to mine the Moon for helium-3, a "clean" non-radioactive fuel for fusion power. Unfortunately, although there is more helium-3 on the Moon than on earth, it is not abundant. It would take roughly about 300 tons of lunar rock to produce one gram of Helium-3. Walsh makes the excellent point, helium-3 is nonrenewable. What do we do with our expensive infrastructure once it is gone? We are just repeating our mistake with fossil fuels.

Fossil fuel extraction on earth has left expanses of land barren and polluted. What about the Moon? We do not fully understand its geology and climate (The Moon has a 'micro-atmosphere.'). We certainly know that humans can transform a planetary environment on a large scale, with both intended and unintended effects. Usually short-term profit takes precedence over long-term consequence, which future generations are left to deal with. The *Beyond Planet Earth* exhibit also examines terraforming Mars. The process would begin by extensive bombing, spraying black dust, building factories to release greenhouse gases, all to raise the temperature to 110° F. Then we introduce plants to provide oxygen. We do not understand our own complex ecosystem and there is no hope of replicating an extensive ecosystem from scratch.

But behind these proposals is the attitude that the Moon and Mars are beneath consideration as environments. If it is out there, we should exploit it. Resource extraction is not our only choice. Walsh believes we have a responsibility as human beings to consider how our choices affect other living beings and even climate and geological processes. Walsh remarks that resource extraction corporations have had trouble meeting ethical standards concerning other human beings. They really are cancers in human form.

Walsh concludes that we should explore space to learn more about our place in the universe. But problems of resource scarcity and never-ending consumption need to be solved here on earth.

At the beginning of Chapter 5 it is stated that eliminating modern technology would not be a good idea because our entire economy and especially our food supply is dependent on modern technology. Thus, civilization as we now know it would collapse. But we can see that it would collapse anyway if we persist in employing modern technology. Once we get to a total human environment with only a tenuous connection to the rest of the environment, we get to a place where, when the disruptive power of modern technology is added to it, the only probable outcome is unmitigated disaster. And climate disaster is only one of any number of possibilities for modern technology to destroy the ecosystem both we and all living things need to survive. The destruction of the ozone layer was another recent real possibility.

If one or more of these disasters happens, they will likely produce population reduction and take care of economic problems and food problems. This is hardly the preferred way of dealing with these problems. However, whatever happens, it is important that no attempt be made to rebuild civilization as we now know it, with its fatal conflicts with the ecosystem.

THE ROLE OF TECHNOLOGY IN THE ECOSYSTEM

We have seen that modern technology is by its nature incompatible with the ecosystem. We have also seen that (some) modern technology is either necessary or unavoidable even if it still has the potential for serious unintended consequences. Two instances:

1. The technology necessary to monitor the earth and its surroundings and help to prevent or mitigate bad consequences.
2. The technology necessary to support biopharmacology to produce new drugs.

I think it is best to minimize any modern technology not included in these two instances. This minimization should be over sufficient time to reduce collateral harm. We also should monitor any modern technology we must use for possible damage to the ecosystem. This is not completely safe because of modern technology's unpredictable consequences, but it may be the best we can do. However, some instances of modern technology are so destructive, either potentially or actually, that they need to be phased out as quickly as possible, without doing extreme collateral damage. These are: Agricultural applications of modern technology, both conventional and biotechnology; and consumer products with no real uses other than those produced by marketing.

These recommendations may seem draconian and wildly impractical. They may be. The reasoning leading to them is correct, so they may be much easier to implement for the fragment of humanity which survives the next fifty to one hundred years. Some authors have provided some ideas of what an economy not based on modern technology might look like—for example Bill McKibben in *Deep Economy* (2007) and E. F. Schumacher in *Small is Beautiful* (1973).

The Internet and World Wide Web are at this point an integral part of any scientific project, including monitoring the earth and its climate. Therefore the infrastructure of the Internet and Web must be continued. All other uses of the Internet and Web should also be included in allowable technology just because of the fact that worldwide communication between individuals in real time is enabled. One of the problems of dealing effectively with climate change is that special interests can appeal to parochial concerns. The more that everyone has access to everything that is going on around the planet, the less likely that special interests can prevail. In 2011 we saw an "Arab Spring". Perhaps before it is too late we will see an "Earth Spring." The fact that we have a genuine worldwide communication system (the World Wide Web was accurately named) is going to make it much easier for anyone to take the point of view of the species.

REFERENCES

Adams, R., & Brown, B. K. (1941). Hydrazine sulfate. *Organic Syntheses*, *1*, 309.

Benson, J. (2011). Genetically-modified, oil-eating bacteria creating dangerous mutant organisms in the gulf. *Naturalnews.com*. Retrieved 4/21/12 from www.naturalnews.com/031090_Gulf_of_Mexico_bacteria.html

Benson, T. (2011). Beginner's guide to rockets. *National Aeronautics and Space Administration (NASA)*. Retrieved 3/26/12 from exploration.grc.nasa.gov/education/rocket/bgmr.html# Brown, L. (2009). *Plan B 4.0*. New York: Norton.

Caliguid, F. (2012, February 6). Big nickel mining firm says rains caused pollution of bay. *Inquirer Mindanao*. Retrieved 3/28/12 from newsinfo.inquirer.net/140773/big-nickel-mining-firm-says-rains-caused-pollution-of-bay

Diamond, J. (2002). *Collapse*. New York: Viking.

Gillam, C. (2012, April 18). Farm group seeks U.S. halt on "dangerous" crop chemicals. *Reuters*. Retrieved 4/18/12 from news.yahoo.com/farm-group-seeks-u-halt-dangerous-crop-chemicals-123359159.html

Herzog, W. (n.d.). *Grizzly man*. [Film] Santa Monica, CA: Lionsgate Films.

Klein, N. (2011, November 9). Capitalism vs. the climate. *Nation (New York, N.Y.)*.

Koplow, D. (2011, February). Nuclear power: Still not viable without subsidies. *Union of Concerned Scientists*.

Krulwich, R., & Abumrad, J. (2008, April 29). Minty e. coli and other bioengineering feats. *NPR*. Retrieved 4/21/12 from www.npr.org/templates/story/story.php?storyId=90014997

McKibben, B. (2007). *Deep economy*. New York: St. Martins Griffin.

Nadeau, R. L. (2006). *The environmental endgame*. New Brunswick, NJ: Rutgers University Press.

Pappas, S. (2012, March 29). Conservatives losing trust in science, study finds. *LiveScience.com*. Retrieved 3/31/12 from http://news.yahoo.com/conservatives-losing-trust-science-study-finds114401347.html

Pfeiffer, E. (2012, April 4). Next great depression? MIT researchers predict 'global economic collapse' by 2030. *The Sideshow*. Retrieved 4/4/12 from news.yahoo.com/blogs/sideshow/next-great-depression-mit-researchers-predict-global-economic-190352944.html

Schumacher, E. F. (1973). *Small is beautiful*. New York: Harper & Row.

Stiglitz, J. (2010). *Freefall*. New York: Norton.

US Environmental Protection Agency. (2011). *Statistics on the management of used and end-of-life electronics*. Retrieved 3/25/12 from www.epa.gov/epawaste/conserve/materials/ecycling/manage.htm

Walsh, J. S. P. (2012, March 2). The final environmental frontier: Space development and its consequences. *Scientific American*. Retrieved 3/21/12 from news/yahoo.com/final-environmental-frontier-space-development-consequences-160000428.html

Warming, G. G. com. (2010). *Measuring climate change*. Retrieved 3/20/12 from www.global-greenhouse-warming.com/measuring-climate-change.html

Wikipedia. (2012). *Energy subsidies*. Retrieved 4/24/12 from en.wikipedia.org/Wiki/energy_subsidies

Wilson, E. O. (2002). *The future of life*. New York: Vintage Books.

ENDNOTES

1 Religions with faith-based doctrines often claim that their set of doctrines is the only correct one, and that other faith-based doctrines are wrong. Therefore, it is probable that none of them is correct. A few religions such as Buddhism do not claim that they have the only correct doctrine and thus might be correct.

2 Naomi Klein has written an eye-opening article on why many super-rich cannot accept climate change. They see (more-or-less correctly) that action to forestall it would entail the destruction of the economic system responsible for their fortunes (Klein 2011).

3 The Montreal Protocol. See the discussion in Chapter 10, "Environmentalism and Sustainability," a detailed plan to achieve sustainability.

4 Mark Lynas supplies most of this information in his *The God Species* (2011), boundary 9. Being a technophile, he does not draw my relatively obvious conclusion.

5 Although even in this case there are questions. Synthetic insulin produced by modified *E. coli* is somewhat more expensive than pig-derived insulin.

6 See Chapter 10, "Environmentalism and Sustainability," "Sustainability with unlimited economic and population growth."

7 In Chapter 9, "Causes of Conflict with the Ecosystem," "Population growth is the main cause," it was argued that increasing human numbers is not intrinsically a good thing.

8 See Chapter 10, "Environmentalism and Sustainability," "Sustainability with unlimited economic and population growth."

9 Besides $50 billion a year in direct subsidies, nuclear industries typically shift "construction costs and operating risks from investors to taxpayers and ratepayers, burdening them with an array of risks including cost overruns, defaults to accidents, and nuclear waste management" (Koplow 2011).

10 Would doing this conflict with our being a species among species? The answer is important but tricky and will be considered in Chapter 13, "Regaining Our Place as a Species among Species."

11 Hydrazine can be synthesized from many common materials such as bleach and ammonia (Adams 1941).

12 A discipline called "social ecology" emphasizes the difficulty in dealing with corporate power. Their solution is to work to change human social structures to eliminate hierarchies. There are two problems with this approach: First, changing human social structures may or may not change the relation of humans to the ecosystem; second, 'hierarchies are always bad' is a dogma of this neo-Marxist doctrine. Complex projects seem to require some type of authority.

13 See Chapter 10, "Environmentalism and Sustainability," "What is Sustainability?"

14 This is a proposal of Senator Frank Lautenberg of New Jersey. See Chapter 2, "Conflicts with the Ecosystem," Toxic Substances.

15 In Chapter 9, "Causes of Conflict with the Ecosystem," Technology First, a New Economics Foundation study is mentioned which had these results.

16 Appropriate species goals will be discussed in Chapter 13, "Regaining Our Place as a Species among Species."

17 At the American Museum of Natural History in 2012 in Washington, DC.

Chapter 12
Human–Animal Relations

ABSTRACT

The proper behavior of humans toward animals is a middle way between human superiority, which allows humans to do whatever they please to animals, and views requiring humans not to interfere with animals at all. Since ethics is principles for social cooperation, only a restricted class of animals—work animals and pets—can have ethical standing. In any case, eco-ethics requires us to treat animals humanely and not as objects.

INTRODUCTION

On the way to regaining our place as a species among species, our relation to our fellow animals deserves separate treatment. Some views about the relations of humans and animals which do not assume human superiority or reject it outright will be discussed. The first of these views, Buddhism, has been around for 2500 years. Another view, veganism, dates from the 1940s. The views of two contemporary philosophers will also be considered: Peter Singer argues with great force, clarity, and consistency in his *Animal Liberation* that we should leave animals alone (2009). Mark Rowlands attempts an entirely different justification for animal rights based on social contract theory in his *Animal Rights* (2009).

A few general remarks: Buddhism includes extensive justification as well as rules of action for its views on human/animal relations. But with the growth of science some Buddhist views on our relation to animals have become more faith-

DOI: 10.4018/978-1-4666-4586-8.ch012

based and less evidence-based. Veganism, by contrast, stipulates rules for behavior with respect to animals without much justification. Singer's views to a great extent provide a justification for vegan rules of behavior toward animals. Singer also holds that humans should leave animals alone. He is in effect promoting another way for humans to be separate from the ecosystem. Finally, Rowland's alternative social contract justification of animal rights turns out to be an exercise in logic chopping. A social contract by definition holds between contributing members of a society. Except in limited cases, animals are not members of society, and no subtle distinctions can make them so.

BUDDHISM

Buddhism is not a theistic religion. The Buddha himself thought that the existence of suffering showed that there was no omnipotent, loving God in charge of the universe. From the beginning Buddhism seems to have been concerned to stay within the realm of the natural rather than the supernatural. Thus, there is an emphasis on karma, or cause and effect, rather than supernatural emanations. And it was the Buddha himself who promulgated the doctrine of no-soul (*anatman*). This is the doctrine that there is no permanent human self, only a stream of experiences.

However, the doctrine of no-soul seems to conflict with another Buddhist doctrine, the doctrine of reincarnation. If every sentient being is re-born as another sentient being with somehow the same identity, each sentient being harbors something permanent enough to survive death. Buddhists call these semi-permanent carryovers, *seeds*. Many Buddhists are not bothered by this seeming inconsistency, but some are. Although I am not a Buddhist, it bothers me[1] (Buddhists generally do not strive for uniformity of doctrine.).

Buddhist reincarnation as a way of survival after death has distinct advantages over the afterlife of Christianity and similar religions. Buddhist reincarnation requires no extra supernatural realm to house survivors after death.[2] Also, since reincarnated people return to this world, there is every reason to preserve and improve this world. In addition, Buddhism requires no fantastic stories such as Jesus and Mary ascending into Heaven. Perhaps those stories did not seem fantastic when people believed in a three-story universe, with Heaven in the sky, mortals living on earth's surface, and death and hell underground. But as soon as people realized the earth was round and that breathable air rapidly ran out as one ascended, the ascension of Jesus and Mary seems preposterous. It would be consistent if Jesus, being God, needed no oxygen mask, but Mary would not have been so fortunate. Also, when Christians restrict the Afterlife to human beings, they affirm a very strong form of unjustified human superiority. Why can't whales go to heaven as well?

Buddhists have respect for all sentient beings. This entails not killing sentient beings and not eating them. One of the forbidden occupations in Buddhism is that of butcher. Respect for all sentient beings is the aspect of Buddhism of most of interest for our current topic of human/animal relations.

The Buddhist doctrine of reincarnation seems to play a large role in justifying respect for all sentient beings. One could view the Buddhist doctrine as a primitive recognition of the laws of inheritance, that sentient beings inherit the properties of earlier sentient beings. Without knowledge of DNA and of the mechanisms of inheritance, as well as knowledge of the fact that acquired characteristics are not inherited, reincarnation is not a bad guess as to what is going on. DNA is a remarkably good candidate for a "seed" that survives individual life. But, unlike the Buddhist seed, it does not contain actual memories or ethical debts. And there seems to be no physical evidence for Buddhist seeds.

Certainly Buddhist reincarnation is closer to the facts than such jaw-dropping claims as that Jesus was born of a virgin mother. We know now that for Jesus to be a male human, someone or something must have contributed a Y-chromosome. If it was the Holy Spirit, Jesus' conception was equivalent to artificial insemination. Or perhaps Mary was a hermaphrodite, with organs of both sexes. It is not clear how this would affect her status as a virgin, though. A person of faith might suggest that God suspended all the rules of biology in this case. But then it is much more likely that the operative deity is not the Creator of heaven and earth, but rather some demiurge (lesser deity) who was mucking around with the mechanisms of life, the intricate creation God had carefully set up. Again, without knowledge of the role of DNA in the continuation of life, people could believe such virgin birth stories.

If the justification of Buddhist respect for all sentient beings requires the acceptance of reincarnation, the justification is not of much use to non-Buddhists. The doctrine of reincarnation as disseminated by Buddhism has additional negative aspects in any case. Buddhists think carnivores are inferior beings which can hopefully be reborn as superior, more enlightened, non-carnivores. In a context of evolution and actual inheritance through DNA, these beliefs represent the imposition of ethics for human beings on the natural world. Like it or not, natural selection has produced a large number of animals that thrive by eating other animals, usually killing them first. In addition, carnivores are usually necessary to keep ecosystems in balance by predation. The current suburban American deer plague is almost entirely due to the elimination of predators of deer such as wolves from the suburbs. Deer, with almost no other natural checks on their population, expand explosively and are forced to extend their food supply to nearly any vegetation within their reach. Besides predators, no other completely effective means of preventing deer destruction has been found (Lemieux and Maynard 1999).

Buddhist treatment of the ethical consequences of reincarnation has other repellent aspects. If things are going badly for you, Buddhists believe that although you have led an exemplary life, one of your (*your)* past lives was probably lived badly and unethically. And you are responsible. Holding you responsible for past lives you have no control over is a good reason for rejecting reincarnation in anything like the Buddhist form. If we have to choose between reincarnation and the no-soul doctrine, the no-soul doctrine is a better choice.

However, we can naturalize the Buddhist idea of respect for sentient beings. Buddhists would have no objection to naturalizing Buddhist ideas: The Dalai Lama himself has said that if any part of Buddhism conflicted with science, he would have to give up that part of Buddhism (Dalai Lama 2005, p. 3). The insight behind the idea of respect for all sentient beings could well be the interdependence of all sentient beings. Science tells us that this interdependence is expressed in DNA changed by the process of natural selection. For Buddhists, this interdependence extends to all living things and also to nonliving things. The result of interdependence is that all explanation is either circular or unending. Thus Buddhists do not accept the idea of an uncaused "big bang" as the beginning of the universe. The same insight about explanation is found in the Antinomies of Kant's *Critique of Pure Reason* (1787) as discussed in Chapter 3, "Science, Modern Technology, and Corporations."

Interdependence means that other beings are not there for our use, so that we cannot dispose of them however we desire. But this does not single out sentient beings for special treatment over other living beings. Nor does it single out living beings for special treatment over non-living things. Perhaps we should say that each type of being deserves the type of respect appropriate to that type of being. For Buddhists, the appropriate respect for sentient beings is the same for human as for nonhuman sentient beings. In particular, all sentient beings have the right not to suffer. Some sects, such as Jainism, extend Buddhist respect to all living things. They do not eat vegetables which must be killed in order to be eaten such as root vegetables, for example, potatoes, carrots, onions, radishes. Also vegetables with a large number of seeds such as eggplant and pumpkin are not eaten (BBC 2012).

The appropriate principle to take away from Buddhism is the idea of appropriate respect for all living things because we are all ultimately related and contribute either directly or indirectly to each others' existence and well-being. This principle will be applied later in the context of human beings as a species among species. Once again, such a principle conflicts with religions with faith-based, that is, non-rational, reasons for human exceptionalism.[3]

VEGANISM

Veganism is the belief that it is wrong for human beings to interfere with animals. As a dietary practice, vegans not only do not eat meat or fish, they also abstain from eating or otherwise consuming any animal products, whether the animal is killed in making that product or not. Thus, vegans do not consume dairy products such as milk, cheese, eggs, butter, or yogurt. Nor do they consume any dish made from such products, such as most baked goods or ice cream. Further, some vegans understand the imperative not to consume animal products to apply to non-edible items, especially leather (Hackett 2013). Presumably silk, the product of silkworm moths, is also prohibited to vegans.

Some of these restrictions seem somewhat draconian, and therefore require a thorough justification. Since vegans regard their choice as an ethical one, their extension of human ethics to animals is what needs to be discussed. As pointed out in Chapter 1, ethics properly speaking is principles for *human* cooperation, grounded either in some version of the Golden Rule, utilitarianism, or a social contract. A short way with veganism would be to declare that its principles are not ethical but rather *moral* principles, principles which hold only for a group with limited membership—perhaps the relevant group is animal lovers.

Vegans would probably vehemently reject this status for their principles, claiming that they apply universally to everyone. However, universality is a necessary condition for being an ethical rather than a moral principle. Many religious sects believe their principles about the wrongness of homosexuality are universal. But since they are not based on anything other than custom, they can't be. If their principles were backed up by the findings of psychology and other sciences, they would be worth considering. But they are not. Veganism is better off than evangelical Christianity. Two potential sources for justification for their beliefs are Buddhism and the animal rights movement. The account of animal rights we will examine is the one eloquently presented by Peter Singer in his book *Animal Liberation* (2009).

As we saw in the preceding section, Buddhism regards all sentient beings as on a par, ethically speaking. Sentience, for Buddhists, is consciousness. For a Buddhist, all animals without exception are sentient beings. These beliefs are intertwined with the doctrine of reincarnation. Although Buddhists regard reincarnation as an empirical matter, the doctrine is not that plausible, for reasons presented in the preceding section.

SINGER AND ANIMAL LIBERATION

By contrast, Peter Singer's discussion of animal rights is resolutely secular. For Singer, ethical concern should extend to any entity capable of feeling pain. It would be consistent with Singer's account to allow the use of animals for various purposes as long as no pain were involved. If this were all to Singer's position, there would be no eco-ethical problem with dairy products and maybe even meat and seafood if the animals were killed painlessly. However, Singer feels that it leads us to the attitude that animals are objects to be treated however we wish (p. 229). Some vegans, by the way, insist that the animal is always harmed in all these cases. The harm may be that these animals are coerced by human beings—this will be discussed shortly. Singer also thinks there are ethical problems in addition to animals being caused pain, very much along the lines of human interference.

Buddhists and Peter Singer have different views about carnivores. Obviously carnivores are non-humans eating meat, which both Buddhists and Singer think is wrong. Buddhism regards predators as having been born as inferior life forms, with eventual good karma allowing them to be reborn as superior life forms which respect sentient beings. Singer's views are based in part on his ethical theory, a form of utilitarianism. Singer acknowledges that animals are not ethical agents. Nevertheless, what matters ethically is only that they can feel pain, not whether they are capable of ethical choices. Being able to feel pain entitles them to equal consideration with human beings (Singer 2009, p. 225).

Singer does consider a version of the Buddhist argument that there is something wrong with the behavior of carnivores or animals that eat meat. For Buddhism, this will be taken care of by reincarnation of inferior carnivores into ethically superior beings who don't eat meat. Singer considers and argues against the idea of eliminating carnivorous species to reduce suffering. He does not bring up the place of predators in ecological systems. Rather, he believes that we should not interfere with animals. As we will see, this principle is an addition to utilitarianism.

Singer argues at length for a third principle, which is to avoid what he calls *speciesism,* a bias toward members of one's own species and a bias against members of other species (2009, p. 6). It is part of speciesism that humans deserve different ethical consideration than animals. He considers the objection that we are still conferring a different status on humans if we require humans not to eat meat but allow animals to do so. His answer is that the important similarity is only that of being capable of feeling pain. And there are of course other differences between human beings and other animals, such as the one just mentioned, the difference in the requirement not to eat meat. Singer also does not mention that there are important ways in which we should be "specieisist." Namely, each species should be concerned for its own continuation, whatever its concern for other species. Singer also does not consider the possibility of species-specific goals. Species goals will be discussed in the next chapter. For now, it is enough to realize that for Singer speciesism is unethical only when it is an illegitimate concern for one's own species.

How good is utilitarianism as an ethical theory ? Singer's version of utilitarianism is known as *hedonistic utilitarianism.*[4] Utilitarianism says that the right thing to do is to produce the greatest good for the greatest number. Hedonistic utilitarianism adds that the only good is pleasure and the only evil pain. Historically, its first main proponent was the English philosopher Jeremy Bentham (1789). Bentham was also an early defender of animal rights. Hedonistic utilitarianism also seems to be implicit in Buddhism, at least in its ethics for human beings. Buddha's Four Noble Truths speak of suffering, the cause of suffering, and an end to suffering through right view, right thought, right speech, right behavior, right livelihood, right mindfulness and right concentration (Buddho Dendo Kyokai, 2009). Therefore, the goal of right behavior is to minimize or eliminate suffering for sentient beings.

In Chapter 1, "Humans as Masters of the World," utilitarianism was criticized because of how it handles the distribution of good. There were four problems: First, if good is produced only if most people follow some rule or practice, you are not producing the greatest good for the greatest number if you break the rule. For example, failing to honor a contract because you could do better otherwise. Second, it is actually impossible to sum goodness over individuals. Third, utilitarianism ignores rights and freedoms. Fourth, as a principle for distribution of good, it allows some to be miserable because of the greater gains of others.

The last criticism is relevant to the present issue. What is the ethical conclusion when animal and human welfare conflict? Consider testing drugs on animals to prevent harm to humans. Singer concludes that most animal testing should not be done because it is not for lifesaving drugs, and even the animal testing of lifesaving drugs should be replaced by alternative methods of determining drug safety (2009, p. 61). Singer's conclusion here does not seem to be utilitarian. Here is the principle which seems to be guiding his conclusion:

(The Noninterference Principle) Humans should not interfere with animals, except for the benefit of the animal (2009, p. 226).

A consistent utilitarian would consider the pain and suffering of all beings involved, human and animal. And if animals could be tested without excessive pain for the prevention of greater human pain and death, that would be OK.

Peter Singer's views on carnivores are also not utilitarian. Singer says that although we (non-meat-eaters) find carnivores behavior regrettable, we have no right to interfere with carnivores being carnivorous. Nor, claims Singer, do we have any right to interfere with any other animal in any way—unless that interference would benefit the animal (Singer 2009, p. 226).

Singer's conclusions here imply that his principle of human noninterference with animals is *not* grounded in the utilitarian avoidance of suffering, but is rather a distinct primary principle, grounded perhaps on feeling or intuition. Looking back at the history of the development of human beings, the domestication of animals was important in human development. Was all of that illegitimate? Leaving aside for the moment breeding and raising animals for food, what about pets or animals used to help humans in various tasks? Someone living in a contemporary urbanized environment could well think that animals are irrelevant to human well-being. All they see are pets—cats, dogs, maybe a few birds, fish, rabbits, snakes, hamsters. All these could be considered unnecessary sources of pleasure for human beings. On a principle of noninterference, all these would go away. However, pigeons, rats, mice, squirrels, and insects such as roaches, ants, and termites are all potential sources of disease. On a strict principle of noninterference, we could not exterminate any of these. Singer does allow, however, that we could reduce their numbers by using technology to reduce their fertility (p. 235).

And in non-urban environments such as pastoral environments or hunter-gatherer environments, animals can work alongside humans. Think of dogs trained to herd sheep, or horses to pull plows. Or barn cats to keep the rodent population down. Or, in urban environments, seeing-eye dogs. The history of domestication of animals goes back possibly 40,000 years for dogs and about 10,000 years for other work animals. Why is any of this wrong? Singer's answer seems to be that we are depriving the animal its freedom to develop naturally. But it would be bizarre to hold the parallel opinion that any interference with any human being is wrong because it deprives them of their freedom to develop naturally. This would make any kind of training wrong. So with both animals and humans, not every interference need be illegitimate. And not every interaction is "interference."

None of what is said here implies anything other than that we should avoid giving pain to anything, animal or human, whenever we can do so without great cost to ourselves. In this connection, the important issue of factory farming will be discussed below. Singer spends a great deal of time reviling the evils of factory farming, which I agree should be almost entirely condemned. But discussing factory farming tends to build an emotional case for Singer's noninterference principle. There are other alternatives besides treating animals inhumanely as material objects or not dealing with them at all.

Suppose we accept utilitarianism just for the moment rather than the much more stringent Noninterference Principle (As pointed out, utilitarianism is also more-or-less the Buddhist view of our relation to animals.). As with utilitarianism for people, the real problem is how to sum up the various pains and pleasures. Do we count each animal's suffering as equal to a human's? Utilitarianism for people yields counterintuitive results unless we consider average amount of pleasure or pain, rather than total pleasure or pain. If we don't consider average utility for people, the Chinese always get what they want because there are more of them. But it is hard to see how to apply averages if we include animals in our utilitarian calculations.

The reason for the difficulty is that utilitarianism makes sense only if applied in the context of a (human) society. But it omits consideration of the reason why a human society is the appropriate context for ethics. As principles of cooperation, ethical principles are relevant only in a context of cooperation, that is, they are relevant for people who share the benefits and burdens of cooperative enterprises. This consideration is critical in attempts to apply social contract ethics to animals.

INCLUDING ANIMALS IN THE SOCIAL CONTRACT

Social contract ethics is more directly concerned with principles holding between people sharing benefits and burdens than utilitarianism. In a limited class of cases, social contract ethics is immediately relevant for our conduct toward animals, namely, when the animal is not wild. This can be because the animal is domesticated or because the animal is captive. Domesticated animals share the burdens of social cooperation. So they have some call on benefits commensurate with their burdens. They are entitled to decent living conditions and some opportunity to do things that give them pleasure. This applies to working animals and beasts of burden. Pets are also entitled to these necessities of life, because we have taken them out of habitats where they can care for themselves. They are dependent on us for the necessities. Universal principle ethics, a foundation for a social contract, would require that we give them care.[5]

Kant, the founder of universal principle ethics, ends up with similar recommendations, but for somewhat different reasons. For Kant, cruelty to animals is a violation of one's duty to oneself to cultivate compassionate feelings, which are ethically significant only with respect to other humans. Kant felt that if animals are killed, it should be quickly and without pain. Also animals should not be made to work beyond their capacity. There should be no painful physical experiments just for speculation, especially when results can be obtained without them. "Even gratitude of the long service of an old horse or dog only belongs indirectly to man's duty to these animals." The duty is always to a human itself (1797, sec. 17). There are two things going on here: For Kant, ethics is strictly for humans and thus one way humans are superior. Second, for Kant, even if our relations with animals are not as beings sharing a cooperative enterprise, the feelings of other sentient beings should still have some weight.

The contemporary philosopher Mark Rowlands attempts to use social contract ethics (also called *contractarianism*) to formulate ethical principles for our relation to animals. Rowlands begins his contractarian defense of animal rights by a flawed attempt to distinguish two kinds of contractarianism, which he calls Hobbesian contractarianism and Kantian contractarianism (Rowlands 2009). Hobbesian contractarianism turns out to be close to what is called *ethics* in this book. And his Kantian contractarianism is close to *morality* as characterized in this book.[6] However, his Kantian contractarianism has nothing whatever to do with Kant and little to do with any social contract.

His mistake concerning Hobbesian contractarianism is that he thinks it is self-interested. But the whole reason for introducing a social contract is precisely that individual self-interest and cooperative interest conflict. In order to obtain cooperative benefits, cooperative interest and not self-interest must prevail.[7] Ethical principles are principles regulating cooperative benefits. Indeed this is how Kant thought of ethics. Rowlands for some peculiar reason thinks Kantian ethics is based on an ungrounded Moral Law known by intuition. But what Kant calls the Moral Law is none other than universal principle ethics founded on his *Categorical Imperative*: Act on principles that could be willed to be universal law. Kant provided several ways of applying the categorical imperative, and one of those is explicitly a social contract. Kant calls it the "Kingdom of Ends:"

By 'kingdom' I understand the systematic union of different rational beings through shared laws. Since laws determine which ends have universal validity, if we abstract from personal differences of rational beings, and also from all content of their private ends or purposes, we can think of a unified whole of all ends in systematic connection . . .This is a kingdom of ends . . . (1785, p. 74).

Kant's suggestion here was elaborated in the 20th century by John Rawls, who was responsible for reviving contractarianism as a viable ethical alternative (1999). But neither Rawls nor Kant thought that an intuited unjustified Moral Law could be the basis for contractarianism. In my terminology, 'Moral Law' rather than 'ethical principle' is the right name for Rowland's principle. For some reason, he thinks it is objective. It is no more objective than any religious or cultural principle that cannot be justified as a cooperative principle.

So Rowlands' justification of animal rights ends up not being on the basis of a genuine social contract at all. He claims that Rawls' social contract, the original position, is merely a heuristic device[8] (Rowlands 2009, p. 137). But for Rawls, we can use the original position to determine our ethical judgements about justice. It is a position we can adopt and reason from. Just as when we are members of a jury, we have no trouble abstracting from information deemed inadmissible to arrive at a verdict. It would be ludicrous to say that jurors abstracting from inadmissible evidence are merely using a heuristic device. So too with Rawls' original position.

Rowlands procedure to admit animals to the social contract is to strip Rawls' social contract of all features that might yield ethical principles for social cooperation. He argues it need not between multiple people and therefore need not be a contract. It is true that *one* person might come up with good judgements, in the same way that *one* juror might come up with the correct verdict. But it misses the fact that what is being agreed on are *social* principles of action, not individual principles. Rowlands also never notices that the point of the original position exercise is to find principles governing distribution of benefits and burdens produced cooperatively and shared by a society.

So it is no wonder that he can include animals in this denatured *social* "contract."

Rawls' contract is a *social* contract holding between members of the same society. Rowlands' "contract" is a heuristic device between animals (human and non-human) deciding what moral principles to adopt. In Rowlands' original position, the fact that one is human is *excluded,* and so is one's rationality. His justification is that these are undeserved attributes. But attributes are deserved or undeserved in the context of social institutions such as employment or other social benefits. Are we competing against penguins or lions for jobs? And excluding rationality in the original position is literally insane. Most of the time, we are in no kind of cooperative relationship with animals. It makes sense, as suggested above, to include working animals in a social contract, just because we are in a cooperative relationship with them.

So this particular attempt to provide a social contract general foundation for ethics concerning non-human animals is a failure. And it is hard to see how any attempt to use a social contract to justify ethical principles concerning animals can get around the fact that a social contract is between participants in a society sharing benefits and burdens. It was for this reason that principles governing our behavior with respect to the ecosystem, should be called *eco-ethical principles* rather than ethical principles.

ECO-ETHICAL PRINCIPLES FOR OUR RELATIONS WITH ANIMALS

Let us return to our naturalized Buddhist basis for eco-ethics, principles for our relations with animals: Since all things are interdependent, other beings are not there just for our use. We cannot dispose of them in any way we desire. No kind of being is privileged over other living beings. Each type of being deserves the type of respect appropriate for that type of being. Human beings deserve the type of respect appropriate for beings of our own species. Other sentient beings deserve the type of respect appropriate for beings with consciousness. Living beings deserve the type of respect appropriate for beings continuing through time by reproducing using DNA. Non-living things deserves the type of respect appropriate to beings which are not alive.

As members of our species, human beings should be treated in accordance with principles which could be universal law for human beings. But this is as far as universal law ethics can go. As we saw just above, universal law ethics is implicitly social contract ethics. You are making laws for beings who share your ability to make and obey laws.

Ultimately we want ethical decisions concerning members of other species to reflect our place among them, as a species among species. There is a certain amount of reciprocity here—we would not want another species to wipe us out arbitrarily, so we should not wipe out other species arbitrarily. There is also some non-reciprocity here—it is legitimate for human beings to wipe out the life forms smallpox virus and polio virus.

Given the nature of the ecosystem, some life forms must sometimes be killed. It is impossible to avoid killing life forms inadvertently. Digesting food kills many microorganisms that we don't even know about. In previous chapters, I also told of my experience attempting to water some plants in my front yard. I severely disturbed a nest of ants, which began furiously running around and rescuing their larvae. I am sure I killed some of them. But I had absolutely no intention of causing any harm. Situations like this occur because the universe is composed of different levels which are largely unaware of each other and whose interests conflict.

There are also many conflicts of interest between living things where one or more of the beings involved are on the same level and very much aware of both sides of the conflict. Here are two: Predator-prey, human-pest. We know that the predator-prey relation is often necessary for a healthy ecosystem and developed through natural selection. The human-pest relation is a different story. From a human point of view, getting rid of pests benefits humans and often harms the ecosystem. Pesticides not only kill insects and unwanted plants, but are harmful to humans as well. Once again, since modern technology ignores the ecosystem, the possibility of human harm is also ignored. After all, humans are part of the ecosystem. But, as Singer points out, technology can find ways to reduce fertility of mammalian pests rather than exterminating them. Reducing insect fertility does not sound like an easy thing to do.

Three other conflicts of interest are:

- Between human space and animal habitat;
- Between humans and animal products; and
- Between humans and animals killed for food.

Since humans ultimately need the ecosystem in ways that cannot be anticipated, it is important that humans leave enough habitat undisturbed so that the ecosystem from which we arose can continue to function. Dave Foreman's estimate of 40 million acres in the US is a decent guess (1991, 187) (40 million acres = 63,000 sq. mi., about the area of Wisconsin.). It is also important not to stint on the amount of land reserved and its freedom from human use. Some human use might be possible, but anything other than any remaining hunter-gatherers or humans under wilderness area conditions would defeat the purpose of such reserved habitat. Definitely no roads, vehicles, or "recreational" use. Exterminating other species does not deserve the name recreation. Other than hunter-gatherers engaging in traditional non-technological pursuits, animals and plants within these ranges should not be subject to management. There might be extraordinary circumstances in which human intervention might be called on to restore a balance, when disaster of some kind has upset the balance.

On the view of many people, there is in general not a conflict when humans use animal products. Cattle, chickens, sheep and silkworms are used for what they produce other than their own flesh. These animals are not killed in producing their products. So, provided that these animals are kept under good conditions, and provided it is not painful for the animal to generate their product, there is nothing wrong with the animal generating the product and our using it. Thus silk and wool are OK eco-ethically. Chickens apparently do not have strong feelings one way or the other about producing infertile eggs. They have been bred to be good at egg laying and seem to regard it as a normal thing to do. So when chickens are kept humanely (free range rather than factory) and are not subject to painful practices such as forced moulting, there is no eco-ethical problem with using eggs. Humane treatment would include letting chickens past egglaying live out their lives. The key to eco-ethical treatment is that insofar as possible, the animal must not be treated as a non-living machine.[9]

Dairy cattle have been bred to produce milk in large quantities, and it is painful for the cows not to be milked. However, cows must be bred once a year and give birth to a calf in order to produce milk. For milk production purposes, the calf is separated from the mother at birth and usually becomes veal. Dairy cows can usually give milk for two to four years and when milk production wanes, are usually slaughtered. Given that the average natural life expectance of a cow is around 20 years, dairy cows are clearly being treated in many countries like milk producing machines. An exception is India, where Hindu belief holds cows to be sacred because they give milk. Many Indian states prohibit the slaughter of cows—aged cows wander the streets. Unless Hindu practices are adopted, the characteristic Western treatment of dairy cows is not eco-ethical. Nutritionally, milk's high protein content is good for growing young animals. But in adults it helps any existing cancers to grow.

There are excellent non-animal substitutes such as soy milk. Vegan friends tell me that there are excellent vegetable-based cheeses, but it is hard to believe that the whole flavor range of animal-milk-based cheese is duplicated. But cheese flavor is not a sufficient reason for treating cattle as machines. However, if animals are not treated as machines nor killed to produce the products, then it can be eco-ethical for humans to continue to collect and distribute these products.

It is true that sometimes the animal is harmed. In which case the eco-ethical course of action would be to shun products produced when the animal is harmed and agitate for an end to such harm.[10] But the vegan objection that the animal is always harmed because we interfere with the animal depends on the principle that it is always wrong to interfere with animals. It was previously indicated why this principle is simply incorrect.

Singer also mentions cruel treatment of sheep in collecting wool in Australia as a reason for not using wool. But surely in getting wool, sheep can be treated in ways that are not cruel. One of the problems with not using animal products is that synthetic replacements are often damaging to the environment. To give up using animal products because the production process is now not humane, condemns both us and the animals to a certain amount of environmental degradation. This seems inappropriate because the welfare of domestic animals can be dramatically improved through human activism.[11]

KILLING FOR FOOD

This brings us to killing for food. Plants will be briefly discussed then animals. As we saw earlier, Jainism calls for not killing plants or animals, so one's diet is limited to plants which are not killed when harvested. All root vegetables are out. Most leafy vegetables are OK, as are nuts, seeds, and seed oil. Traditional Jains were lacto-vegetarians, but became vegans when modern treatment of veal calves was introduced.

Actually, vegans have no problem with root vegetables, so Jainism is now actually stricter than veganism. There is no ethical objection to Jainism or veganism as a religious or limited group practice. But when universality is claimed, we need to have good reasons for that practice. As mentioned at the beginning of the chapter, veganism tends not to justify its practices. So there may be no reason why vegans do eat root vegetables but do not eat dairy products. The Buddhist view that sentience is what matters actually does not eliminate plants. Recent studies have shown that plants are aware of their environment in various ways (Mancuso 2012). Singer discounts claims of plant sentience as unscientific frivolity. He claims that for plants to be sentient, they must have behavior, a nervous system, and the evolutionary value of pain for them must be demonstrable (p. 236).

No one claims plants have a nervous system. Rather, plants obtain information from and give information to the environment by means of hormones. This can include a plant's making the ground around it inhospitable except to plants of the same species. No nervous system, no pain. In any case, Singer has an answer: If they suffer less, it is OK to eat them, given that we have to eat something (p. 236). Actually it seems likely that plants are sentient, but do not feel pain. So no problem for Singer's vegetarianism, but Buddhists seem to be uneasy about sentient plants.[12]

Some of the same considerations for using animal products also apply when the animal is killed for food. The animal should have decent living conditions and should be killed painlessly. Singer believes in general animals should not be killed for food. He mentions a number of permissible exceptions. For example, the ethical principle not to kill animals would not apply to hunter-gatherers who have no choice (I will shortly consider the effect of different stages of human development on the killing of animals.).

For Singer, other exceptions would be:

- When the animal is a predator and immediately threatens human life.
- When it is necessary to eat meat for survival. (The Inuit and Tibetans have no vegetables.)

I would make an addition to Singer's list of exceptions:

- When not eating meat would damage one's health

In other words, Singer's principle of equal consideration for animals does not trump maintaining human health. There is no eco-ethical justification for a human's leading a miserable and possibly shorter life so that some animal can live a longer one. (I am assuming that any meat is obtained by painlessly killing the animal.)

It turns out that of people who adopt a vegetarian life style, about 15% are forced to give it up and include some meat in their diets (Herzog 2011). Here is my daughter Katie's experience. She was very athletic, and the experience she relates happened while she was biking across the US:

[I] started eating a bit of meat during my cross country bike trip because my body just didn't seem to be getting what it needed from a vegetarian diet (granted, this was while traveling through tiny towns in the middle of the US where well-balanced and varied vegetarian foods were not available). But once the trip was over I remained

largely vegetarian until probably 2004 or 2005. At that point, I started getting cravings for meat and it felt that my body was telling me I needed to make a nutritional change...not frequently or in large quantities, but I found I felt better and healthier while eating meat. As you know, we try to eat only free-range, pasture fed meats . . . (Schultz 2012).

The issue here is how much one's health should be impacted to justify eating meat. Does one have to be incapacitated by a vegetarian diet? Or, like Katie, is it enough to feel better when eating meat? It is worth noting that Buddhists, who are vegetarians in principle, make exceptions. The Dalai Lama himself eats some meat for health reasons. And in the higher elevations of Tibet, there are no vegetables. So they count yak as a vegetable.[13]

One way of looking at the issue: Can it be eco-ethically required for me to let my health suffer so that an animal is not killed? Eco-ethics requires that if an animal is killed, it should be killed as painlessly as possible. For me as another animal in the ecosystem, there is no ethical problem eating animals when they have had a decent life and a painless death. I was a vegetarian for a few years, but because of a side effect of an essential medication, I gained a lot of weight very unhealthily. Eating some meat reversed this process. As an animal among animals, if my well-being requires eating meat, I see no eco-ethical objection to doing so. Of course, some meat is obtained under conditions that are not eco-ethical. Factory farming typically treats animals in horrendous ways. (Factory farming is discussed later in this chapter.) In general, if one eats meat, one should eat free range pasture farmed meat that is killed painlessly.[14] Some religions require killed animals to suffer painful deaths. It is up to each individual to decide whether religious rules are more important than an animal's extra suffering.

EATING MEAT IN HUMAN DEVELOPMENT

When we were hunter-gatherers, killing and eating animals was an essential source of protein. Nature, as it were, concentrated the protein for free in the form of prey animals. However, the killing of animals begins to go way beyond their use as food in the megafauna extinctions beginning 60,000 years ago. But for the most part, hunter-gatherers did not have much choice in their diets.

Farmers normally keep food animals such as chickens, pigs, or cattle. These animals are treated well, if conditions permit, and occasionally farmers use an animal for food. Since chickens are used for eggs and cows for milk and milk products, farmers have an incentive to treat these animals well. Even animals used for food are best when treated well before slaughter.

But when we come to civilization, animals are treated as objects. Because food production takes place outside cities, the inhabitants of cities have no direct connection with the animals that become meat. Thus, it is civilization which causes severe problems in human-animal relations.

However, both vegans and Singer seem to believe that human development took a wrong turn when animals were domesticated, and perhaps even when humans developed tools powerful enough for them to kill animals for food. One relevant contrary fact is that human brains began rapidly increasing in size because of the greater availability of high-protein animal meat. Hypothetically, if our brains had never increased in size, we might not be faced with the question of our differing relation to animals.

Vegans are perfectly happy with the disappearance of domestic animals if their policies were universally adopted. They point out that when automobiles replaced horses, the number of horses declined drastically.[15] If indeed their principle seems to be human noninterference of any kind with nonhuman animals, the status of house pets also seems uncertain. We could simply stop breeding cats and dogs by neutering them all, and then we would be left with no "wrong" animal interactions. But this solution to the gap between humans and nature allows us no role in nature whatever. Since we are ultimately part of nature, this solution has to be wrong.

In Chapter 9, it was concluded that having more people is not better. So, the vegan might say, the number of horses or cows or chickens doesn't matter. If these animals fade away, that is not an ethical problem (or moral problem or eco-ethical problem). But it is! If, having created these animals for our own use, we let them go extinct, that is just as bad as any other deliberate human-caused extinction. It is not the animal's fault that we bred them for specific human uses. The conclusion that having more humans is not better allows that there is also a minimum population below which people would do worse. It is the same with human-created animals. We have to keep the minimum number of each species alive that would allow for the species to flourish and for individuals of the species to have a reasonably good life. This much we owe them for their having provided us with support for thousands of years.

Singer has a different take on this issue. He claims that one of the few issues he has changed his mind on, is the issue of whether bringing animals into existence is a positive thing. He now thinks this is a good thing (2009, pp. 228-230). But as we have just seen, merely increasing the numbers of human beings for its own sake is hardly a good thing, and the same holds true for domesticated animals. So Singer's view on this issue has no merit.

As we saw in Chapter 10, "Environmentalism and Sustainability," because of the ecological impact of meat production versus vegetable production, it is probably a good idea to eat 'lower on the food chain'—that is, eat vegetable rather than animal protein. It takes 21 pounds of vegetable protein to produce one pound of animal protein (Singer 2009, p. 165). So eating less meat can yield dramatic savings in farmland and fertilizers. This would be significant in helping the ecosystem.

Recently artificial meat, animal tissue, has been produced directly from stem cells with no animal involved (Collins 2012). This use of technology does not incur any of the objections to killing animals. If the process does not have other negative effects, there seems to be nothing wrong with it. But if we are eating meat for health reasons before artificial meat is developed, we definitely should only eat free-range animals killed painlessly.

FACTORY FARMING

In current western civilization, most meat is produced through factory farming. It is close to impossible for factory farming to satisfy eco-ethical principles for the treatment of animals.

Factory farming is the application of modern technology through corporations to the production of animal or vegetable products. It exhibits all the characteristics of modern technology that make it such a disruptive force. In the case of animal factory farming animals have become the resource to be transformed into products in the most efficient manner possible. As in all uses of modern technology, a new incompatible order is imposed on an older one. To obtain the maximum yield at the minimum expense, modern technology ignores any contrary side effects. In the case of factory farming, this means the maximum number of animals is crowded into the minimum space. The result is enormous pools of animal manure which severely pollute the environment and which factory "farmers" postpone dealing with as long as possible. There is absolutely no concern on the part of the "farmers" (actually corporations) about the well-being of the animals. A certain death rate is acceptable. Chickens are kept in cages no bigger than the chickens themselves, with no room to move. Their beaks are removed to prevent them from damaging each other, and their droppings fall beneath the cages where they are not cleaned up. It is cheaper to put antibiotics in the chicken feed.[16] In the case of cattle, manure and waste can accumulate to depths of a few feet.

One does not have to reject speciesism to believe such practices are inhumane. The term 'inhumane' is interesting. When we treat animals humanely, we treat them in ways appropriate for human beings who are kind and compassionate. In other words, with the same kindness and compassion we bestow on other human beings. It is not uncommon for ordinary people to disapprove of the treatment of animals in factory farms. In 2008, California voters approved Proposition 2, which prohibits the confinement of farm animals in a manner that does not allow them to turn around freely, lie down, stand up, and fully extend their limbs. Voters in other states—Florida, Arizona, Colorado and Oregon—have passed similar laws for pigs and calves (Ballotpedia 2011).

As with many applications of modern technology, factory farming produces unintended 'externalities' which cause harm both to humans and to the ecosystem. Besides the problem of waste disposal, there are serious consequences for the use of antibiotics either to stimulate growth or to obviate the need for cleaning up the animal's environment. Bacteria are evolving quickly, and the use of antibiotics in factory farms has the result that these antibiotics become ineffective both for animals and humans much more quickly.[17] Also, antibiotic-resistant bacteria are evolving in the factory farm environment. Medical practictioners are working to have the use of antibiotics on healthy animals banned. At an absolute minimum, factory farmers, like other corporations driven by modern technology, should pay for any externalities in the environment their operations produce.

Because of the principles behind factory farming, it is difficult to see how it could be transformed into an eco-ethical institution. Built into factory farming is the idea that animals are objects which we can make use of in any way we can. The various propositions to make the life of animals on factory farms more tolerable are certainly a step in the right direction, but they don't address other problems such as the misuse of antibiotics and the accumulation of waste. A dramatic reduction in meat consumption would be one way to obviate the necessity for factory farming.

The justification of factory farming is that it produces the greatest amount of food per dollar. Environmentalist Bill McKibben points out that small-scale farms are more productive in terms of food per land unit than factory farms. And if one factors in environmental damage, it is far from clear that factory farming is more cost effective. McKibben recommends a return to small individual farms for raising vegetable food. He does not explicitly recommend returning to small-scale farms for raising animals, but he gives several salient examples of the greater productivity of such ventures without toxic side effects. His account of the village of Gorasin in Bangladesh is especially telling because the villagers explicitly stopped using the tools of the "green revolution" of pesticides, artificial fertilizers, and genetically engineered seeds and thereby dramatically improved their lives and their productivity (McKibben 2007, pp. 204-206).

This brings up once again the issue of the role of technology in human-animal relations. We have seen a number of negative aspects of modern technology as it relates to animals. But in the course of this chapter, some potentially positive contributions of modern technology to animal well-being have appeared. One was Peter Singer's observation that synthetic materials can replace animal products. Examples would be replacements for leather and fur. As remarked earlier, one would have to examine the ecological consequences of producing these synthetic replacements.

Are there toxic by-products? Since most synthetics are derived from petroleum, the impact on climate change would have to be assessed. It would not do to make life better for some animals at the cost of great harm to all beings, including them.

The second potentially positive contribution of technology to the well-being of animals is synthetic meat, raised from stem cells without having to kill anything. If we need to have the underlying biotechnology anyway, this development seems like a winning proposition for all animals used for meat.[18]

MORE ASPECTS OF OUR ECO-ETHICAL RELATION TO ANIMALS

The discussion of this chapter has dealt with human uses of animals and with humans allowing animals their own space. But anyone who has lived with an animal for an extended period, either as a working animal or as a pet, knows that animals and humans communicate with each other. Such communication is emotional and not linguistic. And, especially with domesticated animals, mutual bonds of caring and respect are formed.

Some wild animals (chimpanzees, gorillas) have been able to engage in linguistic communication with us.[19] Some (whales, elephants) also have their own communication systems which we are trying to understand. Research in these areas should be encouraged as much as possible. The more we can communicate with animals, the less will it be possible to treat them as objects. For how long do you think the Japanese would be able to keep killing whales for their specious reasons if the whales they were about to kill spoke to them in Japanese? Or if the Japanese whalers understood whale communication?[20]

REFERENCES

Avila, J. (2012). Superbug dangers in chicken linked to 8 million at-risk women. *ABC News Blogs*. Retrieved 7/15/12 from yahoo.com/abcnewsblogs/superbug-dangers-chicken-linked-8-million-risk-women-122809803—abc-news-health

Ballotpedia. (2011). *California proposition 2, standards for confining farm animals (2008)*. Retrieved 10/16/11 from ballotpedia.org/wiki/index.php/California_Proposition_2,_Standards_for_Confining_ Farm_Animals_%282008%29

BBC. (2012). *Jainism*. Retrieved 4/3/12 from www.bbc.co.uk/religion/religions/jainism/

Bentham, J. (1781). *An introduction to the principles of morals and legislation*. Retrieved 10/16/11 from www.constitution.org/jb/pml.htm

Borenstein, S. (2012, April 13). See Dan read: Baboons can learn to spot real words. *Associated Press*. Retrieved 4/13/12 from news.yahoo.com/see-dan-read-baboons-learn-spot-real-words-180435856.html

Buddho Dendo Kyokai. (2009). *A teaching a day: Excerpts from the teaching of Buddha*. Berkeley, CA: BDK America.

Collins, N. (2012, April 13). Test tube hamburgers to be served this year. *The Telegraph*. Retrieved 4/13/12 from www.telegraph.co.uk/science/science-news/9091628/test-tube-hamburgers-to-be-served-this-year.html

Cook, G. (2012, June 5). Do plants think? *Scientific American*. Retrieved 6/5/12 from news.yahoo.com/plants-think-110000318.html

Dalai Lama. (2005). *The universe in a single atom*. New York: Morgan Road.

Foreman, D. (1991). *Confessions of an eco-warrior*. New York: Harmony Books.

Hackett, J. (2013). *Vegan - What is a vegan? What do vegans eat?* Retrieved 4/12/13 from vegetarian.about.com/od/glossary/g/Vegan.htm

Herzog, H. (2011, June 20). Why do most vegetarians go back to eating meat? *Psychology Today*. Retrieved 4/10/12 from www.psychologytoday.com/blog/animals-and-us

Kant, I. (1785). *Groundwork of the metaphysics of morals*. Retrieved 3/24/12 from www.earlymoderntexts.com/kgw.html

Kant, I. (1797). *The doctrine of virtue: Metaphysics of morals, part II*. New York: Harper & Row.

Lemieux, N., & Maynard, B. (1999). Deer damage and control. *University of Rhode Island Landscape Horticulture Factsheet*. Retrieved 10/16/11 from www.uri.edu/ce/factsheets/sheets/deermanage.html

Mancuso, S. (2012). *Journal of plant signaling & behavior*. Retrieved 4/10/12 from www.plantbehavior.org/about.html

McKibben, B. (2007). *Deep economy*. New York: St. Martin's Griffin.

Rawls, J. (1999). *A theory of justice* (Revised ed.). Cambridge, MA: Harvard University Press.

Rowlands, M. (2009). *Animal rights: Moral theory and practice*. New York: Palgrave Macmillan.

Singer, P. (2009). *Animal liberation (updated ed)*. New York: HarperCollins.

Thomas, P. (2012, October 23). Talking whale? Captive beluga had learned to mimic human voices. *GrindTV.com*. Retrieved 10/23/12 from www.grindtv.com/outdoor/blog/37509/talking+whale+captive+beluga+had+learned+to+mimic

ENDNOTES

1 One Buddhist attempt out of the inconsistency is to say the seed is not static, it is changing. But that doesn't matter. The seed still has to be *my* seed, and when reincarnated the previous life has to be *my* previous life. If both of these are not true, reincarnation is pointless. Past lives can have no connection with present lives.

2 Later forms of Buddhism add a number of supernatural items possibly in imitation of competing religions. For example, Pure Land Buddhism adds something like the equivalent of a Christian Heaven, with all its existential problems.

3 Chapter 1, "Humans as Masters of the World," has extensive discussion of Judeo-Christian ideas of human dominion over the world.

4 "Hedonism" here means merely "having to do with pleasure and pain." It does not mean compulsively seeking sensual pleasure.

5 See Chapter 1, "Humans as Masters of the World," "Ethics and Eco-Ethics."

6 See Chapter 1, "Humans as Masters of the World," "Ethics and Eco-Ethics."

7 See Chapter 1, "Humans as Masters of the World," "Ethics and Eco-Ethics," especially note 18.

8 Rowlands thinks there is a separate argument for Rawls' principles of justice involving equality of opportunity. Equality of opportunity is an individual right belonging under Rawls' first principle of justice, Greatest Equal Liberty. All individual rights and liberties have to be balanced against each other. Rowlands wants to elevate working out the details of one individual right—equality of opportunity—to the same status as the original position.

9 As will be discussed shortly, in 2008, California voters approved Proposition 2, which prohibits the confinement of farm animals in a manner that does not allow them to turn around freely, lie down, stand up, and fully extend their limbs. Voters in other states—Florida, Arizona, Colorado, and Oregon—have passed similar laws for pigs and calves (Ballotpedia 2011).

10 See note 9 directly above for a successful case of agitation.

11 Again see note 9. It is worth mentioning that this initiative law improving the conditions under which chickens are kept was passed by a majority of California voters. In other words, the policy was not adopted just because of the work of a small group of animal activists.

12 Cook (2012) interviews Dr. Daniel Chamovitz about his work which demonstrates a large number of similarities between animals and plants regarding sentient behavior.

13 The Dalai Lama explained in an interview by Piers Morgan on CNN in May 2012. When he was in Tibet, he ate meat because that was all that was available. When he fled to Dharamsala in India, he immediately switched to a vegetarian diet. However, within a few months he developed severe jaundice and was forced to include some meat in his diet from then on.

14 Buddhists who kill animals prefer to kill large ones, so that less sentient beings are harmed. This makes some sense with reincarnation, but otherwise because smaller animals are more numerous in the ecosystem, this consideration may not matter.

15 My thanks to Neil Anapol for this point.

16 Approximately 70% of all antibiotics in the US is used to prevent having to clean up after animals.

17 See Avila (2012) for a discussion of antibiotic-resistant e. coli bacteria in chicken causing untreatable urinary infections in human women. Of course, the poultry industry says that no study has yet shown the connection. We can speculate whether they will produce their own study with captive scientists or simply deny any causal connection, as the tobacco industry did for so long with cigarettes and lung cancer.

18 How much biotechnology we need to keep in an economy in equilibrium with the ecosystem was discussed in Chapter 11, "The Role of Science and Technology."

19 Writing and reading are purely human inventions dating from only about 7,000 years ago. It is, however, unlikely that writing and reading are based on capabilities that have evolved only in humans. Work with primates suggest that they possess some version of the underlying capabilities. They do not possess capabilities to read and write per se, but they possess similar capabilities to the ones humans exploited when they developed writing and reading. For example, recent work with baboons suggests baboons can distinguish words from non-words. See Borenstein (2012).

20 When this was first written, I thought it was only a fanciful dramatic ending. However, whales can apparently convincingly imitate human voices. See Thomas (2012).

Chapter 13
Regaining Our Place as a Species among Species

ABSTRACT

Civilization may not have that much long-term value for our species because it fosters fairly complete separation from the ecosystem. Ultimately, our survival rests with our being able to use our knowledge to respect the ecosystem and to live within its limits. The most distinctive characteristic of the human species is our ability to accumulate knowledge of our surroundings and act on the basis of that knowledge. Our species goal may be to determine the role of our kind of being in a universe mostly otherwise not conscious of itself.

INTRODUCTION

The relation of humans to the environment is conflicted. We assume we can care for the environment in ways the environment cannot care for itself. Yet we often ignore the fact that we are part of the environment. As pointed out in the Preface, an environmentalist demand that coyotes rather than human hunters reduce excessive deer population, assumes that human beings are not part of nature.[1] Similarly, when animal rights advocate Peter Singer wants humans to have no contact with other animals, he in effect denies that humans are animals.

How do we acknowledge that we are animals among animals in the ecosystem in the face of our great physical superiority over our environment? As we have seen, we have already set in motion changes in the ecosystem as far reaching as those of

DOI: 10.4018/978-1-4666-4586-8.ch013

the stromatolites.[2] We are not in charge, but we have physical superiority. We are already speaking for the ecosystem. There are problems with what we say on behalf of the ecosystem: Firstly, how to say only what will benefit the ecosystem as well as ourselves; and secondly, not to second guess its processes. After all those processes produced us and continue to sustain us as individuals and as a species.

DISTINCTIVE FEATURES OF HUMANS AS A SPECIES

If we are a species among species, what are the distinctive features of our species, which determine our place among species? Neither the cheerleading "Humans, hurray" or the animal rights "Humans, boo" is appropriate. What is appropriate is "Humans, here we are. This is what we are like. What should we be doing with our quite significant capabilities?"

Here is Live Science's list of the top ten things that make humans special: Speech; upright posture; apparent hairlessness; dexterous hands; big ("extraordinary") brains; clothing; fire; blushing; long childhoods; menopause (Choi 2011). This list requires some culling. Clothing and fire are hardly genetic properties of humans. But an animal that is for all practical purposes without fur will likely use its brain and dexterity to make clothing—initially, perhaps, acquiring fur previously attached to another animal. Perhaps lack of fur also made it important for humans to use fire to keep warm. But whatever the motive, both clothing and fire exemplify the human species trait of being able to repurpose natural processes to meet its own needs. And, as we saw earlier, both clothing and fire are early components of a separate human environment.

A long childhood and menopause are also evolutionary traits that function together socially. Long childhoods facilitate "the emergence of human social and cultural complexity", says Jean-Jacques Hublin, director at the Max Planck Institute for Evolutionary Anthropology. "[Long childhood] allows a long maturation of the brain and an extended education period"[3] (Max Planck Society 2010). Both long childhood and, especially, menopause, are examples of evolution acting socially. They benefit the species through benefiting the social group rather than the individual. Thus longer childhood may mean that individuals in the *Homo sapiens* group can outcompete faster-maturing Neanderthal individuals once *Homo sapiens* finishes acquiring more complex skills. Menopause ties in with this: It is much more valuable to the social group to have grandmothers available for the extra child care than for grandmothers to have more children of their own. There are many other examples of social evolution.[4] The prevalence of social evolution shows that the conservative view of both evolution and life as survival of the fittest *individual* is mostly wrong.

The social evolutionary function of blushing has only recently been uncovered. Because it is involuntary, it is a reliable indicator to others that you sincerely regret or are embarrassed by what you did. It therefore forestalls or ameliorates retaliation (Bering 2009).

Speech, upright posture and dexterous hands are clearly important for allowing humans to attain great complexity in their actions. The capacity for speech evolved 350,000 years ago, 100,000 years or so before modern humans evolved. Both our precursor species (some version of *Homo erectus)* and the Neanderthals had the capacity for speech. Speech allows for the complex coordination of cooperative activity, not least of which is an elaborated shared reality. But speech is in itself neutral with respect to our relation to the ecosystem and hence our place as a species among species.

Upright posture and dexterous hands together allow humans flexible use of tools. Upright posture (bipedalism) evolved millions of years before early humans began making tools, and there are a number of somewhat conflicting hypotheses about this evolution. A number of different animals have evolved bipedalism, for example ostriches and kangaroos (Niemitz 2010). However, once early humans began making tools, the upright posture certainly facilitated their use. Darwin hypothesized that the human hand evolved to better make and use tools. His hypothesis has been supported by recent research (University of Kent 2011). The ability to manipulate the environment in complex ways clearly promoted to the separation of human beings from the environment, culminating in the almost totally man-made environments of civilization and modern technology.

The evolution of hairlessness (lack of fur) on humans has been researched extensively by Nina Jablonski. All other primates and most other mammals have fur. There are reasons for thinking the human adaptation of no fur occurred 1 or 2 million years ago, and it was for more efficient cooling (Dreifus 2007). Climate change or the necessity to travel long distances might have called for this adaptation. Its utility can be seen in present-day hunter-gatherers in Africa. Furred animals cool by panting, and they cannot pant and run at the same time. So a gazelle will overheat and have to stop long before a human, who can then simply kill the gazelle (Carrier 1984).

Jablonski also claims that hairlessness and its accompanying cooling system facilitated larger brains in humans. As soon as more effective cooling is possible, the brain can become larger (Dreifus 2007). Elephants, one of the few land mammals also without fur, also have an efficient cooling system and the largest brain of any land mammal.

As soon as the LiveScience article about the distinctive features of our species starts talking about the human brain, it is clear that we are back in human superiority territory:

Without a doubt, the human trait that sets us apart the most from the animal kingdom is our extraordinary brain. Humans do not have the largest brains in the world—those belong to sperm whales. We do not even have the largest brains relative to body size—many birds have brains that make up more than 8 percent of their body weight, compared to only 2.5 percent for humans. Yet the human brain, which weigh[s] only about 3 pounds when fully grown, gives us the ability to reason and think on our feet beyond the capabilities of the rest of the animal kingdom, and provided the works of Mozart, Einstein, and many other geniuses (Choi 2011, p. 6).

So somehow we determine that our brains give us the ability to "reason and think on our feet beyond the capabilities of the rest of the animal kingdom." And we judge this by pointing to the work of such fellow humans as "Mozart, Einstein and many other geniuses." It was pointed out earlier that it is circular to point to human accomplishments in art, science, and technology as justification for human superiority. What meaning can our accomplishments have for non-human species? They did not evolve together with us as part of the complex social Web within which we can communicate with each other and share goals and values. Other intelligent species do not manipulate the environment to anywhere near as great an extent as human beings. Dolphins and other cetaceans have much larger and more convoluted brains than humans but are not using them to manipulate their environment. Their brain size must be useful or else it would be selected away fairly quickly. Complex brains require a lot of energy to maintain. For example, human brains consume at least 33% of the individual's energy. We do not know what cetaceans are using their brains for (Kaiser 1990). And many if not most of our accomplishments are manipulations of the environment.

So which of our species characteristics fits us to be 'a species among species?' That is, a species whose activities are in equilibrium with the ecosystem? Which of: Speech; upright posture; dexterous hands; apparent hairlessness; big brains; blushing; long childhoods; menopause? Some of these improve our ability to coordinate activity with each other: speech, blushing. Almost all of these seem to be neutral with respect to our adaptation to the ecosystem: Speech, blushing, upright posture; dexterous hands, hairlessness, big brains. The only two which are more than neutral with respect to our adaptation to the ecosystem are long childhoods and menopause. The fact that humans are evolutionarily adapted to learn more is essential for human adaptation to the ecosystem. This adaptation must be part of a legitimate species goal: To learn as much as we can about our surroundings. Our legitimate species goal is not to appropriate everything around us in a mindless attempt to expand endlessly.

Two other human characteristics are not mentioned in this list: The human propensity to violence; and the ability to create worlds from a point of view. The propensity to violence will be discussed first.

It was noted earlier that the currently preferred view is that modern humans (*Homo sapiens)* arose once in Africa, then migrated out of Africa about 60,000 years ago to the rest of the earth. *Homo sapiens* then replaced earlier species because it had superior technology. The alternative multiregional hypothesis holds that *Homo sapiens* gradually replaced earlier humans in several areas: Africa, Asia, Europe, and Australia.

This 'Out of Africa' hypothesis has disturbing implications about how *Homo sapiens* replaced earlier species. What could this superior technology have been at this point in human development other than more effective weapons? It is most likely on this hypothesis that *Homo sapiens* simply exterminated the earlier humans. *Homo sapiens* subsequently exterminated all megafauna in sight. Certainly, the subsequent history of *Homo sapiens* has shown a striking tendency to exterminate not only other species, but also large groups of its own species. Genocide has been a regular feature of human history[5]

So what are the consequences for humans as a species among species? If one of the species *Homo sapiens* distinctive traits is its superior ability to kill, the ability of that species to coexist with the ecosystem must be an acquired trait. If, however, the multiregional hypothesis is correct, there is less evidence for the superior ability to kill as a distinctive trait of *Homo sapiens.*

The second characteristic which was discussed earlier is the ability to create representations, which entails the ability to assume multiple points of view. This characteristic is in itself neutral with respect to human separation from the eco-system—it could be used to separate us from the ecosystem or to make us closer. However, the ability to assume multiple points of view is essential to our being able to take a wider point of view than that of human beings striving to get ahead on a planet. Thus we are able to take the point of view of various animals, various plants, and even the ecosystem or the planet itself.

Perhaps this point of view ability is a special human trait. Perhaps animals with the ability to think complete thoughts may also have this trait. However, without language, there seems no way they could demonstrate this point of view trait. Does this trait make us superior, or at least not a species among species? Rather, the trait gives us the ability to consider ourselves as a species among species by allowing us to take the point of view of other species and consider ourselves alongside them.

GOALS FOR THE SPECIES HOMO SAPIENS

Is the best that human beings can do with the considerable talents of their species, to expand mindlessly into their environment in ways that may greatly shorten their tenure on the planet? (This is sometimes mistakenly called growth.). Or are there goals beyond this that we can adopt as a species? Right now, activity incorporating consciousness of our place in the ecosystem tends to be re-active, conservative in the root sense of that word: We *refrain* from obliterating yet another piece of nature to build another suburban tract. We *limit* or *control* population. We *conserve* the environment. Indeed such activities are of the utmost importance until such time as we figure out what we are doing as a species and what we should be doing.

We know as managers that a business without a strategic plan is a business likely to go out of control and therefore out of existence. Yet as human beings with the ability to affect the survival of all life on the planet one way or another by our activities, we do not have a strategic plan that includes the conditions for our own survival. The ultimate ethical question of how we should be using our capabilities is so far unanswered.

Goals usually advocated in this area are all *enabling* ends. If realized, they will free up people to pursue their own special individual ends. But they do not bear on the larger question being raised. For example, Al Gore's (1992) *Earth in the Balance,* includes an excellent summary of the ecological problems which threaten long-term human existence. But Gore's message that we just need to get the economy back on track and avoid wasteful consumption does not get to the root of the ecological problems now facing us.

So, what about the goal of the species as a whole? Growth is not a goal in itself. Expanding heedlessly is not a species goal. Maximizing profits is not a goal. Even arts and sports are not species goals but nice side efforts. Perhaps the closest to a goal of this kind is space exploration. Although a laudable end, it often only amounts to reproducing the aimlessness and lack of direction of our species on other planets.[6] Justin St. P. Walsh's discussion shows the futility of trying to reproduce the exploitation of this planet in outer space (2012).

One reason for attempting to formulate such species-level goals is the fact that our species will not always exist. Although I do not share a "manifest destiny" technologist's belief in the invincibility of the human species or its technology, I finally do hope that the human species has something like the normal two-million-year life span of most species. So what will happen to our accumulated knowledge, our artworks, our buildings, our technology, our systems of communication, our social systems, our ethics, our spiritual connections, and so on? The only thing that prevents human knowledge from dying with its possessors is other human beings who are prepared to carry on the human enterprise. What meaning can it have for non-human species? Philosophers such as Kant and Wittgenstein have noted that our knowledge is in a deep sense human knowledge, tied to the conditions of existence of human beings[7] (Kant 1787; Wittgenstein 1953). So it is problematic to think of passing the torch to other species. It is not even that other species are not intelligent enough. They did not evolve together with us as part of the complex social Web within which we can communicate with each other and share goals and values.

The background goal is preserving whatever is worthwhile in what we have done as a species. But why and for whom? It would be hard for me to accept that in one or two million years (probably a lot sooner unless we start doing more about our relation to the environment) that all that would be left of humanity would be a lot of ruins and all the waste products of technology. The situation would be radically different from other extinct civilizations such as the Maya. There would be no other people, no anthropologists, to pick up the pieces and appreciate what had been going on.

So the question is, what is worthwhile in what we do as a species? And is it still worthwhile if there are no successors to appreciate it? So far as the ecosystem goes, it would be worthwhile if we did not damage it irreparably and left it capable of producing species we could recognize as our successors. Perhaps (hopefully in a million years or so) we could leave a time capsule engineered to be accessible to some future successor species. So our short-term enabling species goal is equilibrium with the ecosystem. Our longer term species goal clearly involves our nature as a self-aware social species.

Modern technology as such *will not* be a positive part of that time capsule we leave for our successors. By modern technology is meant the process of turning everything into resources to be used by other processes of modern technology. However, some of the products of that technology, have an essential role to play in preserving ourselves and our ecosystem. There we mentioned monitoring the earth and threatening asteroids. Also, using I.T., we can model complex systems involving living beings, weather, and climate. Thus we have a much better idea of how our actions affect the other beings we share the ecosystem with. And the fact that we have a genuine worldwide communication system (the World Wide Web was accurately named) is going to make it much easier for anyone to take the point of view of the species for ethical and value judgements.

Author Robert Nadeau may be on the right track about species goals for humans. Nadeau first asks us to suppose there is no extraterrestrial intelligence. Then he says, if we fail to achieve equilibrium with the environment and our species becomes extinct, this would be a tragedy of cosmic proportions,

. . . because consciousness as an emergent property of a universe that had been evolving toward higher levels of complexity for over 13 billion years would cease to exist. In these terms, the universe itself would no longer be aware of its own awareness, and nothing in the vast reaches of interstellar space would have any meaning beyond the brute fact of existence (Nadeau 2006, p. 63).

Although Nadeau has overstated the case for the place of humans in the universe, he is still on to something. By 'consciousness,' he means 'species self-awareness.' Humans are by no means the only conscious species, elephants and dolphins having perhaps the most elaborated consciousnesses. The universe could exist very well without either consciousness or human species self-awareness. But he is right about this: There is something special and awesome about the expanding self-awareness of a species.[8] It is our goal as a species to determine what the proper role of our kind of species is in the universe. Our kind of species is aware of itself and the universe and this is happening through us as social animals. This goal will never be achieved if we go extinct because we cannot coexist with our own life support system.

A useful subsidiary goal would be to develop more communication with other species on earth, especially conscious species. Trying to locate and communicate with species on other planets is likely to be much less fruitful. Species on earth have a lot more in common with us than anything possible for another planet. There is actually some reason to believe this other than mere distance. Even on earth, when conditions are replicated as closely as possible, the same ecosystem does not replicate (Leakey 1995, pp. 167-170). Thus a whole distinct chain of DNA development on some other world will probably not bear that much similarity to ours—certainly much less similarity than other conscious species on our own world bear to us. It is significant that when extraterrestrial human species are portrayed in science fiction, we normally see recognizable human beings with heavy makeup, or humanoids with large slanty eyes. If they figure in the plot, they are able to speak English, but with an accent. At worst, portrayed extraterrestrials are recognizable at least as animals.

WAS CIVILIZATION A MISTAKE?

Modern technology, for all its good points, still must be considered a mistake because it exposes us to the destruction of the biosphere, our life support system. But what about civilization itself? Civilization does not by its nature destroy the ecosystem. Social insects also construct specialized environments for themselves. Bees perform valuable ecosystem functions by pollinating plants. Honey is exploited by humans and other animals such as bears. Ants perform ecosystem functions such as soil creation. Ants also herd and 'milk' aphids, as well as creating and maintaining fungus patches.[9] Termites in dry environments form completely enclosed nests conserving all available water.[10] In tropical settings, they enhance soil fertility, a fact used by tropical farmers.[11]

All of these specialized social insect environments serve ecological functions. Exactly what benefit to the ecology is a human city? Cities seem rather to be ecological liabilities. Few, if any, manage to be in equilibrium with the surrounding countryside in disposing of human wastes. And, with the advent of modern technology, all seem to generate a lot of non-recyclable wastes, for which there is currently no good disposal protocol.

Civilization is, however, the locus for science and scientific research, which we need to continue in order that we ourselves might be preserved. Science is often lumped together with the arts, cultural creations, engineering and technology. Engineering and technology are necessary for science, but art and cultural creations are of value only to humans and pretty much for humans themselves only. Arts and cultural creations usually do not threaten the ecosystem, so there is no reason why they can't be continued as part of humans being a species among species.

However, engineering and (classical) technology are another matter. With the gradual reduction of population which will have to take place, there will no longer be a need for huge urban developments overrunning the natural environment. Probably engineering and classical technology will be needed to repurpose previous concentrated urban environments into facilities that make valuable contributions to the ecosystem. If ants and termites can do it, so should human beings be able to accomplish this.

Again, if the use of modern technology is minimized, it should even be possible to retool some cities so that they are at least ecosystem-neutral. Economic growth will be needed to undo the ill effects of 300-odd years of modern technology and 7,000 years of civilization. Science and exploration can then continue without distraction from frivolous marketing products. Space exploration contributes to our species goal: To determine the place in the universe of a self-aware species descended from animals.

But one can't be confident in giving civilization this relatively clean bill of ecological health. Both farming and civilization entailed dramatic changes in the way humans lived. And both arose relatively recently, farming about 11,000 years ago and civilization about 7,000 years ago. This is simply not enough time to be confident that they are sustainable ways of life for our species for the indefinite future.[12] Especially since both—especially civilization—come into conflict with the ecosystem. Farming's conflicts with the ecosystem are resolvable through sufficient ecological knowledge and a desire not to damage the ecosystem. In earlier chapters, we have seen attempts to redo farming as modern technology, called factory farming. These attempts produce massive conflicts with the ecosystem in the form of artificial fertilizer and its attendant pollution, genetically engineered plants accompanied by more and more disruptive pesticides, and nightmarish waste disposal problems. But there are alternatives such as organic farming which can apparently be as productive as the ecosystem-destroying alternatives. If we go back to farming with sufficient attention to the ecosystem, perhaps there is a precedent. It would be like cetaceans going back to the ocean after living on land. I bet there were dolphin ancestors who were against returning to the ocean.

But we still do not have sufficient justification for the origin of these two ways of life—farming and civilization—so dramatically different from what preceded. If these ways of life had worked well for, say, 50,000 or 100,000 years with all their flaws, there would not be much of a problem. Both arose independently at numerous locations, so it is unlikely their appearance could be the result of chance external factors. Of course, civilization had to wait for farming to develop. There is no other way for a civilized environment that does not produce its own food to come about. Indeed, it seems to be an abundance of food which frees up enough people to start taking up the varied roles of civilization.

When civilization develops into a total environment with only a tenuous connection to the rest of the environment, we get to a place where, when the disruptive power of modern technology is added to it, the only probable outcome is unmitigated disaster. And climate disaster is only one of any number of possibilities for modern technology to destroy the ecosystem both we and all living things need to survive. The destruction of the ozone layer was another recent real possibility.

At the beginning of Chapter 5, "Humans as Hunter-Gatherers," it was stated that eliminating modern technology would not be a good idea because our entire economy and especially our food supply is dependent on modern technology. Thus civilization as we now know it would collapse if we persist in using modern technology. But we can see it would collapse anyway if we persist in employing modern technology.

It is likely that nature together with climate change will produce the drastic reduction in population necessary for human survival within the ecosystem. If this happens, it is important that any attempt to rebuild civilization as we now know it be undertaken with great caution. If we repeat our current mistakes, we will get the same consequences. Unfortunately humans are not always that good at grasping this point. The financial institutions responsible for the 2008 economic collapse are fighting with all their resources to prevent any changes which might make it possible to avoid another collapse in the future.

HUMANS AS A SPECIES AMONG SPECIES

It would be mere playacting if we tried to pretend that we are not conscious of ourselves and that we have knowledge of the ecosystem. The proposal to substitute coyotes for human hunters in order to reduce the deer population is a kind of playacting. So too are extreme animal rights proposals for us to have as little to do

Figure 1. Diagram from Chapter 11

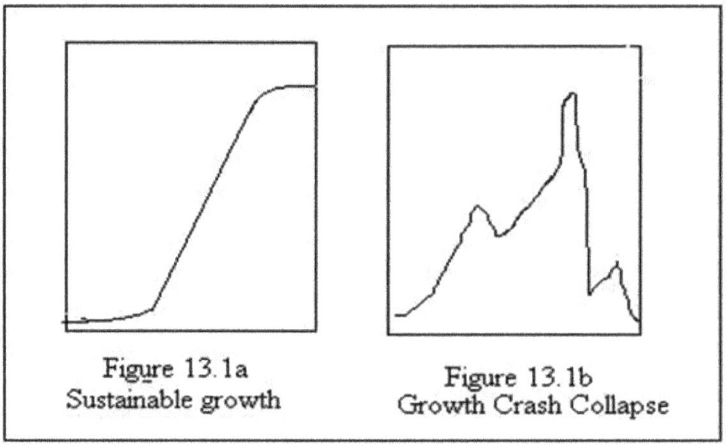

with other animals as possible. As discussed in earlier chapters, E. O. Wilson notes that all biological species expand heedlessly until the environment bites back. The environment bites back through feedback loops such as disease, famine, war, and competition for scarce resources. If *Homo sapiens* ignores the ecosystem, it ignores the feedback loop that could lead to a peaceful and graceful outcome. Conscious restraint in the relation of our species with the ecosystem is that feedback loop. The diagram from Chapter 11 summarizing our choices is repeated in Figure 1.

So ultimately there is a way in which we cannot become a species among species. Our apparently unique self-awareness of our place in the ecosystem cannot be ignored or cast aside. We should neither be misled by our physical superiority nor by a desire not to disturb the ecosystem. We should not be misled by our physical superiority into thinking that all that matters is us humans and our technology. Nor should we be misled into thinking we are not part of the ecosystem, as were those who wanted coyotes to kill the excess deer in suburban Pennsylvania.

Here, then, is a list of what humanity as a self-aware species must do to regain its place as a species among species:

- Members of that species must adopt the attitude that they are not superior to the ecosystem.
- They should use their knowledge of the ecosystem to determine what will benefit the ecosystem as much as themselves; and
- They should not casually alter the processes of the ecosystem unless that is necessary to prevent other demonstrable ecosystem damage.

Within civilization using a market economy:

- All economic environmental externalities must be charged to whatever is producing the externality (such as pollution and environmental damage).

Ultimately to produce a sustainable social order:

- The goal of unending economic growth must be abandoned.
- Resources must not be treated as endlessly available.
- Population should be stabilized at some sustainable number for the planet. 2 billion is a reasonable suggestion. Of course any human-enabled reduction must be voluntary.
- Modern technology itself must be reduced as much as possible.

For other species:

- Sufficient undisturbed habitat reserves must be set aside for animals who are independent of us to live out their life courses.
- For animals involved with us in some way, they must not be treated as objects and should be appropriately rewarded for their service.
- Continue research in communicating with animals.

Unless we adopt these goals, we will continue to treat the world "as a thing." Rather than being a superior species, we will be, in Chomsky's words, a "biological mistake" that no other beings will want to remember. Who exactly were we trying to impress with our seeming superiority? Ourselves?

REFERENCES

Associated Press. (2010, October 18). Coyotes, not hunters, proposed for Valley Forge. *Animal News*. Retrieved 10/20/10 from http://www.bornfreeusa.org/news.php?p=2516&more=1

Bering, J. (2009, September 11). Why we blush: The social purpose of showing embarrassment. *Scientific American*. Retrieved 4/14/12 from www.scientificamerican.com/article.cfm?id=why-we-blush-social-embarrassment

Carrier, D. R. (1984). The energetic paradox of human running and hominid evolution. *Current Anthropology*, *25*(4). doi:10.1086/203165.

Choi, C. (2011, August 22). Top 10 things that make humans special. *Live Science*. Retrieved 1/14/12 from www.livescience.com/15689-evolution-human-special-species.html

Cimons, M. (2011, March 22). Ecosystem function of ants. *US News & World Report Science*. Retrieved 5/2/12 from www.usnews.com/science/articles/2011/03/22/ecosystem-function-of-ants

Dreifus, C. (2007, January 9). Always revealing, human skin is an anthropologist's map. *New York Times*. Retrieved 4/14/12 from www.nytimes.com/2007/01/09/science/09conv.html

Harris, A. (2012). *Social ecology*. Retrieved 4/12/13 from www.thegreenfuse.org/socialecology.htm

Kaiser, D. (1990). *Dolphin consciousness*. Retrieved 1/30/06 from home.onemain.com/~dk1008206/html/dolph1.htm

Kant, I. (1787). *Critique of pure reason*. Riga, Latvia: Hartnoch.

Leakey, R., & Lewin, R. (1995). *The sixth extinction*. New York: Doubleday.

Max Planck Society. (2010). *A long childhood is of advantage*. Retrieved 4/14/12 from www.mpg.de/617475/pressRelease20101111?filter_order=L

Mommer, B. (2003). *Ecological role of termites in dry environments*. Retrieved 5/2/12 from www.biology.iastate.edu/InternationalTrips/1Australia/Australia%20papers/EcolAusTermites%20.htm

Nadeau, R. L. (2006). *The environmental endgame*. New Brunswick, NJ: Rutgers University Press.

Niemitz, C. (2010). The evolution of the upright posture and gait—A review and a new synthesis. *Naturwissenschaften, 97*(3), 241–263. Retrieved 4/14/12 from www.ncbi.nlm.nih.gov/pmc/articles/PMC2819487/

Quinn, D. (1999). *Beyond civilization*. New York: Three Rivers Press.

Sileshi, G. (2009). Integrating ethno-ecological and scientific knowledge of termites for sustainable termite management and human welfare in Africa. *Ecology and Society, 14*(1).

University of Kent. (2011, March 7). Stone tools influenced hand evolution in human ancestors, anthropologists say. *ScienceDaily*. Retrieved 4/14/12 from www.sciencedaily.com/releases/2011/03/110307101504.htm

Walsh, J. S. P. (2012, March 2). The final environmental frontier: Space development and its consequences. *Scientific American*. Retrieved 3/21/12 from news/yahoo.com/final-environmental-frontier-space-development-consequences-160000428.html

Wilson, E. O. (1978). *On human nature*. Cambridge, MA: Harvard University Press.

Wittgenstein, L. (1953). *Philosophical investigations*. New York: Macmillan.

ENDNOTES

1 See Associated Press (2010).

2 See Chapter 2, "Conflicts with the Ecosystem," "Species extinction."

3 This article on long childhood (Max Planck Society 2010) also notes that *Hono sapiens* had a significantly longer childhood than *Homo neaderthalensis*. This opportunity for greater cultural learning gave modern humans a dramatic advantage over the Neanderthals.

4 E. O. Wilson, one of the inventors of sociobiology, explains social evolution well in his *On Human Nature* (1978). Secondary sex characteristics also do not serve to make individuals fitter. For humans, beards for men and breasts for women function solely to attract the opposite sex. Contrary to some feminists, breasts function solely as means to attract men. No other mammal needs them to deliver milk to its young.

5 There are many genocides in history beginning with that of the Midianites, a genocide approved by the Bible. (Numbers 31:18) Genghis Khan often massacred entire civilian populations in the course of his campaigns in the 13th century. More recently 12 million Africans were killed in the course of implementing the slave trade in the 17th to 19th centuries. The Holocaust in the middle 20th century killed 6 million. Subsequent American wars have involved civilian collateral deaths of 1 to 2 million in Vietnam and perhaps 1 million in Iraq. Other genocides include the Armenian genocide by Ottoman Turks at the beginning of the 20th century. Nazis later used this genocide as a model for their own genocide of the Jews. Other later genocides include those in Rwanda by Rwandans, in Bosnia by Serbs, and Darfur by Sudanese.

6 On this planet, a recognized difficulty in using travel to get away from one's personal problems is that one always meets oneself at the new location. In twelve-step programs, this is called a "geographic cure." It does not work for the reason given—the traveler is the cause of most of his own problems.

7 This is the point of Wittgenstein's enigmatic remark that if a lion could talk, we could not understand him (Wittgenstein 1953, p. 223e).

[8] This may be related to what Freud called "one great human unity." See Chapter 7, "Civilization," "Civilization and its Discontents."

[9] The fungus is found only in ant nests, so it counts as cultivated. See Cimons (2011).

[10] See Mommer (2003).

[11] See Sileshi et al. (2009).

[12] Daniel Quinn (1999) has written numerous books detailing why civilization needs to be abandoned. However, his solution, a partial return to hunter-gathering, ignores a number of crucial problems, especially our relation to the ecosystem. Quinn thinks that we should replace the social/economic hierarchical structures characteristic of civilization with "tribal" structures without difference in the status of individuals. Although he mentions civilization's destructiveness to the ecosystem, merely changing a social structure is not going to change humanity's belief in human superiority. Hence, there is no reason to think "tribal" humans will treat the ecosystem much differently from civilized humans. Another social theory called social ecology also holds that the problem with social institutions is hierarchy, and that all major social problems will be at least ameliorated when hierarchical social structures are eliminated. One major difference is that social ecologists believe civilization is still an appropriate institution, just without hierarchy (Harris 2012).

Chapter 14
Extinction

ABSTRACT

Eventually the human race will become extinct, like any other species. Ultimately, we should be concerned with what we can do toward our species goals within the one-to-two million-years lifespan of our species. Our individual life spans are not meaningless even though we know they will end. Therefore, what we do as a species is not meaningless just because our species will come to an end.

INTRODUCTION

It may sound strange to consider extinction as a solution to the problem of the relation of human beings to the rest of nature. But if the human race were to go extinct, that would mean that the problem of its conflict with nature would no longer exist. In any case, since no species lasts forever, extinction is something the human species will someday have to face. We can distinguish two kinds of extinction: Gradual extinction and sudden extinction. Species of living things on this planet have an average life span of two million years. So gradual extinction would be analogous to individual death from old age. The species would simply have run its course in nature, its numbers would diminish, and finally there would be no more human beings, members of the species *Homo sapiens*. Within this scenario, there are two other possibilities: Either there are similar but better adapted species which supersede *Homo sapiens*; or the ecological niche *Homo sapiens* occupies is itself no longer available.[1]

DOI: 10.4018/978-1-4666-4586-8.ch014

The existence of similar but better adapted species—namely us—is exactly what caused the extinction of our predecessor species *Homo erectus, Homo sapiens neanderthalensis,* and perhaps others.[2] The existence of more advanced species among us has been a theme of some science fiction stories. The recent X-men films and stories are an example. It is difficult, however, to see how a separate population of genetically different humans could maintain themselves in the current state of human civilization. With physical communication worldwide, any mutation would be fairly quickly incorporated into our worldwide gene pool. Evolutionary biology tells us that the appearance of a new species superseding an older one requires the physical separation of part of the older species population. This process, called *macroevolution,* can produce a new species in the separated population by natural selection or even simple genetic drift.[3]

The second gradual extinction possibility, the disappearance of our ecological niche, seems most likely as the result of accumulated environmental changes caused by us. For example, a recent dramatic decrease in the fertility of human sperm may have been caused by increased air and water pollution in the environment.[4] If this trend gets dramatically worse, one could see the human population diminishing to zero. Of course, one would expect human beings to take steps to reverse this trend well before extinction took place. But there could well be environmental effects on fertility or other critical factors for species survival whose causes elude us long enough to cause extinction.

Nevertheless, the likelihood of sudden extinction is probably greater. There are also two types of sudden extinction: Those caused by natural processes completely external to human beings and those caused by human action. External causes of extinction are collisions with asteroids or comets, or massive volcanic eruptions. Both have caused mass extinctions in the past. An asteroid collision 60 million years ago in the Yucatan most likely caused the extinction of the dinosaurs (Schulte 2010). And 70,000 to 80,000 years ago the supervolcanic eruption at what is now Lake Toba in Sumatra most likely came close to extinguishing *Homo sapiens.* Some estimate that as few as between 1,000 and 10,000 humans survived (Ambrose 1998). Another example would be the earth's being an unlucky target of a gamma ray burst (Minard 2009). Such events are not predictable.

Probable human causes of sudden extinction are nuclear war and climate change. Proliferation of nuclear weapons among states that are politically unstable raises the threat that the irrationality of their use may not prevent that irrational use. There is also the possibility that some group that is not a state may use them as a terrorist tactic. Although no one denies that nuclear war is a bad thing, this is not true of nuclear energy. In spite of the threat of contamination from malfunctioning nuclear reactors—as was happening at the time of this writing from the tsunami-damaged reactors at Fukushima—economic interests known as the "nuclear power industry" disseminate propaganda dismissing any serious danger. The problem of long-term storage of radioactive waste from nuclear reactors to prevent contamination of the environment also goes unsolved. The problem is that some waste must be sequestered from the rest of the environment for periods on the order of 20,000 years. This is more than double the time period that civilization has existed, let alone in a stable enough form to guarantee actions and policies lasting that long.

Fossil fuel companies have adopted similar tactics to dismiss any dangers from continued release of carbon dioxide in the atmosphere. It is hard to counter economic interests who undermine sound scientific research in an area that conceivably could lead to human extinction. Fossil fuel companies spent over $100 million since 1997 to fund climate change denial. The threat of extinction here is not exactly sudden. The threat is, if anything, worse. Changes in the atmosphere—mainly increases in the concentration of greenhouse gases in the atmosphere—can start processes that can't be reversed but which take long periods of time to manifest. Changes in carbon dioxide levels in the past 150 years do not seem sudden to us. But from the point of view of natural processes which could absorb increased carbon dioxide, they are sudden and irreversible. Such processes normally can require time frames on the order of 100,000 years. It took at least 64,000 years for carbon dioxide levels to decrease from highs of 1,000 ppm at the end of the Mesozoic period to our pre-industrial level of 280 ppm (Pagani et al. 2005). As of June 1, 2012 the CO_2 level had just passed 400 ppm. These numbers have not been tainted by knowledge polluters yet. There are made-up numbers in abundance about the past in climate change denial Websites. In 1850, before industrialization, the level was 280. This works out to an increase of 80 ppm per century. Therefore, it would take about 750 years to reach the level of 1000 ppm that it took nature 64,000 years to undo. There was abundant life during the Mesozoic with its 1,000 ppm carbon dioxide concentration. However, all life had many more years to adapt to the change—64,000 years as opposed to 750. And another real worry is that this speedy an accumulation of carbon dioxide may trigger some form of runaway greenhouse.

We could easily find ourselves in a situation where a delayed response to increased greenhouse gases leaves us with no way to reduce greenhouse gases to levels that permit the survival of most of the life on this planet. If this happens, it will be the slow but certain death of much of what we value. Much worse than an asteroid strike and maybe even worse than a nuclear war.

THE MEANING OF INDIVIDUAL EXTINCTION

So how might we feel about the extinction of the species *Homo sapiens?* Are there parallels with how we feel about individual deaths? Premature individual deaths usually are thought to be tragic, very bad things because potential life is wasted. Even state-sanctioned executions are thought to be sad because of the wasted life ending that way. Of course, people from primitive cultures and religions can at the same time avail themselves of the dubious pleasures of revenge.[5] For most people, such pleasures are not available when innocent people are executed. But those with fascist or totalitarian values believe the state is more important than the individual. US Supreme Court Justice Antonin Scalia claims that there is absolutely nothing wrong with executing innocent people so long as the proper legal procedures leading up to the execution have been followed (Millheiser 2009). Of course Scalia's principle would also justify the Holocaust. The Nazis were very good about following procedures.[6] So there is some justice in calling the claim that innocence is irrelevant, 'fascist.' But since the meaning of that term has been distorted by hysterical right-wing zealots,[7] an equally good description would be 'totalitarian.' The essence of totalitarianism is that the institution is more important than the individuals under it. As such, there is no such thing as totalitarian ethics. Ethics is about rules for cooperation between individuals.

On the other hand, deaths from old age, especially those in which individuals pass away peacefully with no pain, perhaps even in their sleep, after a long, productive, and happy life, tend to cause sadness but are not felt to tragic or very bad things. We may well miss that person's continued company but we don't feel strongly that it would have been better if things had been otherwise. In cases where the last years of the individual's life involved severe pain or severe incapacity, either mental or physical, death may even be counted as a blessing.

The feelings just outlined are usually present even in those who believe in an afterlife. It has always puzzled me why, if one were certain that a better life awaits after death, one would not seek to hasten it. Fringe cults occasionally do this; for example, the San Diego cult Heavens Gate committed mass suicide in 1975. Just like everyone else, most all believers in the afterlife seek to prolong their earthly life as long as possible, so long as one has the ability to live that life reasonably well.

But should we strive to prolong individual lives indefinitely? Again we assume that any such continuation will be lived with close to full capabilities and be happy. But suppose much longer lives prevent significant numbers of future human beings from being born. Is there value lost in preventing human turnover? Surprisingly, scientific progress might be drastically reduced if old scientists did not pass away.

Thomas Kuhn observed in his *Structure of Scientific Revolutions* (1962) that dramatically new scientific theories take hold when old scientists holding old theories die off and younger scientists holding the new theories take over.

Other recent intriguing scientific results are relevant. It is usually assumed that individual death is a malfunction, that if certain biologically-mistaken disease-like degenerative processes are avoided or reversed, life could be prolonged indefinitely. The goal of life-prolonging research is to identify and deal with such aging processes. However, recent research suggests that, rather than death being an imposition on the organism by unwanted processes, death is actually the result of natural selection. Processes leading to death in the organism are coded in the genes selected for that organism.

Before these results are reviewed, we need to recall my discussion of genetic manipulation. Our immediate response to the existence of death genes is, well, so what, let's just alter the genes. Genetic manipulation is a technology which is in general not compatible with the ecosystem. Human needs advanced by technology simply cannot be assumed to take precedence over the ecosystem.

The presence of a death gene was uncovered by Claire Cockcroft and Cynthia Kenyon in the course of their research on nematodes, a small roundworm. Their research showed that there were genes whose function was to produce the aging and death of the organism. Cockcroft and Kenyon raise the question of why such genes might be selected, but don't give much of an answer (2008). For social animals like us, the answer they suggest is that it is useful to have grandparents around to care for grandchildren, but not great-grandparents. Obviously, there is no analogue of this answer for nematodes.

They also may miss an important implication of their research. Indeed their research shows that aging and death are *not* just "the results of a competition between the forces of damage and entropy and repair and prevention." Rather, there is some selective value for death itself (Shuckburgh 2008, p. 201). So in the face of this, Cockcroft and Kenyon conclude their discussion with the desirability of extending human life indefinitely without even an inkling that natural selection probably did not frivolously limit our lifespans. We definitely need to know why the limitation of death was selected before we eliminate that limitation.

The only more-or-less immortal species are one-celled animals which reproduce by fission. With these species, genetic material is simply copied and the organism clones itself. Most larger many-celled animals reproduce sexually. Genetic material from two organisms is combined to form the genetic basis for a new organism. The old ones eventually die. It seems likely that natural selection favored birth and death for more complex organisms. Actually, perhaps it is not so much that birth together with death is favored over immortality. Rather, sexual reproduction seems to require birth/death cycles. And the natural selection reason for sexual reproduction is to repair chromosome breakage.[8] So we can see that immortality for complex beings like us is just one more departure from our foundation in the ecosystem and natural selection. Why, having seen the development of humankind and other species, would you want to opt out of that development? Does it need to be argued that, however accomplished human beings are, they are not completely flawless and therefore could benefit from the continued action of natural selection?[9] Does it now need to be argued why humans cannot effectively take control of natural selection?

THE MEANING OF SPECIES EXTINCTION

So therefore individual extinction is to be expected and dealt with. It is bearable precisely because the species lives on. And, insofar as we are invested in our ancestors and in our children, we are invested in the continuation of our species. But if *Homo sapiens* goes extinct, we no longer have this consolation. Could there be a difference between premature and natural species extinction?

Noam Chomsky reasons as follows:

You could think of it as parallel to premature death for an individual, vs. old age death. If a species goes extinct because it destroys its own habitat, it was a biological mistake. If it goes extinct through normal evolutionary processes or if its habitat is destroyed through other natural processes, it was a "natural" extinction (Chomsky 2003).

Nevertheless, once our species goes extinct, is it any consolation that the extinction happened as a result of normal evolution or the action of natural processes? This is just not parallel to death because of old age, unless there is a successor species. And there is good reason to believe there will not be a successor species to globalized humanity. However, as will be disclosed shortly, it is a bad thing for a species to leave this earth as part of a path of destruction.

Friedrich Nietzsche,[10] in an 1873 essay titled "On Truth and Lie from an Extramoral Point of View," considers how to respond to a universe which no longer contains human beings. Quoted is the essay's striking beginning:

Once upon a time, in some out of the way corner of that universe which is dispersed into numberless twinkling solar systems, there was a star upon which clever beasts invented knowing. That was the most arrogant and mendacious minute of "world history," but nevertheless, it was only a minute. After nature had drawn a few breaths, the star cooled and congealed, and the clever beasts had to die. One might invent such a fable, and yet he still would not have adequately illustrated how miserable, how shadowy and transient, how aimless and arbitrary the human intellect looks within nature. There were eternities during which it did not exist. And when it is all over with the human intellect, nothing will have happened. For this intellect has no additional mission which would lead it beyond human life. Rather, it is human, and only its possessor and begetter takes it so solemnly-as though the world's axis turned within it (Nietzsche 1873).

So how does the species goal proposed in the preceding chapter look from Nietzsche's point of view? In earlier chapters the issue of the meaning of human accomplishments after humans cease to exist was raised. But Nietzsche regards all human knowledge as not worth much, not so much because it will vanish with its bearers, but because it is radically limited and relative. Nietzsche's views on human knowledge and truth verge on a relativism so thorough it is close to complete skepticism. Of course, we can dismiss extreme forms of skepticism and relativism because they cannot have a neutral basis above the skepticism to assert the truth of skepticism or relativism. Therefore, they cannot be true theories because according to them, there is no truth, and if so that includes those theories themselves.

But Nietzsche's relativism is the basis of his attempt to discredit human knowledge. For Nietzsche, because human knowledge is relative, it is not (genuine) knowledge. Buddhism makes a distinction between enlightened knowledge (of emptiness of essential nature) and relative knowledge (of the ordinary world of samsara). For Buddhists, both sorts of knowledge are valid, but enlightened knowledge is higher and more important (Della Santina 1997). For the philosopher Kant, (human) knowledge is relative to some basic conditions of human existence, namely that we are in time and space and connect to the world through our senses. For Kant, knowledge of things in themselves independent of human conditions is impossible (Kant 1781). Kant's account of human knowledge is more plausible than the Buddhist account. If all things do not have essential nature because they are conditioned, we do not have knowledge of the unconditioned as a separate but absent reality. Kant's view that things in themselves are unknowable is much more consistent.[11] So the relativity of human knowledge does not discredit that knowledge.

It does, however, raise the question of whether the species goal outlined in the preceding chapter is worth attaining. It was suggested that humanity's goal as a species was to determine what the proper role of our kind of species is in the universe. 'Our kind of species' is a species aware of itself and increasing portions of the universe around it. From Nietzsche's point of view, this awareness is meaningless because "After nature had drawn a few breaths, the star cooled and congealed, and the clever beasts had to die." In other words, everything having to do with human beings is insignificant compared to the eons of cosmic time. Well, yes. Even the existence of a particular planet or star is insignificant compared to those eons of

Table 1. Life spans of hominid species

Name	First Appearance	Extinct as of	Life Span
	(years ago)	(years ago)	(years)
Australopithecus afarensis	4,000,000	2,900,000	1,100,000
Australopithecus africanus	3,000,000	2,000,000	1,000,000
homo habilis	2,500,000	1,400,000	1,100,000
homo erectus	1,800,000	70,000	1,700,000
homo sapiens neanderthalensis	600,000	30,000	570,000
homo sapiens sapiens	200,000	?	?

Derived from Table 5.2

cosmic time. And the daylong life of a may fly is insignificant compared to a year or a human lifetime. So the question is, what is the appropriate time span for the significance of the human species to be measured against? A plausible answer is, one to two million years, the average life span of a species (see Table 1).

Previous ancestor species lasted at least one million years, with the exception of Neanderthals, who may have merged with us. And even they were around for over 500,000 years. *Homo sapiens sapiens* has been around so far for about 200,000 years. By contrast, farming has been around for about 11,000 years, civilization for 7,000 years, and modern technology for all of about 250 years. We are therefore not doing well as a species if we don't last much beyond 200,000 years—which may happen if the ecosystem collapses around us.

But within a one to two million year timeframe, what is our significance as a species? Previous species could point to successor species they facilitated that were better equipped to handle things. It was previously explained why, in the current globalized circumstances, it is not likely that a new human species can arise. However, if there is a drastic reduction in human population and in mobility, new successor species might be possible.

But even if we zoom in from a cosmic time frame to one to two million years, we still have the problem that most human constructions would be meaningless to any substantially different species. Therefore they can only be valued for their own sakes. This is true of music, visual arts, literature, architecture, performance arts, cinema, whether in the elaborated forms of western civilization or the smaller scale accomplishments of folk art.

And what about our technological and economic achievements? Our achievements in transportation, in weapons and military equipment, in household appliances and tools, in medical areas, factory processes, and financial instruments?

All of these almost entirely have value in accordance with the human ends for which they were created. In some cases, there is a design aspect with visual art value. But such value is always secondary. You do not want a beautiful automobile which blows up when you attempt to start it.[12] And when the human ends of any of these are no longer relevant, their value vanishes.

Nietzsche's cosmic perspective diminishes not only humans and human knowledge, it also diminishes life itself. For life arose about 3.7 billion years ago on earth, and possibly independently elsewhere. Eventually conditions on the earth will not permit life, as when the sun expands out beyond earth's orbit. Of course we may be able to postpone this fate by moving to another habitable world. But this only postpones the inevitable. The entire universe will one day become uninhabitable as the expansion of the universe eventually dilutes all energy to the point nothing can be done (Adams 1997). So there is also a definite time frame within which life exists, whether on earth or elsewhere.

THE ROLE OF OUR SPECIES WITHIN THE LIFE SPAN OF OUR SPECIES

But ultimately we should be concerned with finding out what role our species awareness plays in the universe within our possible one to two million life span. Why are we accumulating knowledge, even it will all go away? We do not regard our individual life spans as meaningless even if we know they will end. So why should we regard what we do as a species as meaningless just because we know it will come to an end? Of course it is way more certain that as individuals we will have descendants than whether there will be successor species to humans.

But perhaps it is enough to follow the implied precept of Claude Levi-Strauss. At the end of his *Table Manners* (1968, p. 508), he says that humanity has no excuse for appropriating the world 'as if it were a thing' and behaving on it with neither decency nor discretion. The implied precept is the inverse: Therefore, in the stay of our species on earth, we should not appropriate the world as if it were a thing, and we should behave with decency and discretion with respect to the world. One purpose of this book has been to spell this precept out in more detail. When we treat the ecosystem with respect, we do not appropriate the world. We cannot treat the ecosystem with respect if we persist in full-scale deployment of modern technology and unlimited economic growth. To behave with decency and discretion is to realize that our role on this planet is limited and that, regardless of our physical power, the rest of the planet does not exist for our benefit. Human superiority and human exceptionalism are ideologies we need to discard.

REFERENCES

Adams, F. C., & Laughlin, G. (1997). A dying universe: the long-term fate and evolution of astrophysical objects. *Reviews of Modern Physics*, *69*(2), 337–372. doi:10.1103/RevModPhys.69.337.

Ambrose, S. H. (1998). Late Pleistocene human population bottlenecks, volcanic winter, and differentiation of modern humans. *Journal of Human Evolution*, *34*, 623–651. doi:10.1006/jhev.1998.0219.

Arendt, H. (1965). *Eichmann in Jerusalem*. New York: Viking.

Chomsky, N. (2003). *Hegemony or survival*. Retrieved 6/24/11 from www.chomsky.info/books/hegemony01.htm

Cockcroft, C., & Kenyon, C. (2008). Surviving longer. In Shuckburgh, E. (Ed.), *Survival: The Survival of the Human Race*. Cambridge, UK: Cambridge University Press.

Colburn, T., Dumanoski, D., & Myers, J. (1996). *Our stolen future*. New York: Dutton.

Colby, C. (1996). *Introduction to evolutionary biology*. Retrieved 9/15/11 from www.talkorigins.org/faqs/faq-intro-to-biology.html

Della Santina, P. (1997). *The tree of enlightenment*. Taipei, Taiwan: The Corporate Body of the Buddha Educational Foundation.

Gonzalez, R. (2012). New research demonstrates how humans are still evolving. *Virpi Lummaa News*. Retrieved 6/2/12 from io9.com/virpi-lummaa/

Heidegger, M. (1954). *Was heisst denken?* Tübingen, Germany: Max Niemeyer.

Kant, I. (1781). *Critique of pure reason*. London: Macmillan.

Kuhn, T. (1962). *The structure of scientific revolutions*. Chicago: University of Chicago Press.

Levi-Strauss, C. (1968). *The origin of table manners*. New York: Harper & Row.

Millheiser, I. (2009, August 17). Scalia says there's nothing unconstitutional about executing the innocent. *Think Progress*. Retrieved 5/9/12 from thinkprogress.org/politics/2009/08/17/56525/scalia-actual-innocence/

Minard, A. (2009, April 3). Gamma-ray burst caused mass extinction? *National Geographic News*. Retrieved 5/9/12 from news.nationalgeographic.com/news/2009/04/090403-gamma-ray-extinction.html

Nietzsche, F. (1873). *On truth and lie from an extramoral point of view*.

Pagani, M., Zachos, J., Freeman, K., Tipple, B., & Bohaty, S. (2005). Marked decline in atmospheric carbon dioxide concentrations during the Paleogene. *Science, 309*(5734), 600-603. Retrieved 6/2/12 from www.sciencemag.org/content/309/5734/600

Schulte, P. et al. (2010). The Chicxulub asteroid impact and mass extinction at the Cretaceous-Paleogene boundary. *Science, 327*(5970), 1214–1218. doi:10.1126/science.1177265.

Shuckburgh, E. (2008). *Survival: The survival of the human race*. Cambridge, UK: Cambridge University Press.

Skakkebaek, N., et al. (1992). Evidence for decreasing quality of semen during past 50 years. *British Medical Journal, 305*(6854). Retrieved 9/17/11 from www.jstor.org/pss/29716905

Williams, G. (1975). *Sex and evolution*. Princeton, NJ: Princeton University Press.

Zimmer, C. (2010, December 22). Siberian fossils were neanderthals' eastern cousins, DNA reveals. *New York Times*. Retrieved 9/17/11 from www.nytimes.com/2010/12/23/science/23ancestor.html

ENDNOTES

1 Another theoretical possibility is that some species we kill for our survival develops an unbeatable defense (Colby 1996). This is highly unlikely to happen to human beings with their physical superiority. Recall the discussion of elephants in Chapter 1, "Humans as Masters of the World."

2 Recently, other immediate predecessors to *Homo sapiens* have been discovered. See, for example, Zimmer (2010) about the species called *denisovans* who lived at the same time as the Neanderthals and early *Homo sapiens*.

3 I am grateful to Andrew Ross for clarification about the mechanism of macroevolution. See also Colby (1996).

4 Niels Skakkebaek and his associates published in 1992 a study "Evidence for Decreasing Quality of Semen During Past 50 Years." In a 1996 book, *Our Stolen Future*, Colborn, Dumanoski, and Myers, argue that industrially-derived "hormone disrupters"—particularly chemicals that behave like estrogen and interfere with fetal development—threaten our ability to reproduce. Of course corporations dependent on the unregulated production of these chemicals

stepped in to discredit these conclusions. See the discussion in Chapter 3, "Science, Modern Technology, and Corporations."

5 Revenge is considered to be a negative thing by both Christians and Buddhists. Muslims and Jews see nothing wrong with it. Heidegger quotes Nietszche with approval as saying 'Revenge is the will's revulsion against time and its "it was" (Heidegger 1954). Thus, revenge is basically an impossible attempt to undo something that has already happened.

6 See Arendt (1965). In Chapter 13, she discusses the Nazi's displeasure with the Romanians, who were killing Jews without using the proper procedures.

7 Glenn Beck formerly of Fox News is an excellent example.

8 See Williams (1975). Also personal communication, Dr. Philip Pack, Professor of Biology, Woodbury University.

9 It was concluded earlier in this chapter that a globalized human species could probably not give rise to entirely species because the conditions for this happening could no longer obtain, namely the physical separation of part of a species population for a substantial period of time. But this does not preclude continuing adaptation of human beings through natural selection. Indeed, many biologists feel this is happening (Gonzalez 2012).

10 My thanks to Venerable Jam Yang for the reference to Nietzche's essay.

11 Although Kant in his ethical writings claims knowability of ourselves as things-in-themselves. this exception to the knowability of things in themselves is not consistent for Kant. In any case, the unknowability of things in themselves is the doctrine which should be followed.

12 The Italians occasionally accomplish this. A good looking Maserati sedan apparently exploded when the test driver attempted to start it.

Chapter 15
Conclusion

ABSTRACT

The principal ideas of the book are summarized. Two stories are told about the future of the planet, a positive story and a negative story.

INTRODUCTION

My principal aim in this book was to reaffirm that humans are a part of nature, and to draw the correct consequences. The book begins by examining some important positions concerning human superiority which hold that humans are not part of nature. Many think humans are superior and that this gives us the right to do what we want to the rest of the environment. But our undeniable physical superiority does not give us "dominion" over the rest of the world, any more than the great power of a volcano gives it the right to destroy cities and towns.

Even though we are part of nature, we can still seriously damage the ecosystem, our support system. A species causing sufficient damage to its own support system can make extinction its role in nature. Modern technological manufacturing regularly produces harmful side effects, either toxic substances or unexpected threats to the entire ecosystem such as ozone layer destruction or global warming. Human economic activity is leading to mass extinction of other species. Humanity is threatening the entire ecosystem. This means that the view of humans as superior is *actually incorrect.*

DOI: 10.4018/978-1-4666-4586-8.ch015

Modern technology dramatically accelerates our conflict with the ecosystem; it is important to understand its nature. The distinguishing feature of modern technology is that it treats everything as nothing but resource, on hand for further use by its own processes. This insight, due to the philosopher Martin Heidegger, explains why modern technology *must* conflict with the ecosystem. It does not recognize the ecosystem as anything other than a set of resources it can exploit. Corporations turn out to be the institution par excellence for advancing modern technology, since their overriding concern is to maximize profits. Thus, everything already reduced to resources is on hand to achieve corporate ends.

I also deal with another type of separation of humans from the environment. This is the view that humans are not part of nature because they are ultimately nonphysical minds or souls or spirits. Such views sometimes lead people believing in these religious doctrines to disparage concern for this world. But it is ultimately very destructive to believe we are not of this world. We know that humans arose on this planet through natural selection and there are no other worlds we currently have access to. Access to any other worlds is by spaceship only.

THE ORIGINS OF HUMAN SEPARATION FROM NATURE

Our immediate ancestor *Homo erectus* made tools complex enough to save and store, thus creating a purely human space. *Homo erectus* later tamed fire and built temporary shelters, again altering the relation of human beings to nature. When our species *Homo sapiens* appeared, our ancestors made and wore clothing, carried out ceremonial burials, began making representations (cave paintings), and exterminated nearly all animals larger than themselves.

But the developments that most separated humans from the ecosystem were farming and civilization. Both are fairly recent developments, farming dating from about 11,000 years ago (6% of our species' time on earth) and civilization about 7,000 years ago (4% of our time on earth). Farming separates us from the ecosystem by replacing the ecosystem with plants and animals of our own choice. But farmers must still respect the environment which allows those plants and animals to thrive.[1] Farming also makes civilization possible by producing a food surplus, so that a number of people are free from work connected with producing food. This separation from the environment enables humans to develop a wide range of activities, pursuits, and occupations, accompanied by many technological innovations. These new activities and innovations are often regarded as demonstrating the superiority of humans. But since almost all these pursuits and innovations benefit humans only, it is only physical superiority which is once more demonstrated.

When modern technology arrives on the scene about 250 years ago, conflicts with the ecosystem accelerate. The main culprits are products and by-products which are either toxic in themselves or cannot be incorporated into the ecosystem—not biodegradable, in other words. With the advent of what could be called postmodern technology beginning about 1900, things get much worse. Postmodern technology largely produces petroleum-based products incompatible with the ecosystem. These products replace products compatible with the ecosystem.

The causes of humanity's conflict with the ecosystem are often said to be at least one of the following: Human institutions misusing technology; technology itself; population growth; economic growth; human attitudes. The principal causes are modern technology together with attitudes of human superiority and human exceptionalism. Population growth and economic growth are secondary causes, but they do make environmental problems worse.

OUTCOMES OF OUR POSITION WITH RESPECT TO THE ENVIRONMENT

How much of our problems relating to the environment could be solved through environmentalism and sustainability? Sustainability is maintaining equilibrium with the ecosystem. Four accounts of how to achieve sustainability are: unlimited economic and population growth; protecting diversity through cooperation with corporations; reducing population and dealing with climate change; and humans adopting the attitude that the ecosystem is primary.

Unlimited economic and population growth is a non-starter. It would have to be achieved through massive applications of modern technology, which would clearly create more environmental problems. Corporations can be trusted only to maximize their profits. Although reducing population and dealing with climate change will help, by themselves they do not address the continuing threat of modern technology to the ecosystem. The final alternative, that humans must come to realize that the ecosystem is primary, is correct. But this realization needs a firmer basis than emotional commitment.

To determine the place of humans in the ecosystem as a species among species, the role of science and technology are considered, as well as our relation to other animals.

The role of modern technology needs to be reduced because of the ongoing threat of catastrophic environmental consequences. Because modern technology needs to be employed to monitor the ecosystem and to deal with potential celestial collisions, some of it must be continued. Parts that don't contribute to the survival of the ecosystem need to be reduced as quickly as possible without causing serious damage.

In evaluating which technological applications to reduce or eliminate, maximin reasoning should be employed, conservative reasoning to choose the alternative which works out best when things turn out for the worst. In any case, economic growth is an illusory goal if environmental costs are not properly included in.

As far as our relations to other animal species, the most important principle from the point of view of eco-ethics is not to treat them as objects. Some animal-rights theorists go overboard and say humans should not have any relations with other animals. This is yet another view which makes humans not a species among species. Humans have had good relations with animals at least since dogs were domesticated. So long as animals are treated well, there is no reason we cannot continue to use their products obtainable without killing them. Even then, clear health reasons should allow humans to eat meat of animals killed humanely. It is also important to continue research into communicating with animals, because it severely reduces the impulse to treat them as objects and kill them indiscriminately.

Determining the role of humans as a species among species involves finding a middle way between cheerleading views of human superiority and negative views of requiring humans to have no contact with the ecosystem. Of the various distinctive characteristics of the human species, only long childhood and menopause bear on our being a species among species. We are therefore adapted to learn more about our surroundings and act on the basis of that knowledge. Indeed, our goal as a species is probably to find out what our role in the universe is, as a knowledge-gathering species.

But even if we minimize modern technology, the long-term value for our species of civilization is problematic in part because it has been around for such a short part of the time of our species' existence and in part because it fosters fairly complete separation from the ecosystem. Ultimately, our survival rests with our being able to use our knowledge to respect the ecosystem and to live within its limits.

Sudden and untimely extinction of the human race is what we hope we can prevent. But, like any other species, eventually the human race will become extinct. How should we feel about this? We can take individual death in stride because we know there will be other humans to carry on our projects and our genetic line. But this will not happen for the species (Because of globalization, it is unlikely a successor species to humanity could arise.). Ultimately, we should be concerned with what we can do toward our species goals within the possible one-to-two million-years lifespan of our species. Our individual life spans are not meaningless even though we know they will end. So what we do as a species is not meaningless just because it will come to an end.

We may be able to avoid sudden and timely extinction of the human race if we treat the ecosystem with the respect it deserves. We cannot treat the ecosystem with respect if we continue to deploy modern technology and unlimited economic growth. To treat the ecosystem with respect is to realize that our role on this planet is limited and that, regardless of our physical power, the rest of the planet does not exist for our benefit. I hope I have made it clear why human superiority and human exceptionalism are ideologies we need to discard.

TWO STORIES

Zog, the extraterrestrial from the beginning of Chapter 1, "Humans as Masters of the World," makes his scheduled return visit to earth 20,000 years from now. Following are two scenarios illustrating what he might find:

- **A Bad Scenario:** Runaway greenhouse;[2] and
- **A Much Better Scenario:** Humans survive as a species among species.[3]

A Bad Scenario

Zog, the extraterrestrial we met at the beginning of Chapter 1, is returning for his next visit 20,000 years in the future. As Zog's ship gets close enough to Earth for our planet to appear as a small disk, Zog could tell something was very wrong. Unfortunately, he had a pretty good idea of what had happened: He had seen it before. But there was still a slim chance.

He called his pilot, Xil.

"Xil, is there any chance we're heading for the second planet, Venus, by mistake?"

Xil replied quickly, "No, sir, that's the third planet they call Earth we're seeing."

Zog felt a burst of sadness. That uniform pale-cream disk could only mean one thing: Earth was now much like Venus. Venus has an atmosphere 99% carbon dioxide, produced by runaway greenhouse—greenhouse effect on steroids—to produce a surface temperature of 800 degrees, with not the slightest trace of water.

Xil said, "You were last here 20,000 years ago. How soon after that visit would runaway greenhouse have started?"

Zog replied, "It was kind of touch-and-go. A few hundred years before my visit they had begun burning large amounts of fossil fuel generating a lot of carbon dioxide extremely quickly. Some of their scientists had figured out the danger, but the humans also had efficient economic arrangements called corporations. These corporations were really good at creating value for themselves but had no concern for human beings as such. So when a governing body completely oblivious to the consequences suddenly gave corporations a lot more power at what was probably the very worst time, the corporations set public policy to preserve and improve their profitability. This resulted in even greater emissions of carbon dioxide."

Zog paused, then continued. "It's hard to say exactly when runaway greenhouse kicked in. During my last visit, there was no consensus on whether runaway greenhouse was even possible on earth.[4] But then in 2030 the polar ice had melted and then by 2050 the melting of the permafrost had doubled the amount of carbon in the atmosphere.[5] Then I left earth." Zog paused to contemplate the rapidly growing pale cream disk.

"I expect with all the methane [from the permafrost] it would not take more than another 1,000 years until temperatures were hot enough for the oceans to boil away. Of course almost all life would be dead by then. Then runaway greenhouse would have occurred. All that water vapor is additional greenhouse gas, the temperature would go up high enough to split the water vapor into hydrogen and oxygen. The light hydrogen escapes into space and the oxygen first combines with carbon in the atmosphere and then 'rusts' or sublimates out any carbon in rocks or anything else on the surface."

"These are major planet wide events. It would probably take 1,000 years at least for the oceans to boil away, and another 5,000 years to lose all that hydrogen. Oxygen being reactive, maybe another 5,000 years to sublimate what was on the surface. So the transformation of Earth into another Venus should take about 11,000 years. Since we're here 20,000 years later, the transformation should be pretty complete by now.[6] I'd like to visit the surface. Do we have the necessary equipment?"

Xil replied, "Yes, we do. Our space suits come with temperature stabilizers that can handle both extremes, from the cold of the absolute zero of space up to about 2000° F. Beyond that point, the space suit would begin to melt. Since we are dealing with a surface temperature on earth of about 700°F, there should be no problem. Shall I get you the suit?"

After putting on his space suit, Zog went into the airlock and closed the door. In a few minutes, a light came on indicating that outside atmosphere now filled the lock. He then pushed the handle on the door to the outside, the door swung open, and he could see earth's current surface.

The light was dim, but he could make out a rolling landscape of orange sand, with a pale cream sky. A strong wind jerked at the outside airlock door and almost tore it off. In the distance he could see some ruins, a large collection of much damaged concrete walls, and some seriously sagging steel skeletons.[7] He knew there would be no remains of any living thing, and the chances of finding human artifacts of any kind—whether drawings, murals or inscriptions—would be low. So he wondered whether it would be worth the effort to walk across to the ruins.

In the end he decided to walk over to the ruins, to acknowledge if even in a small way, that humans had once been the center of attention on this planet. It was slow going through the dry sand, even after he had deployed his Webbed sand shoe.

After a half hour of trudging, he reached the ruins. There was little there besides the partially obliterated concrete walls and the sagging steel skeletons, perhaps one tenth their original height. Even formerly ubiquitous human pollution, including plastic bags, seemed to have disappeared. Perhaps he was seeing the remains of various polluting materials in the multicolored gray and brown streaks scattered across the ruins. Once any carbon they contained had been sublimated out, there would be little structure left. He spent some time circling the various concrete ruins in hopes of finding some human traces. But most paint had contained carbon, so it had sublimated when the free oxygen grabbed all available carbon.

Examining these ruins, it occurred to Zog that the tragedy was not only what humans had managed to do to themselves. The human tragedy was only a small part of what had happened. For humans had managed to take down not only themselves, but over three billion years of the accumulated development of life on their planet. Humans were not only a biological mistake but the biggest biological mistake on earth in over 3 billion years.[8]

Of course, Zog knew that life was not so easily wiped out. In volcanic vents crevices on the ocean floor, there are bacteria which can thrive at temperatures of 300 degrees; perhaps some bacteria might still live at the high temperatures in those vents (Wilson 2002, pp. 5-7). So perhaps in several billion years, there might be a variety of life once again on the surface of the earth. Zog hoped that this life, whatever form it might someday take, would not repeat the mistakes of humanity. As Zog trudged slowly back to his ship, the feeling swept over him that what had happened on earth made life and its development ultimately seem so pointless.

A Much Better Scenario

The extraterrestrial Zog was returning for his next visit to earth 20,000 years in the future. As Zog's ship got close enough to Earth for our planet to appear as a small disk, Zog could tell there had been dramatic changes. Most of the land area of the planet was orange-brown desert. There was water, but it occurred in smaller bodies. There were green areas, but these were concentrated mostly around the northern and southern poles. There was no visible ice at the poles or anywhere else.

Circling around to the dark side of the planet, there were a few areas displaying the glowing lights of cities, almost none compared to what had been there 20,000 years earlier. Zog decided to try to contact whatever human inhabitants that still existed. This was something he had not done on any previous visit. But it was important to him to find out what had happened to the planet.

His ship began sending signals using as many media as possible using as many communication protocols as possible. Since it had been 20,000 years since his last visit, he did not expect any recognizable symbol systems to have survived. He remembered the radio messages earthlings had been transmitting to initiate contact with extraterrestrial beings, and those were some of the messages he was now transmitting to earth. As it turned out, one of those messages was promptly returned. Then for several days, nothing more. Zog and his companions narrowed their search to transmissions similar to the one received. There was clearly a lot of activity on some earthian communication bands. All this activity was routed to their powerful decipherer computer (standard equipment on the interstellar exploration ships of Zog's people). Within a few days, about 75% of the messages were understandable, and Zog had a decent vocabulary he could use to begin communicating with the humans currently inhabiting the planet.

"We have been waiting so long for extraterrestrial contact," the earthling said in his first message. "There are only a few of us still monitoring what we call the SETI equipment. It was only chance that I happened to check this week. We do regular checks only every six months. My name is Rowan."

"That is indeed fortunate," said Zog. "We had planned a visit of perhaps two to three of your months and so we would have been gone before you sent your reply."

"So, have you come to help us escape our follies and difficulties?" asked Rowan.

"Actually not," said Zog. "The role of our species is to try to understand the role of self-aware species in a non-self-aware universe. We determined some time ago that you might eventually be a species that could join us in this endeavor. Because of that goal, it is important for us not to intervene in any significant way in the development of your species."

"But," interjected Rowan, "we are still recovering from the damage done to the planet 20,000 years ago. It has been a slow process and I would have hoped you could help."

"I am afraid not," said Zog. "However, I hope you would still be willing to help us, even if we cannot help you directly. We need every bit of historical information you have, especially concerning that last 20,000 years."

"People did manage to preserve a lot of detail about human activity during that period," said Rowan. "There was a population bottleneck just after 20,000 years ago, so for a while the surviving humans were much more interested in continuing to insure their survival than in keeping good records of what they were up to. This was unfortunately most true during the period when temperature stabilization was being achieved."

"So you don't know exactly how this was done," said Zog.

"We do know that there had been a sudden anthropogenic increase in greenhouse gases, and that there was finally a panic when temperatures began escalating at blinding speeds. In the face of a probable runaway greenhouse, the technophiles of that time took some risky actions altering the atmosphere. They were partially successful in slowing the overall temperature increase, but the variability of temperature greatly increased. Some large population centers endured 160-degree heat for several days. That's what caused the population bottleneck. During that time, the human population decreased to about 500,000."

Zog interjected, "So is the climate more-or-less stable now? What has been your policy toward the technology that caused the climate problem?

"This is where we don't know exactly how things happened. The climate did stabilize, but earth is now a very different planet. It is no longer a water planet, it is a dry planet." [(Abe, et al. 2011)] Rowan continued, "A dry planet is much less susceptible to runaway greenhouse, so this transformation is to some extent insurance against any future runaway greenhouse."

"I was just going to ask," said Zog. "Are humans still practicing the level of technology that led to the climate problem?"

"No," said Rowan. "The negative consequences of the technological solution to runaway greenhouse did the trick. These consequences were several billion dead humans, as well as the extinction of countless other species. It is hard to imagine how the survivors could have dealt with that many corpses.

"Besides, we suspect the conversion to our present earth was done by humans who erased their traces. They would not have wanted us ultimately to get the wrong idea about the reliability of technology. We are careful neither to use nor to develop any technology which impacts the ecosystem in any major way."

Zog remarked, "I am glad some humans are left. My guess would be it will take at least 300,000 or 400,000 years for your resuscitated planet to reach an equilibrium, especially with all the new species you must have."

"We lost most of the ocean species, Rowan said. "The transformation was just too quick for them. But there are some interesting new land species. And there are also a few species we have good communication with. Some of them are quite integrated into human society. My lab assistant is a gorilla, for example."

Zog concluded, "I will be happy to return to this world in 20,000 more years to see what your interesting species has done. On my next visit, I hope I will find that you humans have continued to remain in equilibrium with your ecosystem."

REFERENCES

Abe, Y., Abe-Ouchi, A., Sleep, N., & Zahnle, K. (2011). Habitable zone limits for dry planets. *Astrobiology*, *11*(5). doi:10.1089/ast.2010.0545.

Derrickson, S. (2008). *The day the earth stood still.* [Motion Picture]. Century City, CA: 20th Century Fox Films.

Guidry, M. (1998). *A runaway greenhouse?* Retrieved 5/4/12 from csep10.phys.utk.edu/astr161/lect/venus/greenhouse.html

Hansen, J. (2009). *Storms of my grandchildren*. New York: Bloomsbury USA.

KFC. (2012, January 13). How likely is a runaway greenhouse effect on earth? *Technology Review.* Retrieved 6/3/12 from www.technologyreview.com/blog/arxiv/27495/

Wilson, E. O. (2002). *The future of life*. New York: Vintage Books.

ENDNOTES

1 Factory farming, the application of modern technology to agriculture, greatly reduces this dependence on the ecosystem. See Chapter 12, "Human-Animal Relations."

2 An excellent brief explanation of runaway greenhouse can be found at Guidry (1998). I also need to acknowledge my debt to James Hansen, who uses extraterrestrials to illustrate the consequences of climate change in Chapter 11 of his *Storms of my Grandchildren* (2009).

3 Any other alternatives besides these two would also be catastrophes decimating life on earth. For example, resource collapse; or some other widespread ecological catastrophe which has yet to present itself. Becoming a species among species is the only path to longer term survival for humanity.

4 See KFC (2012).

5 Permafrost has been frozen for over 10,000 years. When it melts, it could release double the amount of carbon presently in the atmosphere in the form of methane which is 20 times more effective as a greenhouse gas than carbon dioxide.

6 My planetary astronomer sources indicate that my times may be too short. A "wet" runaway greenhouse could take on the order of 100,000 years to complete. But for the sake of the 20,000 year timeframe of this story, I have compressed it. The consequences would be the same—the runaway greenhouse would just take much longer to complete (Abe et al. 2011).

7 The basis for this description is knowledge that in runaway greenhouse, all available carbon is connected with free oxygen. Steel contains carbon, so it would be compromised. Concrete (cement) may or may not contain carbon in limestone aggregate, so it would not always be compromised.

8 Noam Chomsky's definition of "biological mistake" discussed in Chapter 14, "Extinction," is used. The notion that humanity's problem is taking down so much of the rest of life with it, features in the 2008 film *The Day the Earth Stood Still.* As discussed in Chapter 7, "Civilization," this promising premise is vitiated by the aliens deciding not to destroy humanity because of Bach, Einstein, and mother love. No self-respecting alien would be swayed by these considerations. Also as discussed, Bach and Einstein are valuable only to other humans. And why would it be a positive thing that a seriously defective species had strong feelings involved with reproducing itself?

Compilation of References

Abe, Y., Abe-Ouchi, A., Sleep, N., & Zahnle, K. (2011). Habitable zone limits for dry planets. *Astrobiology, 11*(5). doi:10.1089/ast.2010.0545.

Achbar, M., Abbott, J., & Bakan, J. (2004). *The corporation.* [Motion Picture]. New York: Zeitgeist Films Ltd.

Adams, G. M. (2013). *Our hidden history of corporations in the United States.* Retrieved 4/11/2013 from reclaimdemocracy.org/corporate-accountability-history-corporations-us/

Adams, F. C., & Laughlin, G. (1997). A dying universe: the long-term fate and evolution of astrophysical objects. *Reviews of Modern Physics, 69*(2), 337–372. doi:10.1103/RevModPhys.69.337.

Adams, R., & Brown, B. K. (1941). Hydrazine sulfate. *Organic Syntheses, 1*, 309.

Adas, M. (1992). *Civilizations of the Americas: The peoples to the north.* Retrieved 1/23/12 from history-world.org/civilizations%20of%20the%20americas.htm

Allan, D. (2011). *Population growth over human history.* Retrieved 8/31/11 from www.globalchange.umich.edu/globalchange2/current/lectures/human_pop/human_pop.html

Amaral, K. (2003). Plastics in our oceans. *Woods Hole Institute.* Retrieved 10/31/11 from www.whoi.edu/science/B/people/kamaral/plasticsarticle.html

Ambrose, S. H. (1998). Late Pleistocene human population bottlenecks, volcanic winter, and differentiation of modern humans. *Journal of Human Evolution, 34*, 623–651. doi:10.1006/jhev.1998.0219.

Ament, P. (2007). Wheel. *The Great Idea Finder.* Retrieved 4/12/13 from www.ideafinder.com/history/inventions/wheel.htm

Ardrey, R. (1977). *The hunting hypothesis.* New York: Bantam Books.

Arena, C. (2004). *Cause for success.* Novato, CA: New World Library.

Arendt, H. (1965). *Eichmann in Jerusalem.* New York: Viking.

Arrow, K. (1951). *Social choice and individual values.* New York: John Wiley & Sons.

Asimov, I. (1954). *The caves of steel.* New York: HarperCollins.

Asimov, I., & Pohl, F. (1991). *Our angry earth.* New York: Tom Dougherty Associates.

Associated Press. (2010, October 18). Coyotes, not hunters, proposed for Valley Forge. *Animal News*. Retrieved 10/20/10 from http://www.bornfreeusa.org/news.php?p=2516&more=1

Avila, J. (2012). Superbug dangers in chicken linked to 8 million at-risk women. *ABC News Blogs*. Retrieved 7/15/12 from yahoo.com/abcnewsblogs/superbug-dangers-chicken-linked-8-million-risk-women-122809803—abc-news-health

Bacon, F. (1620). *Novum organum*. (B. Montague, Trans.). Retrieved 10/14/11 from history.hanover.edu/texts/Bacon/novorg.html

Ballotpedia. (2011). *California proposition 2, standards for confining farm animals (2008)*. Retrieved 10/16/11 from ballotpedia.org/wiki/index.php/California_Proposition_2,_Standards_for_Confining_ Farm_ Animals_%282008%29

Barnofsky, A. et al. (2012, June 7). Approaching a state shift in Earth's biosphere. *Nature*, 486.

Barrow, J. (2006). Glyphosate and cancer. *Australian Plants Online*. Retrieved 4/17/12 from anpsa.org.au/APOL20/dec00-3.html

Barton, M., Riel-Salvatore, J., Anderies, J., & Popescu, G. (2011). Modeling human ecodynamics and biocultural interactions in the Late Pleistocene of Western Eurasia. *Human Ecology*. Retrieved 11/22/11 from www.openabm.org/node/2734

Bartuska, T., Kazime, B., & Owen, M. (1999). *Defining sustainability*. Retrieved 3/14/12 from www.arch.wsu.edu/09%20publications/sustain/defnsust.htm

Batuman, E. (2011, December 19). The sanctuary: The world's oldest temple and the dawn of civilization. *The New Yorker*, 72-83.

Baxter, C. (2011, May 6). Koch industries: Still fueling climate denial. *Greenpeace Blogpost*. Retrieved 11/2/11 from www.greenpeace.org/usa/en/news-and-blogs/campaign-blog/koch-industries-still-fueling-climate-denial/blog/34609/

BBC. (2012). *Jainism*. Retrieved 4/3/12 from www.bbc.co.uk/religion/religions/jainism/

Bednarik, R. (2001). *Beads and the origins of symbolism*. Retrieved 11/27/11 from http://www.semioticon.com/frontline/bednarik.htm

Beliaev, D., Bondarenko, D., & Korotayev, A. (2001). Origins and evolution of chiefdoms. *Revista de Antropologia*, *30*, 373–395.

Benson, J. (2011). Genetically-modified, oil-eating bacteria creating dangerous mutant organisms in the gulf. *Naturalnews.com*. Retrieved 4/21/12 from www.naturalnews.com/031090_Gulf_of_Mexico_bacteria.html

Benson, T. (2011). Beginner's guide to rockets. *National Aeronautics and Space Administration (NASA)*. Retrieved 3/26/12 from exploration.grc.nasa.gov/education/rocket/bgmr.html# Brown, L. (2009). *Plan B 4.0*. New York: Norton.

Bentham, J. (1781). *An introduction to the principles of morals and legislation*. Retrieved 10/16/11 from www.constitution.org/jb/pml.htm

Bering, J. (2009, September 11). Why we blush: The social purpose of showing embarrassment. *Scientific American*. Retrieved 4/14/12 from www.scientificamerican.com/article.cfm?id=why-we-blush-social-embarrassment

Berkeley Plastics Task Force. (2009). Campaign against the plastic plague: Background info. *Earth Resource Foundation*. Retrieved 10/31/11 from www.earthresource.org/campaigns/capp/capp-background-info.html

Bettinger, B. (2007, July 20). Mitochondrial Eve and y-chromosomal Adam. *The Genetic Genealogist*. Retrieved 11/28/11 from www.thegeneticgenealogist.com/2007/07/20/mitochondrial-eve-and-y-chromosomal-adam/

Biello, D. (2012a, June 13). Walking the line: How to identify safe limits for human impacts on the planet. *Scientific American*. Retrieved 6/13/12 from news.yahoo.com/walking-line-identify-safe-limits-human-impacts-planet-120100550.html

Biello, D. (2012b, August). Recycling reality: Humans set to trash most elements on the periodic table. *Scientific American*. Retrieved 8/10/12 from news.yahoo.com/recycling-reality-humans-set-trash-most-elements-periodic-220200185.html

Bilger, B. (2011, December 19). The great oasis. *New Yorker (New York, N.Y.)*, 110–121.

Biology Online. (2009). *Pollution definition*. Retrieved 10/2/11 from www.biology-online.org/dictionary/Pollution

Blanton, R., Feinman, G., Kowalewski, S., & Nicholas, L. (1999). *Ancient oaxaca*. Cambridge, UK: Cambridge University Press. doi:10.1017/CBO9780511607844.

Bolles, E. B. (2008). Neanderthals had speech. *Babel's Dawn*. Retrieved 11/23/11 from www.babelsdawn.com/babels_dawn/2008/03/neanderthals—1.html

Bookchin, M., & Foreman, D. (1991). *Defending the earth*. Boston, MA: South End Press.

Bookstein, J. (2008). *Global warming: The population connection*. [Broadcast]. La Jolla, CA: UCTV. Retrieved 2/4/12 from www.uctv.tv/search-details.aspx?showID=14886

Boren, J., Baker, T., Hurd, B., & Mason, G. (2004). Tanning deer hides and small fur skins. *New Mexico State University*. Retrieved 11/27/11 from aces.nmsu.edu/pubs/_l/l-103.pdf

Borenstein, S. (2012, April 13). See Dan read: Baboons can learn to spot real words. *Associated Press*. Retrieved 4/13/12 from news.yahoo.com/see-dan-read-baboons-learn-spot-real-words-180435856.html

Bower, B. (2011). Stone age paint shop unearthed. *Science News, 180*(11). Retrieved 12/11/11 from www.sciencenews.org/view/generic/id/335172/title/Stone_Age_paint_shop_unearthed

Bradshaw Foundation. (2011). *Cave art of France*. Retrieved 7/24/2011 from www.bradshawfoundation.com/ france/index.php

Brockmann, H. J. et al. (2009). Interspecific behavioral comparisons of hand-reared wolves and dogs. *Advances in the Study of Behavior*, *39*, 87.

Brown, C. (2011, May). El Mirador, the lost city of the Maya. *Smithsonian*. Retrieved 2/24/12 from www.smithsonianmag.com/history-archaeology/El-Mirador-the-Lost-City-of-the-Maya.html#ixzz1nMv460Y0

Brown, L. (2009). *Plan B 4.0*. New York: Norton.

Buddho Dendo Kyokai. (2009). *A teaching a day: Excerpts from the teaching of Buddha*. Berkeley, CA: BDK America.

Burne, J. (2002, January 17). Animal instinct. *The Guardian*.

Caliguid, F. (2012, February 6). Big nickel mining firm says rains caused pollution of bay. *Inquirer Mindanao*. Retrieved 3/28/12 from newsinfo.inquirer.net/140773/big-nickel-mining-firm-says-rains-caused-pollution-of-bay

Carrier, D. R. (1984). The energetic paradox of human running and hominid evolution. *Current Anthropology*, *25*(4). doi:10.1086/203165.

Carson, R. (1962). *Silent spring*. New York: Houghton Mifflin.

Cassidy, D. (1998). Quantum mechanics 1925-1927: The uncertainty principle. *The American Institute of Physics*. Retrieved 9/02/11 from www.aip.org/history/heisenberg/p08.htm

Chirping Bird. (2012). *Plastic ban bag report*. Retrieved 5/27/12 from plasticbagbanreport.com

Choi, C. (2011, August 22). Top 10 things that make humans special. *Live Science*. Retrieved 1/14/12 from www.livescience.com/15689-evolution-human-special-species.html

Choi, C. (2012). Huge ancient civilization's collapse explained. *Live Science*. Retrieved 5/29/12 from news.yahoo.com/huge-ancient-civilization-collapse-explained-123449804.html

Chomsky, N. (2003). *Hegemony or survival*. Retrieved 6/24/11 from www.chomsky.info/books/hegemony01.htm

CIESIN (Center for International Earth Science Information Network). (n.d.). The Montreal protocol on substances that deplete the ozone layer. *CIESIN Thematic Guides*. Retrieved 4/03/09 from www.ciesin.org/TG/PI/POLICY/montpro.html

Cimons, M. (2011, March 22). Ecosystem function of ants. *US News & World Report Science*. Retrieved 5/2/12 from www.usnews.com/science/articles/2011/03/22/ecosystem-function-of-ants

Citizens' Clearinghouse on Waste Management. (2004). *Environmental problems from municipal solid waste*. Retrieved 3/17/12 from www.citizenswasteinfo.org/A559CA/ccwm.nsf/5155d8f53ce25d2785256cc300567828/c050375504a2a75085256e7d005a325f?OpenDocument

Claessen, H. (1978). The early state: A structural approach. In Claessen, H., & Skalnik, P. (Eds.), *The Early State*. The Hague, The Netherlands: Mouton. doi:10.1515/9783110813326.531.

Clark, B. (2011). *Grand challenges of sustainability science*. Paper presented at the 2011 Conference on Resilience, Innovation, and Sustainability: Navigating the Complexities of Global Change. Phoenix, AZ. Retrieved 2/11/12 from www.hks.harvard.edu/var/ezp_site/storage/fckeditor/file/pdfs/centers-programs/centers/cid/ssp/docs/about-us/in-the-news/2011/Clark_Grand_Chal_Sust_Science_110314.pdf

Clottes, J. (2002). Chauvet cave (ca. 30,000B.C.). In *Heilbrunn Timeline of Art History*. New York: The Metropolitan Museum of Art. Retrieved 7/24/2011 from www.metmuseum.org/toah/hd/chav/hd_chav.htm

Co2gether. (2012) Understand climate change. *Co2gether*. Retrieved 1/31/12 from www.co2gether.org/my_co2/understand/

Cockcroft, C., & Kenyon, C. (2008). Surviving longer. In Shuckburgh, E. (Ed.), *Survival: The Survival of the Human Race*. Cambridge, UK: Cambridge University Press.

Colburn, T., Dumanoski, D., & Myers, J. (1996). *Our stolen future*. New York: Dutton.

Colby, C. (1996). *Introduction to evolutionary biology*. Retrieved 9/15/11 from www.talkorigins.org/faqs/faq-intro-to-biology.html

Collins, N. (2012, April 13). Test tube hamburgers to be served this year. *The Telegraph*. Retrieved 4/13/12 from www.telegraph.co.uk/science/science-news/9091628/ test-tube-hamburgers-to-be-served-this-year.html

Cook, G. (2012, June 5). Do plants think? *Scientific American*. Retrieved 6/5/12 from news.yahoo.com/plants-think-110000318.html

Coppinger, R., & Lorna. (2001). *Dogs*. New York: Scribner.

Cribb, J. (2011). Taxonomy: New name needed for unwise Homo? *Nature*, *476*(282). PMID:21850091.

Cummins, J. (2007). Requiem for the honeybee. *Institute of Science in Society Report 24/04/07*. Retrieved 10/3/11 from www.i-sis.org.uk/requiemForTheHoneybee.php

Dalai Lama. (2005). *The universe in a single atom: The convergence of science and spirituality*. New York: Morgan Road Books.

Dawkins, R. (2006). *The god delusion*. New York: Houghton Mifflin.

De Chant, T. (2011). *Population density fostered literacy, the industrial revolution*. Retrieved 11/29/11 from persquaremile.com/2011/11/11/population-density-fostered-literacy-the-industrial-revolution

De Chant, T. (2011, August 17). Hunter gatherer populations show humans are hardwired for density. *Scientific American's Guest Blog*. Retrieved 11/29/11 from persquaremile.com/2011/08/17/hunter-gatherer-populations-show-humans-are-hardwired-for-density/

Della Santina, P. (1997). *The tree of enlightenment*. Taipei, Taiwan: The Corporate Body of the Buddha Educational Foundation.

Derrickson, S. (2008). *The day the earth stood still*. [Motion Picture]. Century City, CA: 20th Century Fox Films.

Derr, M. (2011). *How the dog became the dog*. New York: Overlook Press.

Descartes, R. (1637). *Discourse on method*. Retrieved 10/10/11 from records.viu.ca/~johnstoi/descartes/descartes1.htm

Descartes, R. (1641). *Meditations concerning first philosophy*. Retrieved 9/14/11 from www.wright.edu/cola/descartes/mede.html

Diamond, J. (1987, May). The worst mistake in the history of the human race. *Discover*, 64–66.

Diamond, J. (1997). *Guns, germs, and steel*. New York: Norton.

Diamond, J. (2005). *Collapse: How societies choose to fail or succeed*. New York: Viking Books.

Diamond, J. (2012). *The world before yesterday*. New York: Viking Books.

Diamond, J. (2012). What makes countries rich or poor? *The New York Review of Books*, *59*(10).

Dreifus, C. (2007, January 9). Always revealing, human skin is an anthropologist's map. *New York Times*. Retrieved 4/14/12 from www.nytimes.com/2007/01/09/science/09conv.html

Drew, E., & McGuigan, J. (2008). Prevention of crime: An overview of gated communities and neighborhood watch. *International Foundation for Prevention Officers*. Retrieved 12/09/11 from www.ifpo.org/articlebank/gatedcommunity.html

Earth Talk. (2012). *Do the benefits of recycling outweigh the costs?* Retrieved 3/18/12 from environment.about.com/od/recycling/a/benefit_vs_cost.htm

Ecology Center. (2011). *Plastics task force: Seven misconceptions about plastic and plastic recycling.* Retrieved 10/31/11 from www.ecologycenter.org/ptf/misconceptions.html

Ecology Center. (2012). *Seven misconceptions about plastic and plastic recycling.* Retrieved 3/17/12 from www.ecologycenter.org/ptf/misconceptions.html

Ehrlich, P. (1968). *The population bomb.* New York: Ballantine Books.

Ehrlich, P. (2000). *Human natures.* New York: Penguin Group.

Einstein, A. (1956). *The meaning of relativity* (5th ed.). New York: Barnes & Noble.

Elton, C. (2011). Safety net. *Audubon Magazine, 113*(6).

Emerson, C. (1999). *Aquaculture impacts on the environment.* Retrieved 12/12/11 from www.csa.com/discoveryguides/aquacult/overview.php

Energy Justice Network. (2012). *Dioxins & furans: The most toxic chemicals known to science.* Retrieved 4/17/12 from www.ejnet.org/dioxin/

Fahrenthold, D. (2006, September 29). Wildlife waste is major water polluter, studies say. *Washington Post.* Retrieved 10/3/11 from www.washingtonpost.com/wp-dyn/content/article/2006/09/28/AR2006092801994.html

Farivar, C. (2011). *Ancient humans used stone tools earlier than previously thought.* Retrieved 1/07/11 from www.dw-world.de/dw/article/0,5891406,00.html

Feranec, R. (2011). What became of the megafauna? *American Scientist, 99*(6). Retrieved 12/8/11 from www.americanscientist.org/bookshelf/pub/what-became-of-the-megafauna

Flaherty, R. (1922). *Nanook of the north.* [Motion Picture]. New York: The Criterion Collection.

Flannery, T. (2010). *Here on earth: A new beginning.* London: Allen Lane.

Foreman, D. (1991). *Confessions of an eco-warrior.* New York: Harmony Books.

Foreman, D. (2011). *Manswarm: The killing of wildlife.* Durango, CO: Raven's Eye Press.

Fountain, H. (2012, July 26). Storms threaten ozone layer over U.S., study says. *New York Times.* Retrieved 8/10/12 from nytimes.com/2012/07/27/science/earth/strong-storms-threaten-ozone-layer-over-us-study-says.html

Freud, S. (1930). *Civilization and its discontents.* New York: Norton.

Friedman, T. (2005). *The world is flat.* New York: Farrar, Straus and Giroux.

Ganzel, B. (2003). *The dust bowl.* Retrieved 4/10/13 from www.livinghistoryfarm.org/farminginthe30s/water_02.html

Gascoigne, B. (2001). History of the domestication of animals. *History World.* Retrieved 12/13/11 from www.historyworld.net/wrldhis/PlainTextHistories.asp?historyid=ab57

Gauthier, D. (1986). *Morals by agreement.* Oxford, UK: Clarendon Press.

Gay, M. (2010, December 13). Texas biggest gun source for Mexican drug war. *Huffington Post*. Retrieved 4/17/12 from www.aolnews.com/2010/12/13/report-texas-biggest-gun-source-for-mexican-drug-war/

Gill, N. S. (2000). An introduction to Sumer. *About.com Ancient/Classical History*. Retrieved 5/23/12 from ancienthistory.about.com/od/southasia/a/Sumer.htm

Gillam, C. (2012, April 18). Farm group seeks U.S. halt on "dangerous" crop chemicals. *Reuters*. Retrieved 4/18/12 from news.yahoo.com/farm-group-seeks-u-halt-dangerous-crop-chemicals-123359159.html

Gillam, C. (2012, March 10). Scientists warn EPA on Monsanto corn rootworm. *Reuters*. Retrieved 3/10/12 from news.yahoo.com/scientists-warn-epa-monsanto-corn-rootworm-193441402.html

Goldenberg, D. (n.d.). *Not-so-stupid animal tricks*. Retrieved 1/2/11 from mentalfloss.com

Gombrich, E. H. (1996). *Art and illusion*. Princeton, NJ: Princeton University Press.

Gonzalez, R. (2012). New research demonstrates how humans are still evolving. *Virpi Lummaa News*. Retrieved 6/2/12 from io9.com/virpi-lummaa/

Goodman, N. (1976). *Languages of art*. Indianapolis, IN: Hackett.

Gore, A., & Blood, D. (2012). *A manifesto for sustainable capitalism*. Retrieved 3/17/12 from www.generationim.com/media/pdf-wsj-manifesto-sustainable-capitalism-14-12-11.pdf

Gore, A. (1992). *Earth in the balance*. New York: Houghton Mifflin.

Guidry, M. (1998). *A runaway greenhouse?* Retrieved 5/4/12 from csep10.phys.utk.edu/astr161/lect/venus/greenhouse.html

Hackett, J. (2013). *Vegan - What is a vegan? What do vegans eat?* Retrieved 4/12/13 from vegetarian.about.com/od/glossary/g/Vegan.htm

Hall, M., & Ng, S. (2004). *Overpopulation vs. overconsumption*. Retrieved 2/14/12 from www.odec.ca/projects/2004/ngsi4s0/public_html/indexx.htm

Halweil, B. (2012). Can organic farming feed us all? *Worldwatch Institute*. Retrieved 2/12/12 from www.worldwatch.org/node/4060

Hansen, J. (2009). *Storms of my grandchildren*. New York: Bloomsbury USA.

Hardin, G. (1968). The tragedy of the commons. *Science*, *162*, 1243–1248. doi:10.1126/science.162.3859.1243 PMID:5699198.

Harris, A. (2012). *Social ecology*. Retrieved 4/12/13 from www.thegreenfuse.org/socialecology.htm

Hartman, P. (1996). The stone age, bronze age, and iron age. *The Hartman Website*. Retrieved 1/14/12 from www.naciente.com/essay88.htm

Hawking, S. (2009). Galileo and the birth of modern science. *American Heritage of Invention & Technology*, *24*(1), 36.

Heidegger, M. (1927). *Sein und zeit*. Tűbingen, Germany: Max Niemeyer.

Heidegger, M. (1954). *Was heisst denken?* Tübingen, Germany: Max Niemeyer.

Heidegger, M. (1955). The question concerning technology. In *The Question Concerning Technology and Other Essays*. New York: Harper & Row.

Heimbuch, J. (2010). *Ocean film fest 2010: A killer whale takes down a... great white shark??* Retrieved 11/8/11 from www.treehugger.com/natural-sciences/ocean-film-fest-2010-a-killer-whale-takes-down-a-great-white-shark-video.html

Herzog, H. (2011, June 20). Why do most vegetarians go back to eating meat? *Psychology Today*. Retrieved 4/10/12 from www.psychologytoday.com/blog/animals-and-us

Herzog, W. (2010). *Cave of forgotten dreams*. [Motion Picture]. New York: IFC Films.

Herzog, W. (n.d.). *Grizzly man*. [Film] Santa Monica, CA: Lionsgate Films.

Hiebert, T. (1996). *Rethinking dominion theology*. Winnipeg, Canada: Direction.

Hilton, B. (2012, July 8). Hummingbirds and flowers: A study of co-adaptive relationships. *Howard Hughes Medical Institute*. Retrieved 7/8/12 from www.hhmi.org/coolscience/resources/SPT—FullRecord.php?ResourceId=11

Hirsch, D. (2012). *Nuclear power: Ethical and policy implications*. Retrieved 4/17/12 from www.uctv.tv/search-details.aspx?showID=23504

Hirst, K. K. (2011). Domestication history of sheep. *About.com.Archaeology*. Retrieved 12/13/11 from archaeology.about.com/od/shthroughsiterms/qt/Sheep-History.htm

Hirst, K. K. (2011). *The invention of pottery*. Retrieved 1/18/12 from archaeology.about.com/od/inventions/a/pottery_invent.htm

Hobbes, T. (1651). *Leviathan*. Retrieved 5/20/04 from darkwing.uoregon.edu/~rbear/hobbes/leviathan.html

Humphrey, N. (1998). Cave art, autism, and the evolution of the human mind. *Cambridge Archaeological Journal*, *8*(2), 165–191. doi:10.1017/S0959774300001827.

Huxley, J. (1953). *Evolution in action*. New York: Harper & Brothers.

Infoplease. (2012). *World's ten most corrupt leaders*. Retrieved 2/3/12 from www.infoplease.com/ipa/A0921295.html

Jackson, T. (2008). *The challenge of sustainable lifestyles. In 2008: State of the World*. New York: Norton.

Japan Times. (2000, October 26). Archaeologists find 'oldest' habitation. *Japan Times*. Retrieved 12/4/11 from www.japantimes.co.jp/text/nn20001024a1.html

Jaynes, J. (1976). *The origin of consciousness in the breakdown of the bicameral brain*. New York: Houghton Mifflin.

Johnson, S. (2013). *Acheulean stone tools*. Retrieved 4/23/13 from www.ehow.com/info_8242594_acheulean-stone-tools.html#ixzz2RJbSn6dE

Jones, T. (2007). *Did homo erectus settle down 400,000 years bp?*. Retrieved 11/19/11 from anthropology.net/2007/06/25/did-homo-erectus-settle-down-400000-years-bp/

Joyce, R. (2011). Dressing for the Pleistocene. *Ancient Bodies, Ancient Lives*. Retrieved 12/6/11 from ancientbodies.wordpress.com/2011/01/07/dressing-for-the-pleistocene/

Kaiser, D. (1990). *Dolphin consciousness*. Retrieved 1/30/06 from home.onemain.com/~dk1008206/html/dolph1.htm

Kant, I. (1785). *Groundwork of the metaphysics of morals*. Retrieved 3/24/12 from www.earlymoderntexts.com/kgw.html

Kant, I. (1929). *Critique of pure reason* (Smith, N. K., Trans.). London: Macmillan.

Kant, I. (1790). *Kritik der urteilskraft*. Berlin and Libau: Lagarde and Friederich.

Kant, I. (1964). *The doctrine of virtue: Metaphysics of morals, part II*. New York: Harper & Row.

Katers, R. (2009). *The history of PCBs*. Retrieved 10/28/11 from www.foxriverwatch.com/site_index_fox_river_bay_pcbs.html

Kelly, R. (2005). The evolution of lethal intergroup violence. *Proceedings of the National Academy of Science, 102*(43), 15294–15298. Retrieved 12/22/11 from www.ncbi.nlm.nih.gov/pmc/articles/PMC1266108/?tool=pmcentrez

KFC. (2012, January 13). How likely is a runaway greenhouse effect on earth? *Technology Review*. Retrieved 6/3/12 from www.technologyreview.com/blog/arxiv/27495/

Kirby, P. (2011, February 15). Religion the ultimate tyranny. *Washington Post*. Retrieved 4/27/12 from onfaith.washingtonpost.com/onfaith/panelists/paula_kirby/2011/02/religion_the_ultimate_tyranny.html

Klein, N. (2007). *The shock doctrine*. New York: Henry Holt.

Klein, N. (2011, November 9). Capitalism vs. the climate. *Nation (New York, N.Y.)*.

Knud-Hansen, C. (1994). Historical perspective of the phosphate detergent conflict. *Conflict Resolution Consortium, University of Colorado*. Retrieved 10/31/11 from www.colorado.edu/conflict/full_text_search/AllCRCDocs/94-54.htm

Koplow, D. (2011, February). Nuclear power: Still not viable without subsidies. *Union of Concerned Scientists*.

Kramarae, C., & Spender, D. (Eds.). (2000). *Routledge international encyclopedia of women*. Florence, KY: Taylor & Francis.

Kristula, D. (2001). *The history of the internet*. Retrieved on 4/3/08 from www.davesite.com/Webstation/net-history.shtml

Krulwich, R., & Abumrad, J. (2008, April 29). Minty e. coli and other bioengineering feats. *NPR*. Retrieved 4/21/12 from www.npr.org/templates/story/story.php?storyId=90014997

Kuhn, T. (1962). *The structure of scientific revolutions*. Chicago: University of Chicago Press.

Kuijsten, M. (2006). *Reflections on the dawn of consciousness*. Henderson, NV: Julian Jaynes Society.

Lai, L., & Morrison, K. (2013). *Nuclear energy fact sheet*. Retrieved 4/13/13 from www.wagingpeace.org/menu/issues/nuclear-energy-&-waste/nuclear-energy-fact-sheet.php

Lautenberg Press Office. (2010). *Lautenberg introduces safe chemicals act to protect Americans from toxic chemicals*. Retrieved 10/29/11 from lautenberg.senate.gov/newsroom/record.cfm?id=323863

Leakey, R., & Lewin, R. (1992). *Origins reconsidered*. New York: Anchor.

Leakey, R., & Lewin, R. (1995). *The sixth extinction*. New York: Doubleday.

Lee, R. (1980). *Lactation, ovulation, infanticide, and women's work: A study of hunter-gatherer population regulation*. Retrieved 5/22/12 from tspace.library.utoronto.ca/bitstream/1807/18007/1/TSpace0104.pdf

Leeming, D. A., & Leeming, M. A. (2009). *A dictionary of creation myths*. Oxford, UK: Oxford University Press.

Lemieux, N., & Maynard, B. (1999). Deer damage and control. *University of Rhode Island Landscape Horticulture Factsheet*. Retrieved 10/16/11 from www.uri.edu/ce/factsheets/sheets/deermanage.html

Lerner, M. (2000). *Spirit matters*. Charlottesville, VA: Hampton Roads Publishing.

Levi-Strauss, C. (1962). *The savage mind*. Chicago: University of Chicago Press.

Levi-Strauss, C. (1968). *The origin of table manners*. New York: Harper & Row.

Lewandowsky, S. (2011). Acceptance of science and ideology. *Shaping Tomorrow's World*. Retrieved 11/9/11 from www.shaping-tomorrowsworld.org/ideologyScience.html

Lindemans, M. F. (1998). Xipe totec. *Encyclopedia Mythica*. Retrieved 11/27/11 from www.pantheon.org/articles/x/xipe_totec.html

Lynas, M. (2011). *The god species*. Washington, DC: The National Geographic Society.

MacIntosh, H. S. (2007). *What's the dirt on phosphate-free soaps?* Retrieved 11/1/11 from www.treehugger.com/culture/ask-treehugger-whats-the-dirt-on-phosphate-free-soaps.html

Malik, H. (2012). *Managerial accounting - Profit planning*. Retrieved 1/27/12 from hafeezrm.hubpages.com/hub/Managerial-Accounting-Profit-Planning

Mancuso, S. (2012). *Journal of plant signaling & behavior*. Retrieved 4/10/12 from www.plantbehavior.org/about.html

Markel, H. (2010, October 15). How the word 'scientist' came to be. *NPR*. Retrieved 10/15/11 from www.npr.org/templates/story/story.php?storyId=127037417

Mason, M. (2013). *Early history of Napster*. Retrieved on 4/13/13 from www.moyak.com/papers/napster-history.html

Max Planck Society. (2010). *A long childhood is of advantage*. Retrieved 4/14/12 from www.mpg.de/617475/pressRelease20101111?filter_order=L

McAuliffe, K. (2010, September). If modern humans are so smart, why are our brains shrinking?. *Discover*. Retrieved 1/27/12 from discovermagazine.com/2010/sep/25-modern-humans-smart-why-brain-shrinking/article_view?b_start:int=2&-C

McBroom, P. (1999). Meat-eating was essential for human evolution, says UC Berkeley anthropologist specializing in diet. University of California, Berkeley. Retrieved 11/27/11 from berkeley.edu/news/media/releases/99legacy/6-14-1999a.html.

McFadyen, J. (2012). What is the immigration and nationality act? *About.com Immigration Issues*. Retrieved 3/13/12 from immigration.about.com/od/immigrationlawandpolicy/f/INA.htm

McKibben, B. (1989). *The end of nature*. New York: Random House.

McKibben, B. (2007). *Deep economy*. New York: St. Martin's Griffin.

McKibben, B. (2011). *EaArth*. New York: St. Martins Griffin.

Mill, J. S. (1848). Principles of political economy. *Library of Economics and Liberty*. Retrieved 2/28/12 from www.econlib.org/library/Mill/mlP.html

Mill, J. S. (1863). *Utilitarianism.* Retrieved on 5/10/04 from //etext.library.adelaide.edu.au/m/mill/john_stuart/m645u/

Millheiser, I. (2009, August 17). Scalia says there's nothing unconstitutional about executing the innocent. *Think Progress.* Retrieved 5/9/12 from thinkprogress.org/politics/2009/08/17/56525/scalia-actual-innocence/

Minard, A. (2009, April 3). Gamma-ray burst caused mass extinction? *National Geographic News.* Retrieved 5/9/12 from news.nationalgeographic.com/news/2009/04/090403-gamma-ray-extinction.html

Mommer, B. (2003). *Ecological role of termites in dry environments.* Retrieved 5/2/12 from www.biology.iastate.edu/InternationalTrips/1Australia/Australia%20papers/EcolAusTermites%20.htm

Moravec, H. (1988). *Mind children.* Cambridge, MA: Harvard University Press.

Morris, A. (2010, April). Climate change in Foley, J., "Boundaries for a Healthy Planet. *Scientific American.*

Mote, V. (1975). Air resource management in Soviet Russia. *Proceedings of the American Philosophical Society, 119*(3). Retrieved 10/30/11 from www.jstor.org/pss/986670

Mountain Justice. (2012). *What is mountain top removal mining?.* Retrieved 1/31/12 from mountainjustice.org/facts/steps.php

Murchison, K. (2007). *The snail darter case: TVA versus the endangered species act.* Lawrence, KS: University Press of Kansas.

Murnau, F. (1927). *Sunrise: A song of two humans.* [Motion Picture]. Century City, CA: Twentieth Century Fox Film Corporation.

Nadeau, R. L. (2006). *The environmental endgame.* New Brunswick, NJ: Rutgers University Press.

Nagel, E., & Newman, J. R. (1958). *Gödel's proof.* New York: New York University Press.

National Geographic News. (2011, May 18). Ancient fertile crescent almost gone, satellite images show. *National Geographic News.*

New Economics Foundation. (2009). *The happy planet index 2.0.* Retrieved 3/4/12 from www.happyplanetindex.org/public-data/files/happy-planet-index-2-0.pdf

News, B. B. C. (2000, March 1). World's oldest building discovered. *BBC News Sci/Tech.* Retrieved 12/4/11 from news.bbc.co.uk/2/hi/science/nature/662794.stm

Niemitz, C. (2010). The evolution of the upright posture and gait—A review and a new synthesis. *Naturwissenschaften, 97*(3), 241–263. Retrieved 4/14/12 from www.ncbi.nlm.nih.gov/pmc/articles/PMC2819487/

Nietzsche, F. (1873). *On truth and lie from an extramoral point of view.*

Nolan, C. (2010). *Inception.* [Motion picture]. Burbank, CA: Warner Bros.

Norman, J. (2013). The earliest hearths. *From Cave Paintings to the Internet.* Retrieved 4/11/13 from www.historyofinformation.com/index.php

Norris, J. (2009, June 25). Sugar is a poison, says UCSF obesity expert. *UCSF News.* Retrieved 1/31/2012 from www.ucsf.edu/news/2009/06/8187/obesity-and-metabolic-syndrome-driven-fructose-sugar-diet

Northwest Fisheries Science Center. (2009). *Domoic acid poisoning.* Retrieved 10/2/11 from www.nwfsc.noaa.gov/hab/habs_toxins/marine_biotoxins/da/index.html

NRDC (National Resources Defense Council). (2005). *Pollution from giant livestock farms threatens public health.* Retrieved 12/12/11 from www.nrdc.org/water/pollution/nspills.asp

O'Neil, D. (2012). *Early transitional humans.* Retrieved 4/11/13 from anthro.palomar.edu/homo/homo_1.htm

Olesen, A. (2006, July 22). Stephen Hawking warns about global warming. *USA Today.*

Pagani, M., Zachos, J., Freeman, K., Tipple, B., & Bohaty, S. (2005). Marked decline in atmospheric carbon dioxide concentrations during the Paleogene. *Science, 309*(5734), 600-603. Retrieved 6/2/12 from www.sciencemag.org/content/309/5734/600

Pappas, S. (2012, March 29). Conservatives losing trust in science, study finds. *LiveScience.com.* Retrieved 3/31/12 from http://news.yahoo.com/conservatives-losing-trust-science-study-finds114401347.html

Parry, W. (2011). Bid to rename Homo sapiens called unwise: Most scoff as science writer says Latin term for 'wise man' no longer deserved. *MSNBC on Science.* Retrieved 11/27/11 from www.msnbc.msn.com/id/44182081/ns/technology_and_science-science/t/bid-rename-homo-sapiens-called-unwise/#.TtQZpLIr1dQ

Partch, J. (2004). *A brief history of type.* Retrieved 4/12/13 from graphicdesign.spokanefalls.edu/tutorials/process/type_basics/history.htm

Pennisi, E. (1999). Did cooked tubers spur the evolution of big brains? *Science, 283*(5410), 2004–2005. doi:10.1126/science.283.5410.2004 PMID:10206901.

Pfeiffer, E. (2012, April 4). Next great depression? MIT researchers predict 'global economic collapse' by 2030. *The Sideshow.* Retrieved 4/4/12 from news.yahoo.com/blogs/sideshow/next-great-depression-mit-researchers-predict-global-economic-190352944.html

Pruetz, J., & Lindshield, S. (2011). Plant-food and tool transfer among savanna chimpanzees at Fongoli, Senegal. *Primates, Online First™.* Retrieved 12/7/11 from www.springerlink.com/content/?k=(au%3a(pruetz)+OR+ed%3a(pruetz))

Putnam, H. (1981). *Reason, truth and history.* Cambridge, UK: Cambridge University Press. doi:10.1017/CBO9780511625398.

Quine, W. V. (1982). *Methods of logic* (4th ed.). Cambridge, MA: Harvard University Press.

Quinn, D. (1999). *Beyond civilization.* New York: Three Rivers Press.

Rafferty, J. (2009). Zhoukoudian. *Britannica Online.* Retrieved 4/11/13 from www.britannica.com/EBchecked/topic/114721/Zhoukoudian

Rawls, J. (1999). *A theory of justice* (Revised ed.). Cambridge, MA: Harvard University Press.

Relaxnews. (2011, October 28). Iceland to help France save trees from global warming. *Relaxnews.* Retrieved 11/1/11 from news.yahoo.com/iceland-help-france-save-trees-global-warming-101350209.html

Reports, N. (2009).. . *Climatic Change.* doi: doi:10.1038/climate.2009.95.

Rockström. Johan, et al. (2009). Planetary boundaries: Exploring the safe operating space for humanity. *Ecology and Society, 14*(2). Retrieved 2/11/12 from www.ecologyandsociety.org/vol14/iss2/art32/

Rosenfield, I., & Ziff, E. (2012). Review of Shepherd's Neurogastronomy. *The New York Review of Books, 59*(11).

Rowlands, M. (2001). *The nature of consciousness*. Cambridge, UK: Cambridge University Press. doi:10.1017/CBO9780511487538.

Rowlands, M. (2009). *Animal rights: Moral theory and practice*. New York: Palgrave Macmillan.

Russell, E. (2011). *Evolutionary history*. Cambridge, UK: Cambridge University Press.

Sahlins, M. (1968). Notes on the original affluent society. In Lee, R. B., & DeVore, I. (Eds.), *Man the Hunter* (pp. 85–89). New York: Aldine Publishing Company.

Santayana, G. (1896). *The sense of beauty*. New York: Scribner's & sons.

Sapien, J. (2007, April 26). Superfund today: Massive undertaking to clean up hazardous waste sites has lost both momentum and funding. *The Center for Public Integrity*. Retrieved 10/30/11 from projects.publicintegrity.org/Superfund/report.aspx?aid=851

Scherer, G. (2004, October 27). Christian-right views are swaying politicians and threatening the environment. *Grist*. Retrieved 11/18/11 from www.grist.org/article/scherer-christian

Schulte, P. et al. (2010). The Chicxulub asteroid impact and mass extinction at the Cretaceous-Paleogene boundary. *Science, 327*(5970), 1214–1218. doi:10.1126/science.1177265.

Schultz, R. (2006). *Contemporary issues in ethics and information technology*. Hershey, PA: IGI Global.

Schultz, R. (2009). *Information technology and the ethics of globalization*. Hershey, PA: IGI Global Press. doi:10.4018/978-1-60566-922-9.

Schumacher, E. F. (1973). *Small is beautiful*. New York: Harper & Row.

ScienceDaily. (2010). Conscious choice of food can substantially mitigate climate change, research finds. *ScienceDaily*. Retrieved 3/18/12 from http://www.sciencedaily.com /releases/2010/06/100628075744.htm

Seabrook, J. (2012, March 26). The song machine. *New Yorker (New York, N.Y.)*.

Service, E. (1975). *Origins of the state and civilization*. New York: W. W. Norton.

Sewell, A. (2012, June 18). Environmental group demands full San Onofre review. *LA Times*. Retrieved 7/3/12 from latimesblogs.latimes.com/lanow/2012/06/san-onofre-friends-of-the-earth-embargoed-until-7-am-monday.html

Shuckburgh, E. (2008). *Survival: The survival of the human race*. Cambridge, UK: Cambridge University Press.

Siebert, C. (2006, October 8). An elephant crackup? *New York Times Magazine*. Retrieved 9/01/11 from www.nytimes.com/2006/10/08/magazine/08elephant.html?pagewanted=1

Sileshi, G. (2009). Integrating ethno-ecological and scientific knowledge of termites for sustainable termite management and human welfare in Africa. *Ecology and Society, 14*(1).

Singer, P. (2002). *Animal liberation*. New York: HarperCollins.

Singer, P. (2009). *Animal liberation (updated ed)*. New York: HarperCollins.

Skakkebaek, N., et al. (1992). Evidence for decreasing quality of semen during past 50 years. *British Medical Journal, 305*(6854). Retrieved 9/17/11 from www.jstor.org/pss/29716905

Solecki, R. S. (1975). Shanidar IV, a Neanderthal flower burial in northern Iraq. *Science, 190*(28), 880. doi:10.1126/science.190.4217.880.

Specter, M. (2012, May 14). The climate fixers. *New Yorker (New York, N.Y.)*.

Spinoza, B. (1677). *Ethics*. Retrieved 12/6/11 from users.erols.com/nbeach/spinoza.html

Stanford University. (2007). Harmful environmental effects of livestock production. *ScienceDaily*. Retrieved 11/3/11, from http://www.sciencedaily.com/releases/2007/02/070220145244.htm

Stevens, W. (1959). *Collected poems*. New York: Vintage Books.

Stiglitz, J. (2010). *Freefall*. New York: Norton.

Swenson, K. (2010). *The bible and human 'dominion' over animals: Superiority or responsibility?* Retrieved 9/15/11 from www.huffingtonpost.com/kristin-m-swenson-phd/the-bible-and-human-domin_b_681363.html

Tainter, J. (1988). *The collapse of complex societies*. Cambridge, UK: Cambridge University Press.

Taylor, R. (1974). *Metaphysics*. Englewood Cliffs, NJ: Prentice-Hall.

The free dictionary. (2011). *Definition of corporation*. Retrieved 11/1/04 from legal-dictionary.thefreedictionary.com/Corporation

Thomas, P. (2012, October 23). Talking whale? Captive beluga had learned to mimic human voices. *GrindTV.com*. Retrieved 10/23/12 from www.grindtv.com/outdoor/blog/37509/talking+whale+captive+beluga+had+learned+to+mimic

Thoreau, H. D. (1854). *Walden*. Retrieved 1/1/12 from www.online-literature.com/thoreau/walden/

Toedt, D. C. (2004). *Serious inconsistencies in the new testament writings: Six reasons for skepticism about the traditionalist account*. Retrieved 11/12/11 from www.questioning-christian.com/2004/10/troubling_incon.html

Tolkein, J. R. R. (1954). *The lord of the rings*. London: Allen & Unwin.

UN Food and Agricultural Organization. (2005). *General situation of world fish stock*. Retrieved 11/2/11 from www.fao.org/newsroom/common/ecg/1000505/en/stocks.pdf

United Nations Environmental Programme (UNEP). (2011). *Recycling rates of metals: A status report.* Retrieved 8/12/12 from www.unep.org/resourcepanel/Portals/24102/PDFs/Metals_Recycling_Rates_110412-1.pdf

University of Kent. (2011, March 7). Stone tools influenced hand evolution in human ancestors, anthropologists say. *ScienceDaily*. Retrieved 4/14/12 from www.sciencedaily.com/releases/2011/03/110307101504.htm

US Environmental Protection Agency. (2011). *Statistics on the management of used and end-of-life electronics*. Retrieved 3/25/12 from www.epa.gov/epawaste/conserve/materials/ecycling/manage.htm

von Uexkull, J. (2007, February 21) *The post corporate world.* Paper presented at the 6th annual Frank Kelly Lecture. Santa Barbara, CA: University of California.

Wachowski, A., & Wachowski, L. (1999). *The matrix.* [Motion picture]. Burbank, CA: Warner Bros.

Walsh, B. (2011, October 4). Who's bankrolling the climate-change deniers? *Time*. Retrieved 11/4/11 from http://www.time.com/time/health/article/0,8599,2096055,00.html

Walsh, J. S. P. (2012, March 2). The final environmental frontier: Space development and its consequences. *Scientific American*. Retrieved 3/21/12 from news/yahoo.com/final-environmental-frontier-space-development-consequences-160000428.html

Warming, G. G. com. (2010). *Measuring climate change*. Retrieved 3/20/12 from www.global-greenhouse-warming.com/measuring-climate-change.html

Waugh, R. (2012, February 8). Right-wingers are less intelligent than left wingers, says study. *Daily Mail Online*. Retrieved 4/17/12 from www.dailymail.co.uk/sciencetech/article-2095549/Right-wingers-intelligent-left-wingers-says-controversial-study—conservative-politics-lead-people-racist.html

Weiss, J. (2011). *FAQ: Bottlenose dolphins*. Retrieved 11/8/11 from www.fieldtripearth.org/article.xml?id=1496&ordinal=3

Welker, C. (1985). The social structure of primates. *Anthropologischer Anzeiger, 43*(2), 97-164. Retrieved 5/26/12 from www.ncbi.nlm.nih.gov/pubmed/3927823

Wieland, C. (2006, July 6). Mitochondrial Eve and biblical Eve are looking good: Criticism of young age is premature. *Creation Ministries International*. Retrieved 11/27/11 from creation.com/mitochondrial-eve-and-biblical-eve-are-looking-good-criticism-of-young-age-is-premature

Wikipedia. (2011). *Extinction*. Retrieved 10/17/11 from en.wikipedia.org/wiki/Extinction

Wikipedia. (2012). *Energy subsidies*. Retrieved 4/24/12 from en.wikipedia.org/Wiki/energy_subsidies

Wikipedia. (2012). *Planetary boundaries*. Retrieved 3/4/12 from en.wikipedia.org/wiki/Planetary_boundaries

Wilford, J. N. (2009, May 7). Homo floresiensis. *New York Times*. Retrieved 7/29/2011 from topics.nytimes.com/topics/news/science/topics/archaeology_and_anthropology/homo_floresiensis/index.

Williams, G. (1975). *Sex and evolution*. Princeton, NJ: Princeton University Press.

Wilson, E. O. (1978). *On human nature*. Cambridge, MA: Harvard University Press.

Wilson, E. O. (2002). *The future of life*. New York: Vintage Books.

Winton, R. (2011, April 6). Giants fan had feared for safety. *Los Angeles Times*. Retrieved 10/14/2011 from articles.latimes.com/2011/apr/06/local/la-me-dodgers-beating-20110406

Wittgenstein, L. (1953). *Philosophical investigations*. New York: Macmillan.

Wolf, Z. B. (2010, February 11). Inhofe uses blizzard to refute global warming. *ABC News*. Retrieved 4/17/12 from abcnews.go.com/blogs/politics/2010/02/inhofe-uses-blizzard-to-refute-global-warming/

World Commission on Environment and Development. (1987). *Report of the world commission on environment and development: Our common future*. Retrieved 3/16/12 from www.un-documents.net/wced-ocf.htm

Worldwatch Institute. (2008). *2008 state of the world*. New York: Norton.

Young, J. (2002). *Heidegger's later philosophy*. Cambridge, UK: Cambridge University Press.

Yunus, M. (2003). *Banker to the poor*. New York: Public Affairs.

Zimmer, C. (2010, December 22). Siberian fossils were neanderthals' eastern cousins, DNA reveals. *New York Times*. Retrieved 9/17/11 from www.nytimes.com/2010/12/23/science/23ancestor.html

About the Author

Robert A. Schultz received his Ph.D. in philosophy from Harvard University in 1971 with a dissertation in ethics under John Rawls. In 1980, he became Data Processing Manager at A-Mark Precious Metals, in Beverly Hills, CA. He was Professor and Chair of Computer Information Systems at Woodbury University from 1989 until he retired in 2007. He currently teaches an online course in IT and Ethics for the AIM Masters Program at the University of Oregon. He has published in the areas of database design, IT and ethics, and the philosophy of technology. He has also published two books with IGI Global on topics in IT and ethics.

Index

Q

R

S

T

U

V

W

CPSIA information can be obtained
at www.ICGtesting.com
Printed in the USA
BVHW01*0340300318
R8586200001B/R85862PG511661BVX1B/1/P

9 781466 645868